ONCE A GRENADIER . . . .

Other Books by Oliver Lindsay CBE F.R.Hist.S.

The Lasting Honour: The Fall of Hong Kong 1941

At The Going Down of the Sun:
Hong Kong and South-East Asia 1941–1945

A Guards General:
The Memoirs of Major General Sir Allan Adair Bt (Editor)

*DEDICATED TO ALL GRENADIERS,
SERVING AND RETIRED,
THEIR WIVES AND FAMILIES.
THIS IS THEIR HISTORY.*

Contents

Part 2 1945–1995

APPENDICES

The Second World War ended 50 years ago and, but for a few minor conflicts, we are supposed to have been at peace ever since. However, this history makes it quite clear that there has been little peace for the British Armed Services. The British infantryman in particular, has had to face an extraordinary range of difficult and challenging tasks, each of which has demanded a very high level of resourcefulness and adaptability.

I think it would be fair to say that the regiments of Foot Guards have been taxed harder than most. The frequency and speed with which they have had to exchange tunic and bearskin for combat kit and the modern instruments of war is a recurring theme of this book. Over the last 50 years more Grenadiers have been to more parts of the world, engaged in a greater variety of training, adventurous activities and operational deployments than their predecessors could have imagined.

This book shows just how well Grenadiers have succeeded in reconciling their well-deserved reputation for ceremonial and public duties in London with the mastery of the most advanced techniques of the contemporary military art. As might be expected, they discharged whatever duty was given them with the same thoroughness and attention to detail for which the Regiment is justly famous.

This is a history of human achievement under every kind of stress and it emphasises the strength and durability of the very special human relationships that the Grenadier Guards have fostered for so many generations. Colonel Oliver Lindsay has written a fascinating account of a half century of dedicated service. I am quite sure that those thousands of Grenadiers, whose story this is, will find that he has captured the spirit, ethos and way of life of one of the world's most renowned military organisations.

Introduction

This History deals with the tasks and armed conflicts in which the Grenadier Guards took part from the victory over the Nazi tyranny in May 1945 to the relative peace of 1995. This book seeks to be unique, for controversial Regimental matters are aired, and I hope that it will be of interest, not only to those who have served and are now serving in the Regiment, but also to their wives and families too. However, the principal readers I have in mind are those in the 21st Century who will wonder what Grenadiers and the Army accomplished in every quarter of the globe over many difficult years.

The story is largely related by Grenadiers themselves. Part 1 deals in a chronological order with the 1945–1976 period. Part 2 covers a broader canvas over the entire 50 years, including such matters as adventurous training, and the Guards Parachute Company and G Squadron SAS.

Chapters on similiar events would make the History dull, repetitive and much too long. I have therefore had to be very selective when it comes to describing soldiering in Germany, Northern Ireland and London District, not referring to similar events twice. I have chosen the tours in question, not because they were more or less successful than others, but instead because they provide an interesting contrast, and I have a personal knowledge of some of them.

I alone am responsible for all errors and omissions.

The chapters do not need to be read in a chronological order. The reader may prefer instead to pick out, for example, the dramatic events in Germany (Chapters 1, 5, 12, 26), or to follow in turn those relating to operations in Northern Ireland (14, 17, 23, 30), or to read of the great ceremonial events in London (6, 9, 29).

The History does not seek to boast. Not everything was perfect. Nevertheless the story reflects some glorious deeds, great loyalty and proud endeavours which bind together with a golden thread all those who served in the Grenadier Guards during fifty years of peace and wars.

Throughout this time the Regiment has been wonderfully supported, first by Princess Elizabeth as Colonel of the Regiment, and then as Colonel-in-Chief. Moreover since March 1975 the Duke of Edinburgh has been Colonel of the Regiment. Therefore the Grenadiers have greatly benefited from the support by and close connection with the Crown. All ranks have been enormously proud to serve as soldiers of the Queen.

OLIVER LINDSAY

Illustrations

The author is indebted to those who have given permission for photographs in their possession to be reproduced in this book. The frontispiece is reproduced by kind permission of Tim Graham; plates 35, 38, 50 and 51 were provided by Roger Scruton; plate 45 by SJ Kingsley-Jones and 52 by Patrick J Stocks. In addition a number of Grenadiers have contributed photographs. I am most grateful to them all.

Maps

Glossary of Abbreviations and Terms

For those unfamiliar with the Grenadier Guards and the Army

ABBREVIATIONS

BAOR – British Army of the Rhine
GPMG– General Purpose Machine Gun
GOC – General Officer Commanding
LMG – Light Machine Gun
RUC – Royal Ulster Constabulary
UDR – Ulster Defence Regiment

Appointments (as in 1996)

The Major General	–	The Officer Commanding the Household Division who is also the General Officer Commanding London District.
The Lieutenant Colonel	–	The officer, serving or retired, who commands the Grenadier Guards. Until 1989 he was a serving officer, holding the rank of Colonel. Subsequently the appointment has been filled by an officer senior to the Regimental establishment, selected by the Regimental Council, who carries out the duties in a part-time capacity.
The Senior Major	–	The Second-in-Command of a Grenadier Battalion
The Captain	–	The officer in the rank of Major who commands The Queen's Company

The establishment of units has varied considerably. Often units were below strength. In a typical infantry battalion in the UK in 1971, three sections (each of up to 10 men) formed a platoon and three platoons with a small platoon headquarters made up a rifle company of which there were three. In addition a battalion had a headquarter company and a support company of mortar, anti-tank and assault pioneer platoons. The establishment of a mechanized battalion in BAOR was 712 officers and men with a total of 126 vehicles. A mechanized brigade in 1971 contained three mechanized battalions and one armoured, artillery and engineer regiment with supporting units, and numbered 4407 officers and men.

Ranks

RSM	–	Regimental Sergeant Major
CSM	–	Company Sergeant Major
CQMS	–	Company Quartermaster Sergeant
CSgt	–	Colour Sergeant
Sgt	–	Sergeant
L/Sgt	–	Lance Sergeant. A rank equivalent to Corporal, unique to the Foot Guards
L/Cpl	–	Lance Corporal. Foot Guards Lance Corporals wear two chevrons
Gdsm	–	Guardsman

PART ONE

1945 – 1976

CHAPTER 1

THE AFTERMATH OF BATTLE

1945

On 4 July 1945 1st Battalion Grenadier Guards entered Berlin, the ruined capital of Germany. Burned-out tanks littered the roads. Dark crowds of Germans lined the pavements, clustered in knots on the uneven rubble. The smell of death and dust was everywhere. The Grenadiers were among the very first British troops to enter the City. They had come a long way over the last five dangerous years.

*   *   *

In 1940 all three regular Battalions had formed part of the British Expeditionary Force. The campaign, a great operational victory for the German army, culminated in the evacuation from Dunkirk to England. Throughout, the Grenadiers, as at Corunna 132 years before, were judged by all to have set a fine and much needed example of discipline amidst the chaos. No weapon was lost and the Grenadier Battalions kept their cohesion. The traditions were being maintained. In June 1944 the 1st and 2nd Battalions landed in France once more, forming parts of the Guards Armoured Division commanded since 1942 by Maj Gen Sir Allan Adair, Bt DSO MC. The Regiment had expanded to six Battalions. The 4th, reformed in 1940, was detached in 6 Guards Tank Brigade as an independent formation to support infantry divisions. They fought through France, Belgium, Holland and Germany.

The 3rd Battalion participated in the Tunisian Campaign 1942 – 1943 before fighting up Italy 1943 – 1945 and into Austria where their exploits will be recounted in the next chapter. The 5th Battalion had also fought in Tunisia and Italy. Its hardest trial came in the Anzio beachhead between

Germany, 1945-95

22 January and 8 March 1944. The 6th had formed part of 201 Guards Brigade in the 8th Army, advancing in the Western Desert in March 1943 after the great victory of El Alamein. Their way was barred by a strong position at Mareth in Tunisia. The attack went in by night across a large minefield. Fourteen officers were killed and 280 casualties received in this their first battle. Admiration for the courage and skill of the Grenadier performance in appalling conditions was expressed by all, not least by the Germans themselves. The 6th Battalion, first to land in Italy, went from success to success in the bitter struggle northward up the peninsula, but had no finer hour than its first battle among the mines, crossfire and darkness of Mareth. Both the 5th and 6th Battalions had reinforced the 3rd

Battalion after their official disbandment in May 1945 and December 1944 respectively.

Grenadier casualties throughout the war had amounted to 4,422. The Battle Honours added to the Colours were Dunkirk 1940, Mareth, Medjez Plain, Salerno, Monte Camino, Anzio, Mont Pincon, Gothic Line, Nijmegen and Rhine.

The spirit of comradeship, loyalty to the Regiment and pride in its achievements were to prove a great inspiration to the new generations of Grenadiers who followed those who had fought with such gallantry in the Second World War.

By Victory in Europe Day, 8 May 1945, all the Battalions were taking over new responsibilities. The 1st Battalion (commanded by Lt Col PH Lort-Phillips DSO) was moving to Stade, near the west bank of the Elbe close to Hamburg; the 2nd Battalion (Lt Col JNR Moore DSO) to a small part of Freiburg on the Elbe estuary; neither Battalion was to be there long; within two weeks they were south of Bremen; the 3rd Battalion (Lt Col PT Clifton DSO) was approaching Austria, while the 4th Battalion (Lt Col Lord Tryon DSO) was moving to villages south-east of Kiel. Four days earlier Field Marshal Montgomery had received the surrender of all German forces in north-west Germany, Holland and Denmark.

The Grenadiers in Germany were responsible for an area as large as some British counties. Squadrons and companies were therefore sub-allocated their own districts. There were innumerable difficulties. The first priority was the disarming and demobilization of the German soldiers. Their weapons were quickly gathered into various dumps scattered all over the countryside. But collecting the Germans into groups and demobilizing them became a nightmare. The job proved so complicated that a large stretch of the Baltic coast was set aside as a mammoth prisoner-of-war cage. Nearly a million prisoners were poured into it.

Few of the German civilians in the Battalions' areas had seen the war at first hand. Patrols in Churchill or Sherman tanks were therefore sent through the villages providing 'Displays of Strength.' Local Burgomasters were ordered to arrange for the handing in of all weapons and to prepare lists of Nazi party members. Houses were searched. Hours were set aside for dealing with the local civilians. Hoards of Germans would assemble each day for an interview seeking passes, perhaps, to visit a neighbouring village or to listen to the denunciation of a fellow-citizen. Gradually the Germans learnt the meaning of total defeat.

A most urgent problem was helping the displaced persons. In the 4th Battalion's area alone there were at least 3,000 Russians. They had led a miserable existence, divorced from any semblance of a civilized society. Most nationalities had been sent home, apart from Poles and Yugoslavs.

It had been decided that more infantry would be required now that the war was over. It was appropriate therefore that the Guards should revert to their traditional infantry role. But first a 'Farewell to Armour' parade should be held on 9 June at Rotenburg airfield near Hamburg.

The sun was shining when Field Marshal Montgomery, the Commander-in-Chief, arrived. Over 200 tanks painted battleship grey were driven across the airfield, traversing their turrets, their commanders saluting, and then, over the crest of the hill, they were gone. It was now the turn of the infantry. Five columns of Guardsmen marched to the front of the stand where Montgomery addressed them: 'I have been terribly impressed with everything I have seen. . . . In this realm of armoured warfare you have set a standard that will be difficult for those who follow after you to attain. You have achieved great results. When the Guards do a thing, they do it properly – they give a lead to everyone else. You will long be remembered for your prowess in armoured warfare . . .' He ended by paying a warm tribute to the character and ability of Maj Gen Allan Adair whose memoirs, *A Guards' General*, describe these days in detail.

The Guards' newspaper which reported Montgomery's speech contained a postscript: 'Hitler married before Berlin fell, and may still be alive, said Marshal Zhukov.' Many wondered what was happening there. The 1st Battalion was about to discover for they were required for the Victory Parade.

On 24 June they moved to the Brunswick area where they were welcomed by 7 Armoured Division, nicknamed the Desert Rats. An intense programme of spit and polish commenced; 400 gallons of paint made the vehicles immaculate; the green webbing became white.

On 4 July the Battalion entered the Russian Zone at Helmstedt, driving beneath an enormous red flag adorned with hammer and sickle. The Russian soldiers examined the passes upside down, cadged cigarettes and politely saluted as the convoys drove on. At Magdeburg, a mass of chaotic rubble, the Division was held up by the Russians for three hours at 'Friendship Bridge.' Below in the Elbe unclaimed corpses protruded from the water to the delight of the Ukrainian and Mongolian soldiers who pointed them out with glee. At last, on the move again, the convoys passed ramshackle Russian lorries which contained everything from red-cheeked girls in sheepskins sitting on sacks of flour, to half a dozen pigs and a grand piano.

On the Berlin autobahn burned-out tanks littered the road to Potsdam in increasing numbers. They drove towards Spandau, passing German graves with curved steel helmets stuck awkwardly on rough wooden crosses. At last, after the nine-hour drive, they approached the saluting base. The 11th Hussars led the way before the 1st Battalion marched past

Maj Gen LO Lyne, the Divisional Commander, who was flanked by cameras and military police.

The Battalion's destination, the Hermann Goering barracks, proved to be a grim, half-destroyed building swarming with flies and without sanitation. The Russians had broken all the basins, lights and furniture. Their horses had been stabled in the Officers' Mess. It would take 1,000 civilians over three weeks to make the barracks habitable.

Two days later Grenadiers watched the Union Jack being run up a flagstaff at the foot of the German memorial to the victory in the Franco-Prussian war. The ceremony took place in a drizzle of rain and seemed to awe the Germans. The flag would fly, it was said, as long as British troops remained in Berlin. Nobody anticipated that they would still be there nearly 50 years later.

On 21 July the British Victory Parade was held on the Charlottenburger Chausee, the long, tree-lined avenue in the heart of the German capital. Churchill, Prime Minister for a further five days, was accompanied by Eden, Attlee and four Field Marshals – Montgomery, Brooke, Wilson and Alexander. Flags and bunting decked the stands along the route. The sun was shining. Churchill was driven slowly past the long line of guns, tanks and infantry, reviewing the 10,000 men on parade. The inspection over, the march past began. The infantry was led by the 1st Battalion. As they drew level with the Prime Minister the massed bands opposite the saluting base broke into 'The British Grenadiers.'

Berlin was scarcely recognizable to the few Grenadiers who knew the city before the war. The Tiergarten, originally a deer park, was reduced by the urgency for firewood to a desert of tree stumps. Many of the statues were scarred by shell fire and headless. The River Spree was choked to death and decay, an open sewer with stagnant and scum-bound waters. Three in every four houses were smashed or gutted. In the British Zone less than ten per cent of the buildings were undamaged. The streets were disfigured with tram lines torn up, craters and piles of wreckage.

Eight weeks before, the Germans had locked themselves into houses and cellars and waited for the next explosion and the hurrying footsteps of the advancing Russians. Food supplies had broken down and the water was foul. There was no gas, electricity or transport. After the surrender, the Russians had been through buildings like a whirlwind, smashing windows, bayonetting chairs, slashing pictures and ripping out electric fittings. The Russian reign of terror only ended when troops from the West arrived. Their initial deliberate and calculated terrorism did them incalculable harm politically. Some of the Russian units in the Berlin area were the least disciplined and crudest – filthy, down-at-heel and intent upon looting and rape. The Germans feared them, knowing that some German units had behaved no better in Russia. The appalling behaviour of the Red Army was played down. The pretence was made in some quarters: 'Good Uncle Joe, rather left wing, of course, but all right.' The Guardsmen saw through this, regarding them as animals. They got 'Uncle Joe' out of their system extra-ordinarily fast.

The Germans regarded the troops from the West more as liberators from the Russians than as an occupying power. The Germans appeared to the Grenadiers as neither sullen nor resentful. Some cheered and clapped when the Battalion had first driven into Berlin. The atmosphere was unreal but the terror continued in the Russian Zone.

Sgt FJ Clutton MM vividly recalls hearing from the Russian Zone women's screams at night and occasional bursts of machine-gun fire. Women tried to get into the Grenadiers' barracks for safety to escape the Russians. Two months earlier he had participated in the liberation of Sandbostel concentration camp near the River Oste. All but three of his platoon in the Kings' Company had become casualties. He was horrified by the conditions in the camp – the open pits filled by skeletons and the terrible smell of decaying bodies. After Victory in Europe Day Sgt Clutton's platoon was lodged in a large, attractive house near Stade. He gave choco-late to a pretty, young German girl who was extraordinarily pale because her father, a doctor, had kept her hidden in the attic throughout the war, convinced that she would otherwise have ended up in a Nazi 'Stud Farm' having to breed perfect children for 'the master race'.

'Each morning in Berlin every company Sergeant-in-Waiting went down to the barrack gate to collect 100 women each to scrub out the barracks', Clutton recalls. 'The women's cards were stamped at the end of the day which entitled them to extra rations; there was no lack of volunteers. The Guardsmen in the cookhouse gave them extra crusts as well. Despite the fraternisation ban, a lot of German women wanted to marry Guardsmen to get out of Berlin. Much later some did and lived happily every after. When we were at Stade earlier, German women paraded by the river in swim suits and high-heeled shoes in the hope that we would go out with them and give them food. Many women preferred a life of promiscuity to rubble cleaning and semi-starvation. Quite apart from women, there was the black market: bully beef, soap, sardines, chewing gum and tobacco could all be sold to Germans.'

Grenadiers visited the Reich Chancellery where something of Hitler's evil presence still lingered. The building was pitted by Soviet mortars and rockets.

Gen Sir Andrew Thorne, a Grenadier Battalion Commander in 1917, and Military Attaché in Berlin before the war, had written an article on the battle for Gheluvelt. Hitler had also fought there and had the article translated into German. A copy was found in the Chancellery amidst other documents, files and books in damp disorder all over the floors alongside broken glass and medals rusting in heaps. Slogans and messages had been scrawled over the walls; electric wires dangled from the smashed ceilings. Maj KEM Tufnell MC removed a Christmas card signed by Hitler and a picture of Mussolini.

The Victory Parade over, the 1st Battalion returned in August 1945 to Bonn. The 2nd and 4th Battalions were still in Germany, while the 3rd Battalion was having an interesting and exciting tour in Austria.

CHAPTER 2

AUSTRIA:
THE NAZI ROUND-UP

MAY – AUGUST 1945

While bonfires burned in Britain to celebrate Victory in Europe Day, 8 May 1945, 3rd Battalion Grenadier Guards crossed the border from Italy into Austria, still then part of the German Reich. The road to Villach was excellent and undamaged but the sights which greeted the Grenadiers were both strange and unbelievable to those who had not seen the Germans' defeat in North Africa. Hundreds of fully armed Germans were standing or sitting by the road completely indifferent to the Grenadiers. Austrian civilians threw bunches of lilac at the Battalion's vehicles. They seemed resigned to a firm but not oppressive military government. They were puzzled, humbled, slightly nervous but not over-awed. British, Dominion and allied prisoners of war were seen on the road. They had been well treated by the Austrians and employed on local farms. Prisoners of other allied nations had suffered severely.

On the first evening in Villach an emaciated figure, dressed in thin, striped overalls, walked into 1 Guards Brigade HQ and introduced himself as Louis Balsan, head of a department in the French Ministry of Finance. For two years he and 600 other French had been interned in a concentration camp on the Loibl Pass where they had been mercilessly bullied, starved, overworked and many of them shot or allowed to die for want of medical attention. On the previous day the German guards had fled and the prisoners had seized a train which they had managed to bring to a station on Villach's outskirts. They were given a whole lorry load of captured food and British uniforms. Soon afterwards they were returned to France – except for Balsan who insisted on remaining behind to bring his tormentors to justice. He had brought with him detailed records of German crimes, copied at night from the SS records in the concentration camp. He was made a temporary Lieutenant in the British army; unrecog-

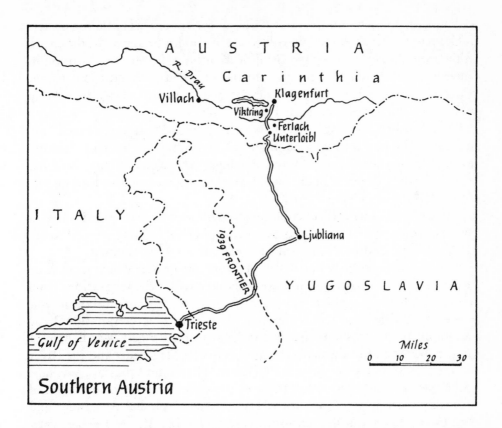

Southern Austria

nizable in his battle dress and dark glasses, he visited the vast prisoner of war camps, looking closely into the face of each man before occasionally pointing out an individual. He was never wrong.

On arrival at Villach the Grenadiers' convoy came to a stop outside a hutted barracks which was still occupied by German soldiers of 1/112 Artillery Regiment. The German commander was ordered to clear out while all his transport and military stores were enthusiastically commandeered and distributed to various Battalion departments. The Commanding Officer Lt Col PT Clifton DSO, ordered No 1 Company (Maj EC Russell), No 3 Company (Capt GGH Marriott MC) and No 4 Company (Maj WJL Willson DSO) to establish road blocks while Support Company (Maj ECWM Penn MC) sent detachments to the two main railway stations. Villach was the key town in the railway network of Southern Austria; in the next 48 hours some 5000 assorted refugees passed through Maj Penn's hands for sorting out and escorting to the various Prisoner of War cages. On the following day, 9 May, Lt PR Freyberg MC, commanding the anti-tank platoon, decided that it was time to put a stop to the increasing rail traffic and cut the line with a demolition party. Guards

11

and duties increased rapidly. Capt PATG Meyrick sent his Battalion drivers in German trucks as mobile patrols to catch and lift prisoners to the cages. In the evening No 2 Company (Maj SE Bolitho MC) sent one platoon to supervise 450 drunk Hungarian soldiers who had revolted against their officers; the second platoon was despatched to guard a large fuel dump while the third guarded Brigade Headquarters.

Fortuitously a Swedish YMCA worker appeared with three British ex-internees who had formed a band and played to the Grenadiers. Another interesting character was an American singer whose charm had raised her from scrubbing floors at Buchenwald concentration camp to a leading role in the Berlin State Opera.

The Commanding Officer took on strength an excellent Alfa-Romeo, a cinema projector and a trailer generator capable of lighting a small town. He later commandeered an open Mercedes and a motor launch.

A very serious problem had developed with Yugoslavia which grew more grave daily. Tito's partisans and army claimed territorial reparations from Italy and Austria, for which they were not prepared to wait until the matter was discussed at a peace conference. Tito was a fiercely nationalistic Communist who was sufficiently confident to challenge Great Britain. 1 Guards Brigade moved east to the area between Klagenfurt and the River Drau some 35 miles north-west of Trieste. They found that every village had its section and every small town its platoon, company or battalion of Yugoslav partisans. Negotiations between Field Marshal Alexander and Tito broke down and the tension was considerable. It rose further when the Yugoslavs issued a proclamation in Klagenfurt promising the annexation of Carinthia to Yugoslavia and the end of British military government in Austria. The Yugoslavs started to remove large loads of German arms at the Drau bridge to the frustration of the Grenadiers who had orders not to interfere.

The Grenadiers' task had changed from combat duties to administrators – feeding, watering, administering and caring for thousands of people. There were refugees to be sorted out, Yugoslav regular and partisan forces to be deterred, and prisoners to be guarded.

The Battalion was responsible for manning a Prisoner of War camp under command of Lt JW Harkness. The camp required a Grenadier company to guard it. No one in the Battalion was allowed within the perimeter and its internal administration was left entirely to the German commander, Oberst Von Seerler, and his staff. He had great difficulty exerting his authority over the troops of other nationalities although they had fought with the Germans. They included 5,000 Russian deserters from the Red Army, 9,000 Yugoslavs and Slovenes who claimed to have passes signed by King Peter of Yugoslavia.

The situation deteriorated further when Tito's soldiers drove at speed past the camp and opened fire on German officers. Fortunately they were poor shots. A young SS Sergeant told a Grenadier that Nazi werewolves were still at large in the camp. The informer quickly identified fifteen tough young NCOs from various SS units including one girl dressed in SS uniform. They were originally part of a sabotage group and were found to have a cleverly concealed wireless set which had been in contact with Berlin.

There was very welcome news on 20 May when the Battalion heard that an agreement had been signed which would lead to the evacuation of Yugoslav forces from Carinthia the following day. Part of the agreement necessitated returning to Tito those Yugoslavs who had fought against him. 'It was not a pleasant task to break up their domesticity, to slaughter their horses for food, and send their families to a sterner captivity,' reported Capt N Nicolson, the 1 Guards Brigade Intelligence Officer, ' but we had pledged our word to Tito, and the reception given to them by his men encouraged us to hope that their fate would not be too harsh.'

The return of those Russians who had fought alongside the Germans became a matter of great controversy and led to the largest libel damages in British legal history.

Lt GW Lamb and others were aware that a Russian corps with many women and children had surrendered to the British to escape from Stalin's tyranny. 'I saw them piling weapons throughout a whole day. They were sent on to the Welsh Guards for guarding but some slipped away into the hills seeking freedom. We knew that British prisoners were in Poland and it seemed fair enough to swap them for the Russians. So when we received the orders that they should be returned, we complied.'

It was now time to round up the many isolated and desperate German soldiers and other undesirables for whom the war had not ended. Some were hiding in mountainous, inaccessible country adjacent to the Austria –Yugoslav frontier. It was therefore decided that a cavalry platoon should be formed, commanded by Lt PG Hedley-Dent and assisted by Lt CMH Murray. Plenty of Grenadiers volunteered, although few had been astride a horse before. So many German units, and also the Cossack Division, had surrendered that there was no shortage of horses. One German officer told Maj Willson that he wanted to bring horses over from Yugoslavia. Maj· Willson pointed out an excellent grass area, expecting several thousand at most, to be told that the German had 53,000. Field Marshal Alexander unexpectedly inspected the cavalry platoon to the consternation of Lt Hedley-Dent who, not sure of cavalry drill and on a frisky horse, dismounted and gave his mount's reins to the Divisional Commander to hold while he accompanied the Commander in Chief.

'Operation Ratting' occupied several long and tiring days. Its aim was to find the remaining enemy. Questionnaires in German were given to civilians before their houses were searched. Boats on the River Drau were impounded or sunk, arms were collected and the location of every vehicle noted. Rain made the operation most unpleasant but 200 arrests were made, including five company commanders of the Prince Eugen SS Division who had walked the whole way from northern Greece. They were interrogated by the Battalion Intelligence officer, Lt the Hon JRB Norton. The cavalry platoon and Grenadiers on foot scoured the 4,000-foot peaks with great determination. The views beyond the frontier posts were staggering. Visitors from the Army of Occupation much admired the wild, beautiful country. One of them, the Adjutant of 3rd Battalion Welsh Guards, strayed across the frontier while collecting rare flowers and was promptly arrested by a woman political commissar.

The severest test for the cavalry platoon occurred on 4 June. A Yugoslav recce group was seen in the morning watching a Grenadier post at the frontier. That evening all the Old Etonians celebrated the day in keeping with their tradition. Those present included the Army Commander, and Commander-in-Chief, although he was an Old Harrovian. An enormous dinner and much wine were followed by fireworks, arranged by Maj the Hon DSTR Dixon. However, just before midnight a message was received that Tito's forces had crossed the border. The Battalion orders group at 1.30 am was a less alcoholic affair. The cavalry platoon was ordered to advance at dawn and secure the bridge over the River Drau and the emergency scheme was put into effect.

Morning found the whole Battalion sitting on the banks of the Drau enjoying the glorious weather and cursing the Tits, as Marshal Tito's forces were now called. At 3pm it was confirmed that there had been no invasion. A Wykehamist at Brigade HQ was held responsible for the false alarm.

As tension subsided the only major military commitment was for a company at Unterloibl to man road blocks and frontier posts. The rest of the Battalion relaxed to enjoy for the first time some of the pleasures of peace. Grenadiers found themselves in a lovely part of the world, one of the favourite holiday resorts of many. The billets were the villas and hotels formerly occupied by senior Nazis. 'Our pleasures had been their pleasures,' noted Capt Nicolson, 'speed-boating on the lakes, waterskiing, sunbathing, riding, racing, rowing, shooting, swimming, climbing and fishing. All that was lacking was the companionship of civilian families. Even that was offset by the prospect of leave to England and the not long to be delayed day of demobilisation.' There was now time for cricket and football. Those who preferred eating and drinking patronized NAAFI bars

1. Marshal Zhukov, accompanied by Field Marshal Sir Bernard Montgomery, inspects a Guard of Honour of the 1st Battalion in Berlin, July 1945. The Reichstag stands in ruins behind them.

2. Officers of the 2nd Battalion at Caterham, 1948.
   Back: 2/Lt M H Burrill; Lt R M O de la Hey; Lt N Hales Pakenham Mahon; Lt P J C Ratcliffe; Lt R M V Porcelli.
   Centre: Lt M T Middleton-Evans; Major J H Wiggin; Major W G S Tozer; Capt (QM) B H Pratt; Capt P G A Prescott; Capt N W Alexander; Capt J A Fergusson-Cuninghame; Lt R A G Courage.
   Seated: Capt F J Jefferson; Col E H Goulburn; Lt Col G C Gordon Lennox; The Colonel; Maj P W Marsham; Maj E C W M Penn; Lt Gen Sir Frederick Browning.

3. Operations in Malaya, 1949, destroying squatters' houses near the Batu caves. From right to left Major J C Trotter, Lieutenant Colonel T F C Winnington and Major R C Rowan with two Guardsmen of the 3rd Battalion.

4. The Sergeants' Mess, 3rd Battalion, Chelsea Barracks, 1949.

around the lake called the Wörther See. Grenadiers ate well, obtaining eggs and chickens from locals.

The policy of non-fraternization, difficult to enforce and even more difficult to resist, was the only disadvantage to an otherwise enjoyable, restful and entertaining change from the long and monotonous months of battle in Italy.

Everyone was attracted by the fair-haired pretty Austrian girls, particularly those in the flimsiest of clothing doing the harvest in very hot weather. Italian girls had been very different and closely guarded by their mothers. The Austrian girls regarded the British Army as their liberators – either from the Germans or from the Yugoslavs and Russians who would certainly have raped them, if not kidnapped them and shot their parents. This led to some clandestine fraternization among the Guardsmen, although it was strictly forbidden for all ranks. Few young Austrian men had yet returned from the war.

No 4 Company HQ in Ferlach discovered that a German breeding farm had been established nearby. Blond well-built German women had been sent there to conceive Hitler's 'master race'. However, all had dispersed. The Company Commander optimistically put up a notice by Ferlach's swimming pool saying that it was out of bounds to civilians except for Austrian girls between the ages of 17 and 30.

Grenadier NCOs were much in demand as drill instructors for neighbouring units. D/Sgt H Wood DCM had a shock when sent to 7th Battalion The Rifle Brigade. His 'step sharper' was not necessary as they hurtled round the square at 180 paces to the minute.

On 23 June orders were received for the impending departure of 1 Guards Brigade for England en route to the Far East to fight against the Japanese. The news became public when the Battalion was celebrating Lt J Tayleur's success in the local hunter trials with his gelding Leave to Speak. Everyone welcomed the prospects of returning home but some were uneasy about the final destination.

A farewell parade was held on 29 June on a football ground near Viktring; 1,500 men were on parade despite the small space. Gen McCreery congratulated the Grenadiers on their outstanding turn-out and drill. At the final parties which followed it was sad saying goodbye to close friends in the Ayrshire Yeomanry, 16/5 Lancers and to the other Regiments who had supported the Grenadiers so well in past battles.

The Battalion left the following morning for Fano in Italy on the Adriatic Coast. Two days later they reached their destination which turned out to be a stubble field – an unwelcome change from the handsome Austrian schlosses. A strong gale arose and blew down most of the tentage. The medical officer, Capt M Knowles, tried to save the Officers' Mess marquee and held onto a guy rope when an extra gust caught the tent, lifting his sixteen stone into the air. The following morning dawned clear and fine. The Battalion settled down to intensive bathing, for the sea was close by. There was little else to do in the heat and dust. There were a few film shows and NAAFI canteens in Fano or Pesaro, but these were soon filled by hundreds of other troops including Americans from 15th Air Force who spent hours bartering with the Guardsmen for Luger pistols in return for GI watches or sunglasses.

To everyone's frustration and impatience the Battalion remained at Fano for a month. But on 1 August Support Company climbed into ten Liberator bombers to start the return to England. Eight hours later they were flying over the White Cliffs of Dover – all except the Company Commander's plane which broke down, leaving Maj Penn in Marseilles.

Within a fortnight the whole Battalion was in England. They now really felt that the war in Europe had definitely come to an end.

CHAPTER 3

PALESTINE 1945 – 1948

COUNTDOWN TO CATASTROPHE

In August 1945 Japan surrendered. The 3rd Battalion, having returned from Austria in July, was at Hawick in Scotland. Plans to send them with 1 Guards Brigade across the Atlantic to America for the war against Japan were cancelled. Almost everyone heaved a sigh of relief. The news of the Brigade's imminent departure for Palestine instead met with a mixed reception. Some even wished that the Japanese had held out just a little longer! Jungle training was replaced by instruction on cordons and searches. But, as usual, there was inadequate time to train thoroughly for the unexpected.

The Battalion, commanded by Lt Col PT Clifton DSO, sailed from Southampton on 11 October 1945, being seen off by the Lieutenant Colonel Commanding the Regiment, Col JA Prescott, and Lt Col RH Bushman. The French liner *Champollion* was comfortable while the food and weather were both excellent. On arrival in Haifa the Battalion moved after a few weeks to Camp 260 on the Mediterranean coast seven miles north of Acre, close to the Lebanese border. By now they had received plenty of lectures on Palestine's historical background.

In 1919 Britain was entrusted with the Mandate for Palestine after the British Army, with the help of Arabs, had defeated the Turks. The League of Nations supported the British proposal in the Balfour Declaration to establish in Palestine 'a National Home for the Jews', although only 55,000 Jews lived there amidst 600,000 Arabs. Jewish immigration drove the Arabs to rebellion which was put down by seventeen British battalions in 1939. Most Jews cooperated with the British during the Second World War but there were occasional guerrilla attacks against the police and Army.

British policy in 1945 was to promote the well-being of Jews and Arabs alike under the Mandate and to prepare the country for self-government while supervising limited Jewish immigration without prejudicing the

17

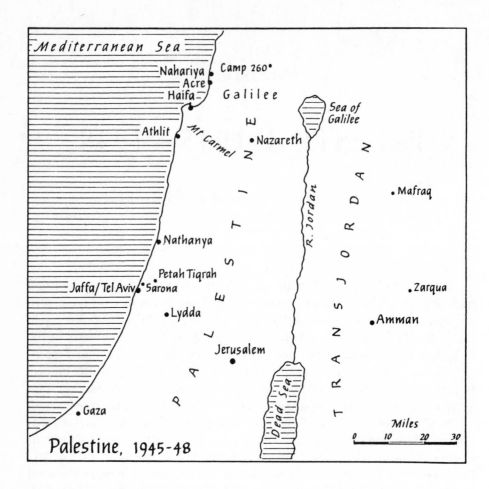

Palestine, 1945-48

Arabs' civil and religious rights. The Arabs declared that they had never accepted the Balfour Declaration and were vocally violent and bitter; the Arab League was getting into its stride as a major force in Arab politics. Jewish sabotage and terrorism were being carried out by the Hagana, the Irgun Zvai Leumi, and the Stern Gang who were the worst of the killers. Not all supported the terrorists. In 1945 many Jews and Arabs were ostensibly getting on well together despite increasingly active political agitators.

The Grenadiers' primary role in 1 Division in Palestine was therefore internal security, while forming part of the Middle East strategic reserve. An early visitor to the Battalion was Field Marshal Lord Gort, the High Commissioner, who had won the Victoria Cross when commanding the 1st Battalion of the Regiment in 1918. It was his brave decision to withdraw the British Expeditionary Force in 1940 to the Channel ports which undoubtably saved so many of his men. Lord Gort was leaving Palestine a very sick man and died from cancer in 1946.

On 13 November the British Government announced that Jewish immigration would continue at 1,500 a year, which satisfied nobody, and that an Anglo–American committee would seek a lasting solution. President Truman was pressing for 100,000 Jews to be admitted at once. The following day there was serious rioting in Jerusalem and Tel Aviv. The police could not cope; shots were fired and the Army was called in to help. But all remained quiet in the north.

Camp 260 was very dilapidated and rather crowded; it was tented and had no electricity, relying instead on pressure lamps. The Royal Engineers started to erect a camp nearby for an armoured car regiment; the Battalion lived amidst a welter of stones, concrete mixers and Basuto, Italian and Indian labourers.

The company of Basutos was commanded by a British officer and a CSM. The natives were all fed up as they had been promised they could go home when the war was over. According to the Transport Officer, Capt GW Lamb, on Christmas Day the Grenadiers apparently received a panic signal from the officer asking to be rescued; he was in his tent surrounded by angry Basutos who were said to have eaten his CSM for Christmas dinner and they were now after him!

In the spring a severe storm blew the roof off the Battalion's orderly room, scattering documents and files everywhere. Some were found the next morning hanging on thorn bushes almost a mile away. The Adjutant, Capt WS Dugdale MC, found the storm a good excuse for failing to reply to some of 1 Guards Brigade's more troublesome letters.

Operational commitments were not yet very demanding. They included mounting one snap road check by day and night, each of at least two hours, every four days. In addition each week two night patrols were sent out and observation posts were manned. Occasionally when curfews and road restrictions were ordered, road blocks and patrols were stepped up. The snooping patrols sought to show the Jews and Arabs that British troops were alert and present. Force could be used when necessary to repel an attack; offenders could be shot as a last resort if a felony was being committed and there was no way to capture the offender.

The Jews were active in the south. On 15 April 1946 they murdered seven soldiers of the Parachute Regiment who were in a tent in Tel Aviv. Twelve rifles were stolen. Two months later terrorists kidnapped five unarmed British officers, but all were released, the last two only after the General Officer Commanding had commuted two Jews' death sentences to life imprisonment.

On 10 June three trains were blown up south of Haifa. The following week two Arabs near the Grenadiers' camp saw a party of up to forty men approaching a railway bridge. The Arabs were suspicious, although the party claimed in Arabic to be British soldiers and some were in British uniform. The Arabs warned the seven policemen guarding the bridge, who opened fire, killing the Jewish terrorist bren gunner. A lucky shot exploded one of the terrorists' gelignite charges, killing up to five others. The Commanding Officer ordered one troop of D Squadron King's Dragoon Guards, in support of the Battalion, to patrol while the Grenadier emergency platoon which was always on standby established section posts. The Pioneer Platoon, accompanied by police, cleared mines towards the bridge. In follow-up operations that night the Grenadiers and police found the body of a dead student, as well as weapons and ammunition. The following morning the Battalion cordoned off Mesuva, finding a wounded man and arresting others. Such incidents in North Palestine were unusual. Cordon and search operations, on the other hand, were frequent, not only in Palestine, but in many other trouble spots in the years to come, notably Cyprus.

The Jewish intelligence system was better than our own: previous reconnaissance of a village would be noticed, thereby making air photographs important; telephone lines were insecure and locals, employed in all camps, were thoroughly untrustworthy. Soldiers were therefore often woken at midnight, hastily briefed and despatched in three tonners by 3 am to ensure that the cordons were in position around the village before dawn. An outer cordon stopped other villagers from interfering while the inner cordon prevented people escaping. Road blocks diverted traffic away. Barbed wire cages were next erected, together with tents for the interrogators, women and children. While soldiers formed up to start the search, the Commanding Officer and senior policeman went to the Mayor's house to explain that terrorists were believed to be in the village which was to be searched. The Mayor, called a Muktar in Palestine, was asked to provide reputable citizens to accompany each search party to see that no unnecessary damage was done. The search party often consisted of an officer, NCO and ten men with a Palestinian policeman. Everyone was then escorted to the cage apart from pregnant women and children. A thorough search for people and weapons was a long and laborious business lasting at least six hours. Royal Engineers concentrated on gardens with mine detectors.

Finds of weapons were photographed. Their hiding places were ingenious. For example Grenadiers found some weapons concealed in a dismantled, hollow maypole. Others were found beneath a wooden window-sill from where the shutters hung. At the cages sexes were segregated. Wired-in passages led to the interrogation tent in which Palestinian police screened everyone. A medical officer was on hand, together with an interpreter, and an army water cart provided drinking water. Soldiers with fixed bayonets stood on guard. Suspects were kept in a special 'wanted' cage. On conclusion, the Mayor was asked to sign a certificate stating that no damage had been done. Soldiers were searched by their officers in order that future claims of looting could be refuted. As the last troops left with the suspects, the curfew was lifted and life returned to normal. Not all operations went to plan. Sometimes the younger inhabitants refused to go to the cage and had to be carried there forcibly; sometimes the soldiers were stoned. All operations required careful planning and some proved worthwhile. On one occasion two policemen were kidnapped from a swimming pool. The Battalion's search found both them and the kidnappers.

There were frequent film shows in the Grenadiers' camp and a repertory company was formed within the Division. Cricket, football, table tennis and water polo were played. A gramophone was available which led to frequent recitals of classical music chosen, surprisingly, by the Guardsmen. A fortunate few found it possible to travel widely. Due to the time difference, one could have one lunch in the camp followed by another in Beirut which was renowned for its 'flesh pots'. Some afternoons were spent swimming and skiing in Lebanon.

When the situation deteriorated in June leave was sadly curtailed. The last of the leave parties returned from Cyprus and Beirut. Thereafter many never left the camp. Life in the Officers' Mess became fairly monastic with occasional bridge, poker and reading. The Jewish cafés in neighbouring Nahariya were usually out of bounds. The village had been partly built by Jews who had escaped from Germany before the war. They seemed to get on well with the local Arabs.

In June 1946 a vast search and arrest operation, codenamed Agatha, involved most of the 100,000 British troops in Palestine. 600 weapons were found but complete surprise was not achieved. The Irgun retaliated by planting explosives under the King David Hotel which housed government and military offices, murdering ninety.

Jewish politicians refused to participate in a London conference unless their detained leaders were released. In October the British Government suspended the policy of general searches to try and get political co-operation. The new British High Commissioner, Gen Sir Alan Cunningham, refused to give the Army a free hand, occasionally agreeing only to a curfew and road restrictions. He also felt that keeping the peace was a police rather than a military responsibility.

According to Maj CB Frederick, Commanding No 1 Company: 'Most people in the Forces thought that the British Government and the High Commissioner were very weak. We were never allowed to be tough and I do not believe that we did a lot of good.' Some soldiers did not appreciate the realities of 'police actions'.

Cunningham had commanded the Eighth Army in the desert campaign before being sacked by Auchinleck in 1941. Field Marshal Montgomery, now Chief of the Imperial General Staff in London, was exasperated by the Labour Government's decision to release from detention the leaders of the terrorist campaign. More and more restrictions were placed upon the troops in Palestine regarding their activities in the maintenance of law and order. Meanwhile over 100 British soldiers and British members of the Palestinian police had been killed. 'Over two per day,' recorded Montgomery in his memoirs. 'Yet the Army was not allowed to take appropriate action against its assailants. Offensive action was not permitted. We had, in fact, surrendered the initiative to the terrorists. I reported my views to Whitehall saying that: "The whole business of dealing with illegal armed organisations is being tackled in a way which is completely gutless, thoroughly unsound and which will not produce any good results."'

There were the usual changes in Officers' postings. Maj Kinsman and Captains Van Der Woude, Corbett and Rollo, after a short threat of compulsory deferment, were released. They were followed by the departure of Maj PE Scarisbrick and Captains PN Whitley and the Hon JRB Norton MC.

Lack of offensive action freed battalions to train for more conventional war. Intensive exercises in Transjordan's desert provided an enjoyable change to internal security operations. One exercise at Brigade level was described as the biggest undertaken by the British Army since the war. Another was modelled on Operation Market Garden which had sought to liberate Holland in September 1944. Mafraq was substituted for Eindhoven and Zarqua for Nijmegen. While in Transjordan the Battalion provided a Guard of Honour for King Abdullah. The day was blanketed by a dust-storm covering the lunch with sand, but the parade went well.

On the Battalion's return to Palestine they were soon involved in preventing illegal immigration once more. On 1 November they cordoned Haifa docks at 4.30 am while 1300 miserable, pathetic Jews from Europe were unloaded from the *San Dimitrio* which had a thirty degree list to starboard. The Jews were embarked into other boats and sent to a spartan detention camp in Cyprus. On another occasion they were sent back to Hamburg amidst appalling, anti-British publicity.

On 25 November the Battalion cordoned the docks again; 4000 illegal immigrants in *Lohita* were determined to resist deportation and drove the Royal Artillery boarding party off the ship. No 2 Company had to assault up the gangway wearing respirators and steel helmets. Under cover of tear gas and a hose, order was reimposed.

Arresting Jews in such a manner troubled some, but legitimate orders had to be obeyed. These Jews had survived the holocaust and Nazi concentration camps such as that at Sandbostel described in Chapter 1; they were coming to their 'promised land', financed and aided by United Nations Refugee Relief organizations in leaky, filthy and unserviceable boats. 'You are worse than the Nazis,' one Company Commander was told. But British servicemens' sympathies often lay with the Arabs who had lived in Palestine since biblical times and wanted to keep their land. Illegal immigrants had already set up twenty-eight new Jewish settlements.

Standing by to intercept illegal immigrants as their ships approached Palestine became a frequent Battalion commitment, although it was primarily a Royal Navy and RAF responsibility to intercept them well before. If the ship had been beached the Navy shone a searchlight beam vertically for one minute and then depressed it at least six times slowly onto the suspected position. The Battalion was then responsible for cordoning the area, preventing interference from inland and manning observation posts and guards. On one occasion, for example, the Battalion stood by near Tel Aviv while the 1st Battalion Welsh Guards were also on call. The latter earmarked a rescue party of fifty swimmers in life jackets to supplement a Royal Engineer rescue squad.

By December 1946 insurgent attacks had driven the police from the streets, forcing them to patrol in armoured cars, further alienating them from the public and their sources of information. But Britain's role continued to be to look after Palestine under the Mandate.

In January 1947 the Battalion, over 800 strong, moved closer to the action in the turbulent Lydda District. The camp there was initially shared with the RAF all of whom had sheets whereas the Guardsmen only had blankets. The RAF later moved elsewhere.

3rd Battalion Coldstream had a minor disaster in their camp at Sarona. A party of 'thugs' as the terrorists were referred to, entered the camp disguised as Post Office workers and parked their van, packed with concealed explosives, beside the police telephone exchange. They then set off the time fuse and escaped. A few moments later the exchange was completely destroyed.

On 2 March 1947 the Battalion placed road blocks around Tel Aviv to control movement. Gdsm D Roberts was fatally injured near Tel Aviv by a 4/7 Dragoon Guards armoured car which was blown up by a mine and landed on him.

The occasional execution of several terrorists often led to disturbances. At 3.0 am on 16 April, for example, the Battalion, now commanded by Lt Col HRH Davies, took up security positions to be imposed following Dov Groner's execution that day at 4 am. No 1 Company (commanded by

Maj Frederick) stationed at Petah Tiqrah maintained continuous patrols for the next forty-eight hours while Support Company (Maj RH Heywood-Lonsdale MC) did the same at Ramat Gan and Ra'anana. No 3 Company (Maj RC Rowan) manned three road blocks while No 2 Company (Maj HWO Bradley) guarded Brigade Headquarters. A Squadron of 12th Lancers was in support. The defence of the Battalion's camp was entrusted to the Senior Major (Maj D Dixon) and Headquarter Company (Maj WJ Hacket Pain).

Battalion order groups at strange hours became more frequent. Those present, apart from the above, included the Adjutant (Capt AN Breitmeyer), the Transport Officer (Lord Montagu of Beaulieu) and RSM H Wood DCM.

One who missed this operation was Lt RF Birch-Reynardson who was sent with a 1 Guards Brigade shooting team to Iraq. Most of their vehicles failed to complete the 560 mile journey each way without breaking down.

After a very hectic five months in the Lydda and Tel Aviv area, the Battalion moved twenty miles north on 30 June to a tented Brigade camp near the Jewish town of Nathanya (now called Netanyua), the centre of the diamond industry. A week later the Jews started mining roads, blowing up several vehicles. To reduce this danger the Battalion mounted frequent patrols and ambushes. On one occasion tanks and flame throwers were brought in to reduce an orange grove to scrub to prevent the terrorists' mining operations.

On 29 July three Jews were hanged in Acre Jail which was of particular concern because Jewish terrorists had kidnapped Sergeants Martin and Paice of the South Palestine Field Security Section on 14 July in Nathanya. 1 Guards Brigade was reinforced by two battalions and an armoured regiment to carry out a thorough search. Martial law was imposed and between 14–27 July a tight cordon was maintained round the city of 15,000 people. Daily searches led to the capture of 18 wanted people and economic pressure was brought to bear.

On 1 August the Mayor of Nathanya reported the location of the Sergeants' bodies; the area was said to be mined. Capt SC Tilleard and Lt OP Stutchbury were driven down a sandy track through orange groves to an eucalyptus wood. The bodies could not be seen and so the two Grenadier officers, now joined by Royal Engineers, the press, local officials and police, all formed in extended line, started to comb the wood. The bloody bodies were found five yards apart hanging from trees. On their chests were pinned typewritten sheets in Hebrew which listed the charges, the judgement, and then read: 'An appeal of the convicted for mercy was rejected. The sentence was carried out on 30.07.47.'

The Royal Engineers swept a path up to the bodies for the American and

Jewish press, who were clamouring to take photographs. Capt JFL Denny arrived in a Brigade Headquarters jeep while the Royal Engineer officer started sawing at the rope above the right hand body. There was a sudden, large explosion. The body which had fallen on the mine completely disintegrated; the other was found horribly mangled twenty yards away. The explosion started a small fire; the press stampeded away. Miraculously the Royal Engineer officer was not seriously hurt.

In August seventy suspects, including three Mayors, were arrested by the security forces, but with that offensive the counter-insurgency campaign directed against terrorists rather petered out. While the British Government and United Nations deliberated the future of Palestine, the Jews and Arabs started to fight each other rather than the British. Between 8 August and 30 September 1947 there were more than twenty-five incidents of communal violence, compared to only thirteen attacks on the security forces.

On 7 August the High Commissioner learnt that in view of potential difficulties in Egypt (related to the decision to remain in the Canal Zone) there would be no large scale reinforcements available for Palestine. Battalion strengths dropped; fortunately companies of the 1st Battalion joined the 3rd Battalion well before the latter's return to England. But the Welsh Guards were reduced to 200 men. Moreover further reductions in the size of the armed forces, accompanied by reduced defence spending, meant that it would not be possible to impose martial law on Palestine as a whole: the security forces were insufficient to maintain law and order. Finally, Palestine was no longer regarded as a worthwhile base area. The British had no viable option but to withdraw. Insurgent terrorism had played a role in persuading the British Government to relinquish the Palestine Mandate. Unwilling and unable to impose a solution by force, embarrassed by the violence and plagued by economic crisis, the British Government sought an honourable exit from the Mandate.

In November 1947 the United Nations decided on partition for Palestine. Although the Jews now amounted to one-third of the population the UN decision gave them half the country (Israel). They were hysterically jubilant while the Arabs were stunned with disbelief.

The British declared that they were not prepared to enforce any policy and, in the absence of an agreed settlement, they would leave Palestine: the Mandate would end on 15 May 1948, but not before, because 260,000 tons of stores had first to be evacuated.

Few sympathized with Britain. 'The Jews are so emotional and the Arabs so difficult to talk to that it is impossible to get anything done,' announced President Truman. 'The British of course have been exceedingly uncooperative.'

Inter-communal violence escalated. On Christmas Day 1947 most servicemen, including the Battalion, remained uncommitted, confined to their camps while Arabs and Jews murdered each other throughout Palestine. Over 60,000 Arabs sought refuge elsewhere. Fifty years later the refugee camps are still there.

Gdsm F Sergeant was then serving in Palestine: 'At Nathanya we had a sandbagged Bren machine-gun position guarding the camp's main entrance. Covering the rear was a watch tower, another Bren, and a search-light which took three minutes to warm up. One night Arabs attacked through the orange groves to the rear. The searchlight did not work and the Bren jammed. They got to the guardroom just as Cpl Phillips was about to change the sentries. After a brief exchange of fire he was hit in the shoulder and later lost his arm through gangrene. The Arabs got away. Our time was spent mostly on road blocks stopping and searching everything: transport was two universal carriers and three ton trucks.'

On 12 January 1948 the Battalion moved south to the Jaffa–Tel Aviv area. Musketry fire could be heard by day and night. Grenadiers were frequently shot at. 'Two platoons were sent to Manshiya, a village about a mile north of Jaffa,' continues Gdsm Sergeant. 'One platoon guarded a petrol line and my platoon a Palestinian police station called Apak. Snipers were very active, mainly coming from the Terhan Bey mosque which, being so high, had a good all round field of fire. They shot at anything that moved around our crossroads. Three of us were always on duty amidst the sand-bags on the roof of our police station, armed with No 4 rifles, a Bren gun and a rifle with telescopic sights. One day while off duty Gdsm Ken Taylor and I crossed the road; we had a little money to spend at the small Jewish shop. We asked our sentry to fire a couple of bursts to keep the Arabs' heads down. It worked all right. On leaving the shop, believe it or not, we completely forgot about the snipers. We got to the middle of the road, side by side, when I remembered them. "Snipers," I shouted to Ken and ran like hell, but he was hit and died on the police post's steps. Sgt W Grant, a Guards Armoured Division World War 2 veteran, looked after his body. The platoon wanted to go in and sort out the mosque but we were told that because it was a place of worship we could only shoot at it. This lack of free hand didn't go down well. We were quickly sent to Jaffa to guard a railway track. Arabs attacked us mainly at night with a few grenades and lots of sniper fire: nothing serious. I believe we were called "Trigger Happy" by the Sherwood Foresters who relieved us.'

'We took over one post from the Arab Legion at the General Officer Commanding's house. Our orders were to open fire on anyone seen on a track in a deep ravine which ran through barbed wire past the back of the house. Early one morning I saw an Arab on the track leading a mule carrying two containers. I fired a couple of short Bren-gun bursts. I never knew what happened to the Arab; the mule turned tail. The following day we received a message from the GOC, Lt Gen GHA MacMillan, commending us on our vigilance, but could we in future let his milkman through. We later moved to Mount Carmel where I was mainly responsible for escorting pregnant Jewish women to maternity hospitals. I enjoyed my stay in Palestine.'

It was planned to move the population of the various prisons, mostly captured terrorists, from one jail to another so that they would be in the appropriate areas when partition came. Divisional HQ kept postponing this, fearing the reaction of the public when it saw the unusual convoys. When it could be put off no longer Maj Frederick was finally given the job with armoured cars and three tonners. 'It was supposed to be a dead secret,' he noted, 'and the lorries had canopies to conceal the nature of the cargo. We called first at the central prison in Jerusalem and took on board the Stern Gang. I was surprised to find how many of this murderous fraternity had evidently been captured over the past year or two. There may have been about 20. They were individually and collectively chained.

'Within a few hundred yards all the canopies had been slashed to ribbons and the prisoners, clearly visible, were exchanging badinage with passers by. There were not too many of these as we took the least frequented route possible. Those that we did pass showed surprisingly little interest and certainly not the violently anti-British reaction that had been feared. But then the bulk of the Jewish population, while wanting to be rid of the British, probably did not hold a particular brief for the more extreme measures of the terrorists.

'We took these people to Athlit where they were greeted with rapture and song by female prisoners. For the next three or four days we took similar convoys back and forth without any trouble.'

One Company Commander regarded his tour as easily the best soldiering he ever had: 'Excellent climate, lovely wild flowers, a spice of danger, but a lot of fun – riding, shooting and trips to Jerusalem and the Holy Places.' In the final months the dangers escalated. Capt GRM Sewell, the Assistant Adjutant, became so exasperated by one Jewish sniper that he took an anti-tank team and demolished the entire position.

On 4 April 1948 the 3rd Battalion moved to Victoria Barracks, Windsor, being replaced in Nathanya by the 1st Battalion, commanded by Lt Col CMF Deakin. The change over of Battalions took place one Company at a time over a three-month period to avoid a whole Battalion of inexperienced troops coming in together.

On the 1st Battalion moving to Haifa, the Jews attacked and captured the town. During the battle No 3 Company (Capt P Freyberg) remained with Battalion Tactical Headquarters on Mount Carmel, the King's Company (Maj AMH Gregory-Hood MC) at a former convent called Carmelite House, while No 2 Company (Maj J Pelham) was divided between the oil refineries and railway marshalling yards. The Battalion's role was to secure the lines of communication to the docks for embarkation. They were guarded by Royal Marine Commandos. The official policy was unchanged – to remain neutral between Jews and Arabs. For three years the British had been ruling in Palestine without any apparent policy, amid turbulence, vilification, assassination and kidnapping. The tolerance and patience of the servicemen were remarkable.

The final withdrawal was under way. In the north two platoons of the Irish Guards made a fighting withdrawal from the north-east of Galilee while surrounded by Arab invaders from across the border. The situation was approaching conventional warfare.

British Comet tanks smashed Jewish and Arab road blocks; houses of both sides were blown up, batteries of Royal Artillery opened fire if necessary, while rifle companies tried to impose a curfew in Haifa. L/Sgt PR Clarke and Gdsm F Howlett were killed in action, as was the Battalion's Greek Cypriot interpreter, Sergeant Dimitriadis. He was ambushed by Arabs who threw grenades into the scout car before machine gunning it. Lt WM Miller, L/Sgt F Allen, Cpl J Phillips, Gdsm M Fairiey and J Jakes were wounded in different actions. Beside the roads stones marked each kilometre. The Stern Gang made a false one, filled it with explosive, disguised it with plaster of paris and blew up a passing Grenadier foot patrol. The Commanding Officer of 1st Battalion Coldstream Guards, Lt Col J Chandos-Pole, was wounded by a Jewish sniper while leading a convoy of ambulances into Haifa to evacuate Arab casualties.

The higher command directed that no further attempt should be made to prevent inter-communal fighting in Haifa. The 1st Division continued to concentrate instead on keeping the roads open to the docks and maintaining some degree of order around their camps. The 1st Battalion's sandbagged positions in buildings were spattered by bullets. The Brigade Commander, Brig JNR Moore DSO, who had taken over from Brig EH Goulburn DSO – both Grenadiers – decided to pull back some of the exposed company positions. The withdrawal by day in three tonners was difficult and at night more so because both sides fired at the vehicles not knowing who they contained. The King's Company was sent to guard Haifa airport, with a Squadron of 4/7 Dragoon Guards in support. The Jews were believed to have bribed three of their tank drivers to hand over their tanks. Capt MS Bayley discovered that the three had driven off at night bursting through the wire. He and his CSM followed in the pitch darkness and found one tank in a ditch; a lone Jewish sentry who came towards them was arrested. Lt PNS Frazer's platoon equipped with carriers surrounded the area. The following day the RAF's Typhoon aircraft hunted unsuccessfully for the missing tanks to shoot them up. Rumour had it that the drivers were later murdered; the Jews probably found the two remaining tanks useful against the Arabs.

1 Guards Brigade were the last troops to leave Palestine. Supported by a squadron of 4/7 Dragoon Guards, the Chestnut Troop Royal Artillery and two Royal Marine Commandos, they carried out their final withdrawal on 30 June in a series of bounds covered by the RAF and Royal Navy.

Maj Gregory-Hood felt that his withdrawal was hilarious: 'The Company was in an isolated position above the docks. Nobody told us that we were to pull out. I was with Michael Bayley wearing a dressing gown sunbathing on the flat roof. The Jews started shooting at us with machine guns. We tried to hide behind a small parapet which was being chipped away by bullets. The Jews shot out the vehicles' tyres so we drove to the docks on flat ones. The Arabs were delightful, rather inefficient but extraordinarily nice. But it was the Jews whom I admired; they were extremely hard-headed and very efficient.'

Shortly after 1 pm the motley collection of ships in the anchorage steamed out of Haifa. The Union Jack had been lowered on shore for the last time, thus ending thirty years of British occupation and of British labours for Palestine. Little had been achieved. The result of being driven out was to weaken Britain's overall strategic position in the Middle East, and that of the Western World generally in the struggle between East and West. By this time India had her independence and the British forces were on the way to being driven out of Egypt. The importance of remaining in Libya was emphasized to Ministers.

Immediately Israel proclaimed itself, it was attacked by Egypt, Syria, Iraq and Jordan. Israel survived. A million Arabs fled from Palestine, settling in Jordan, Lebanon, Syria and the Gaza strip. Many of the refugees felt that their return would only be achieved through the elimination of Zionism. Israel and the Arab world were at war again in 1956 and 1967. Israel hung onto her 1967 conquests including the West Bank (taken from Jordan), the Gaza strip and the Sinai (taken from Egypt) and the Golan Heights (taken from Syria). A fourth war followed in 1973 and led to a peace treaty between Israel and Egypt. The creation of a State of Israel engendered a semblance of Arab unity and created lasting enmity among some Arabs towards Britain and America.

And so it was that the 1st Battalion saw its last of Palestine and sailed for Libya.

CHAPTER 4

THE MALAYAN EMERGENCY

1948 – 1949

On 16 June 1948 the High Commissioner in Malaya, Sir Edward Gent, declared a State of Emergency. Violent crime was rife. The British had lost prestige following their defeat by the Japanese in 1941 and the subsequent blunders of the British Military administration did little to help to restore it. The British Government had decided on a new constitution which reduced the Rulers to mere religious figureheads and their state governments to provincial administrations in which Chinese were to have equal rights with Malays. The Chinese numbered 38% of the five million in Malaya. Every Malayan Ruler and politician had boycotted Sir Edward Gent's installation. In this critical situation the Malayan Communist Party exerted itself and turned to direct action. The arms caches were opened and distributed to the Communist cadres who as the Malayan Peoples Anti-Japanese Army (MPAJA) under Chin Peng had, with British support by Force 136, provided the opposition to the Japanese during the occupation. These cadres had remained largely under cover and now changed their name to the Malayan Races Liberation Army (MRLA). Attacks on isolated planters and on the rubber estates and tin mines increased. The intelligence experts warned repeatedly that a plan for armed revolt was developing. Gent was summoned to London for discussions. Approaching London his aircraft collided with another and he was killed.

Nicol Gray, formerly a Royal Marine, the Inspector General of Police in Palestine, with whom the 3rd Battalion had served, was appointed as Commissioner of Police in Malaya. He was told that he could have any troops he wanted and apparently promptly asked for the 3rd Battalion Grenadiers again. The War Office replied that the Battalion was long overdue for home service; apart from two months in England in 1945, they

30

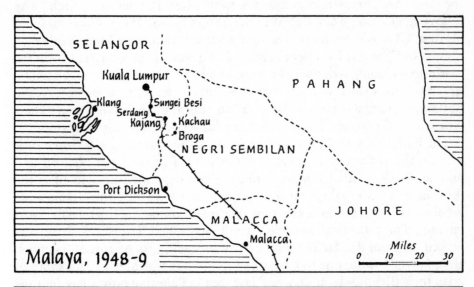

Malaya, 1948-9

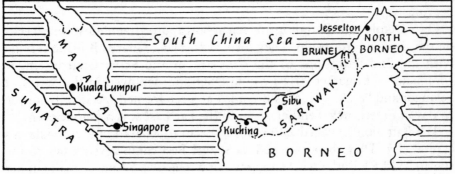

had been abroad since 1942. However, the Commissioner insisted he should have the troops he knew and trusted.

The 3rd Battalion at Victoria Barracks, Windsor, was mobilized on 14 August. 150 reinforcements joined them including Maj DW Fraser who had been commanding the company of trainees at the Training Battalion at Pirbright. He was told to take his three senior platoons with their instructors to form a fourth rifle company for Malaya. Some Grenadiers came straight from the Guards Depot after only 12 weeks' training.

The newly arrived Commanding Officer, Lt Col TFC Winnington, told the Battalion's Padre, the Rev W Temple Bourne and the Adjutant, Capt AN Breitmeyer, to find out how many compassionate cases ought to be left behind. 'I recommended fourteen men who had genuine compassionate reasons for staying with their families,' recalled the Padre. 'Apart from these not a single man ratted apart from the Doctor. On the last night in

31

the Mess we threw him many times through the service hatch. The following morning we marched to the station behind the Regimental Band playing "The old grey mare, she ain't what she used to be" and then a racy version of "The girl I left behind me." There were quite a lot of girls on the platform including half the debs from London. I have never been kissed by so many girls, sisters, wives and mothers in my life.' Prior to departure, the Battalion practised gas drills, weapon training, including firing their weapons and mortars, and parades in field service and marching order.

The Battalion sailed from Liverpool on 10 September 1948. Some officers on the voyage read Spencer Chapman's *The Jungle is Neutral* and another book entitled *Jungle Jottings* which explained how to fight the Japanese. On 15 September it was announced that all the Army's conscripts would have to serve an extension of three months due to international tension. The following day the Battalion reached Tripoli. A draft was picked up from the 1st Battalion. They brought gifts of grapes, peanuts, gin and whisky. But due to a heavy swell nobody was able to get ashore to greet their friends whom they had last seen in Palestine only a few months previously.

On 7 October the Battalion arrived at Singapore after almost four weeks on the crowded troopship. They formed part of 2 Guards Brigade serving with the 2nd Battalions of the Coldstream and Scots Guards, under command of Brigadier MD Erskine DSO and deployed in the States of Selangor and Perak. 2 Guards Brigade had been formed specifically for the Malayan Emergency. Sending them to the Far East was a unique event.

The Battalion travelled 200 miles by train to Kuala Lumpur, Malaya's capital city. They then marched to Sungei Besi camp seven miles to the south. RSM RE Butler was among the first to arrive. Seeing a small gunner with a rifle sitting on a huge mound of tentage, he asked him where the camp was. 'I'm sitting on it,' replied the gunner. There was no building, no electric light, no hardstanding and an unattractive water supply from a great disused tin mine which was said to be the biggest man-made hole in the world. The following day the heavens opened, causing a large number of tents, including the Officers' Mess, to collapse amidst the mud and water. However, the camp was surrounded by pleasant, fertile, green hills on a semi-circular ridge. The tin mine became a lake which later proved popular for swimming.

Before leaving England the Brigade Commander had told the Battalion that their objective was 'to restore confidence and round up bandits. There will be much publicity and so you must do well. Being such a young lot you have much to learn.' On 13 October the Battalion split into sections and did a route march carrying loaded rifles.

They were, like some other units, quite untrained for jungle warfare, although some officers and senior NCOs had attended a course on arrival in Malaya.

The GOC Malaya, Maj Gen CH Boucher, lectured the Battalion. This was followed by a good police briefing on local operations. The police in the very early days were regarded as fairly inactive in the area. They were poorly equipped and their morale was low. Fortunately they were reinforced by some tough, experienced British police who had served in Palestine. Needless to say these were not popular with the old hands who were in any case divided between those who had escaped imprisonment and those who had suffered imprisonment at Changi jail.

It was assessed that the Malayan Races Liberation Army (MRLA) or Communist bandits, as they will continue to be called, consisted of 3,000 to 4,000 armed men supported by 250,000 Min Yuen supporters in the towns and villages. Soon the figure was raised to 5,000. Against them, in late 1948, were some 10,000 police, three Malay, six Gurkha and seven British battalions. The Communists held the initiative and were raiding from their jungle bases almost at will.

The Government needed extra powers to cope with the Chinese squatters who were the main source of support and supply for the bandits, providing food, drugs, information, recruits and money, largely by coercion and extortion. Compulsory registration of the entire population was started. Identity cards were issued bearing both photograph and thumbprints. In November the new High Commissioner, Sir Henry Gurney, introduced regulations permitting deportations to China. Even so, the bandits' attacks continued – setting fire to a railway station, ambushing a train, throwing hand grenades into a newspaper office, killing ten British soldiers, four of whom were swimming, and bayonetting to death three labourers while burning buildings.

British, Gurkha and Malay servicemen did their share of burning too. Patrols finding empty Chinese villages in isolated areas, which may have been sympathetic to bandits, set the shacks on fire. On other occasions mass arrests were made prior to deportation. Notes were left in Chinese reading: 'To the squatters of this area. Some of you may have been away at your daily work when your families, your wives and children were evacuated from here. You may want to know where they are now. You can rejoin them by reporting to the nearest police station. To help you, arrangements can be made for you to accompany your families to China.' Such a policy led to greater, rather than less, support for the bandits who became more aggressive. A raid on the camp at Sungei Besi was expected. The Commanding Officer was among those who slept with a pistol beneath his pillow.

Excitement was caused at 3.0 am on 23 October when a bullet was fired through the camp's water tank, but apparently sentries were firing at thieves. Later that day a company clerk went absent with his rifle, leaving behind communist literature. Majors JC Trotter, HWO Bradley and Capt AN Breitmeyer out shooting green pigeon, saw him dashing across a road and persuaded him to return. The Sergeant of the Guard was somewhat startled seeing the officers armed with shot guns returning with the absentee.

It was shortly after this that the Adjutant, Capt Breitmeyer, was woken from a siesta, after a Sunday curry tiffin in Kuala Lumpur, by the Chief Clerk, ORQMS Euling: 'I think that you ought to see this, Sir,' he said as he handed him a signal which read 'Breitmeyer posted to Paris as ADC to FM Montgomery. Hand over to Sewell forthwith and arrange priority passage.' So Capt Breitmeyer departed with alacrity, after over 2½ years in the appointment, to more salubrious surroundings, and Capt GRM Sewell took over for the remainder of the Battalion's tour.

The MRLA had their own newspaper, *Vanguard*. Issue 17 claimed that: '600 red-hair pigs (British soldiers) who arrived here from Palestine two months back are camping on a hillock near a European mine in S.Besi. According to an MP in London the majority of these soldiers have not handled rifles before. They are sent almost daily to the jungle and rubber estates for training. During the rainy season in the tropical countries, there is a strong sunshine in the forenoon and torrential rain in the afternoon. This kind of change of weather greatly affects those who have not been acclimatized to tropical countries. Due to non-acclimatization and excessive drinking of whisky every one of them has fallen sick of a tropical fever. A clerk in the nearby tin mine said that he had seen about ten soldiers being carried away for burial daily. He added that the epidemic was continuing. The people said that the British soldiers deserved such punishment because they had arrested and burned down houses in many places.'

Had the 'clerk' been watching one parade, attended by the High Commissioner, he would have seen no less than fourteen Grenadiers faint on parade, but they were scarcely buried! The torrential rain did indeed prove a problem. When the Lieutenant Colonel Commanding the Regiment, Col EH Goulburn, visited the Officers' Mess on an inspection the officers had to eat standing up in their mess dress because the roof was leaking.

By December operations were well under way. They mainly consisted of searching Chinese squatter areas, cordoning villages for police searches and patrolling the rubber estates and the jungle.

The Battalion was responsible for supporting the police in the southern half of Selangor which included Kuala Lumpur. One Company, sometimes reinforced by extra platoons from Sungei Besi, was located in the Chinese

Communist school at Kajang. This town was south of the capital astride the main north/south road and railway. It was the headquarters of the Police District. The police there were led by Inspector Ironside with an assistant called Mole.

Most of the information came from the police but there was, initially, a tendency for the police to act on it themselves, rather than involve the Company Commander. This resulted in some abortive operations which might have been successful if properly coordinated. However, this improved because Kajang was a 'hot' area with a bandit gang led by a bearded Chinese called Lieu Kon Kim who was terrorizing both the tappers on the rubber estates and also the local Chinese population, especially the squatters illegally settled on land adjacent to the jungle fringe.

The gangs lived in camps in the jungle usually alongside a stream. The camps were well constructed with living huts, cookhouses and parade grounds. They were protected by a series of outlying sentries covering all the ground. They gave warning of approaching patrols by pulling long ropes of jungle creeper.

The Kajang gang were supported by the Min Yuen Communist organization in the urban areas. These extorted money and food for the bandits who in turn provided the killer squads to terrorize the population.

The police often sent 'Chinese detectives,' interpreters and informers with the Grenadier patrols. A dozen Dyak trackers from Borneo proved very helpful in following tracks left by bandits in the jungle. The Dyaks were small, tattooed all over and invested their savings in gold teeth. They were invariably grinning and surprisingly often made a lot of noise in the jungle. Their English was practically non-existent. In anticipation of an inspection by Brig Erskine, they had been coached by some Guardsmen as to what to say when the Brigadier arrived. On the afternoon of the inspection they were paraded outside their tents and were asked to tell him how well they were getting on. Believing they were saying all the right things, they cheerfully bombarded Brig Erskine with a shrill volume of coarse 'f's and b's' about the food and conditions which had the Brigadier and his party astonished and absolutely dumbstruck.

Short patrols of several days were gradually increased to a week or more. A patrol of 15 men was organized with a reconnaissance group with one Bren gun. It led and had a rifle group in the centre behind them and a Bren gun in the rear.

The order of dress was jungle green bush jacket and trousers tucked into rubber-soled jungle boots laced up at the front, half way to the knee. Patrols wore the webbing introduced before the war – large green packs and front pouches. They carried a poncho cape from which two men could construct a reasonable bivouac. Besides having grenades and ammunition, patrols carried rice and dried fruit, tied in waterproof bags to the mens' belts.

The rifle was the No 5 short barrelled .303 and, besides the Sten gun, an Australian sub-machine gun called the Owen gun was introduced.

Communication was by man-pack 68 set back to Company Head-quarters, but the range was limited on the move and communications were only established when one made camp by setting up a wire aerial.

Movement in the rubber estates was easy, but in the jungle the leading man often had to hack his way with a parang or matchet, making very slow progress. Occasionally one could follow a path made by animals or the bandits, in which case one had to beware of being ambushed.

Rivers were sometimes an obstacle, as was marshy ground. Swamps stank of decaying leaves while thick creepers hung in dense jungle. The hundreds of fallen trees and rotting logs were the worst obstacle: some were submerged, only half-buried beneath the filthy, foul-smelling water.

Hornets stung, while leeches squeezed their way through any opening and clung to the skin until they fell off, gorged with blood, or were persuaded to give up their hold when touched with a lighted cigarette. Red ants fell from the trees and bit, as did the whining, stinging mosquitos. It was surprising that there was not more sickness.

A series of exchanges with the Royal Navy took place. Three Guardsmen spent a week on HMS *Consort* while a Petty Officer and three ratings went into the jungle which they did not enjoy. One fell sick. RSM Butler met the RAMC Doctor in Kuala Lumpur who looked after the Battalion. 'Ah Grenadiers!' complained the exasperated Doctor. 'I've got a Grenadier in my hospital who must be mad because he thinks he's a sailor. He came in suffering from exhaustion after a jungle patrol with other Grenadiers and now says that he must get back to his ship.' RSM Butler had great diffi-culty in persuading the Doctor that the sailor was telling the truth.

Jungle sores were treated by painting the affected areas with what looked like a colourful purple dye. Sgt Chappel who ran the Battalion medical centre painted his patients bright colours: they began to resemble Indians. Officers, NCOs and Guardsmen received strange hues of scarlet, green, purple and mauve, but he discreetly gave the Warrant Officers only a light yellow shade.

In Kuala Lumpur there was a NAAFI, canteen, a swimming pool, two cinemas, a cheap 'Happy World' amusement park, two Chinese theatres, an outdoor stage, a large dance hall and several brothels. Life in England was so austere, it was marvellous getting delicious Indian, Malay and Chinese food. The Coliseum was a favourite haunt of the officers.

Views differed among the Grenadiers as to the friendliness or otherwise of the British civilians in the District. The Commanding Officer found the

British in Kuala Lumpur neither particularly pro the Army, nor anti the Grenadiers. All seemed to him to be oblivious to the Emergency which was threatening to strangle the economy of their country. He asked the Lake Club's committee if his officers could join as honorary members and was told 'no'. The Golf Club was more sympathetic and all their facilities were used instead. Protocol was preserved in the Lake Club. Dinner jackets were appropriate despite the Sten guns and revolvers carried by the planters. Some residents were still suffering from the after-effects of Japanese imprisonment and were hoping to restore pre-independence imperialism. The planters on their threatened estates were much more hospitable.

Due to the humidity and dampness of the Malayan climate, part of the white silk of the Union Jack of the Regimental Colour had rotted away. Capt LAD Harrod gave his silk shirt, a perfect match, to Sgt PR Harvey, the Master Tailor, who repaired the Colour which was laid up in Manchester Cathedral two years later.

While in Malaya, rather to his surprise, Capt Harrod became the Battalion Quartermaster (when Maj H Lucas had to retire owing to ill-health). He produced a ration which could enable patrols to operate for seven days on patrol without resupply.

The Battalion's area of operations was so close to Kuala Lumpur that it was possible to be on a dangerous patrol one night while enjoying an excellent meal and film the following day. Kuala Lumpur seemed unaffected by the Emergency. Capt JW Scott's diary summarizes contrasting events: '30 Oct 1948 to races for day. Golf Club wonderful. Watch polo. 5 Nov: men working hard, keen and interested. 6 Nov: Inkerman Day celebrated. 8 Nov: KL very cheap for shopping. Heavy mess bills. 10 Nov: Reveille 0140 for first bandit operation. Two platoons putting out stops another two will be driving south east of Serdang. Operation a flop. Back at midday. Swimming pool in KL. How lucky we are to be near civilisation. 24 Nov: 50 Grenadiers with 40 men of 4th Hussars and 30 police escort 22 manacled prisoners to Malacca. Escort of two Spitfires for road move. 1 million dollars spent on their camp by the seaside. 90 Grenadiers earmarked for demob. 6 Dec: cordon a village by 0400 to prevent tappers getting out. No shooting but police detain two wanted men. 30% of the men have skin trouble of some sort. 18 Dec: In evening while dining with police hear of information of 30 bandits in Broga area. Had all hoped for Sunday rest as due to search Broga Forest on Monday. However must go.

'Sun 19 Dec: Broga operation a success. Heard shots from Gavin Anderson's group. Some Guardsmen went to ground. I fired from rubber tree killing two myself. Quite close. Under fifty yards. All very quick. Very exciting, but damned lucky. 3 Jan: capture terrorist who was asleep. Evelyn

Fitzherbert's Sten jams and 5 got away. 6 Jan: Maj Gen Boucher promises that troops will move from the edges of the squatter areas and begin their pursuit deeper into the jungle. Large sweep of 400 police and 8 platoons. Nothing seen. Set fire to two houses. Back in evening for cinema show. John Richardson killed one and wounded a second who is being chased by Dyaks.

'14 Jan: rumours of Jap commanding patrol. Suddenly hear Evelyn's lot firing. Investigate with Jock. Two shots near us. Take up the hunt. Soon pass one bandit down. Put a round in his head. Later hear not so badly wounded before, and they'd have got him to talk. Taken away and operated on as he was. Find a camp for 5 just deserted. Destroy. Long Day. 17 Jan: Jock Rowan goes out burning – the squatters move to towns, but will find it difficult making a living there. Getting them out of the countryside is the main target. 26 Jan: Football team suffering from sickness and operations. 31 Jan: area creeping with bandits but not always recognizable.

'7 Feb: Francis Crossley got a known killer with his section. 11 Feb: Electric light arrives in camp, everywhere but the officers' own lines. Bn responsible for training special constables. 19 Feb: News of ER's success: killed a man as he was pulling a pin out of a grenade. 1 Mar: Have not spoken to any girl out here other than a Chinese in the Eastern nightclub. Kings Own Yorkshire Light Infantry Bn has been abroad for 27 years east of Bombay. 7 Mar: Heard Lt John Farrer was killed – due to go to the Depot in 10 days and had got a good bandit 14 days earlier – shot going round his sentries. He was challenged three times and then shot through the head.' Lt Farrer was later mentioned in despatches for his earlier successes – one of some half a dozen operational awards the Battalion was to receive in Malaya.

On 11 March the RAF attacked a camp overlooking the Sungei Jeloh squatter area where 600 squatters had been evacuated under new emergency regulations. They had been assisting the terrorists with food and money. Grenadiers followed up the air strike and found nothing of interest except a few long-deserted huts.

The following day a Grenadier patrol commanded by Capt DW Hargreaves was ambushed in the Sungei Jeloh valley near Kajang, 15 miles south of Kuala Lumpur. His platoon of twenty men was driving up a track which was found to be impassable due to logs blocking the way. The bandits opened heavy automatic fire when the vehicle stopped. Three Guardsmen were killed immediately. Grenades were thrown at them and the bandits fired 2 inch mortar bombs.

'The platoon, having jumped out, fired back, but it was very difficult to see the enemy.' Hargreaves recalls. 'Leaving someone with the Bren gun firing overhead from the vehicle, I charged up the hill with three men, although the scrub was waist high. I could see the flashes of the enemy's fire. I was hit in the right shoulder, the bullet missing my backbone by one eighth of an inch. Behind me three more Guardsmen and Cpl JP Chriscoli MM who had fought at Mareth were hit. The Corporal and one of the Guardsmen died of their wounds. Sgt Black, my platoon Sergeant, made an improvised stretcher and got an ambulance. I was fairly conscious. On getting to hospital I was made to hold a drip in my hand while left in the boiling sun.' It was thought that the bandits numbered up to fifteen although only two were seen. Blood was found in their position which was carefully prepared with protective wiring and well dug in. It was subsequently learnt that one of the enemy had been killed. The dead Guardsmen, J Ryan, A Martin, J Hall, V Herrett were buried with Cpl J Chriscoli in Kuala Lumpur. The two wounded Guardsmen were T Stanley and G Hilman.

Maj AG Heywood, commanding No 2 Company, describes two operations while deployed in the Kajang area four miles south of Sungei Besi. 'One day Robert Heywood-Lonsdale, G2 Intelligence at HQ 2 Gds Bde, came to lunch. Midway through the meal a message arrived from the police saying that they had some "hot" information; a large communist gang of at least twenty members were on a rubber estate nearby. Within minutes a plan had been made: the few Grenadiers available in camp were to be taken in transport to a point north of the estate to try to establish a cordon between it and the primary jungle to which it was thought the bandits would withdraw. The police jungle squads would, after a suitable delay to allow us to get into position, act as beaters to drive the bandits, as we hoped, into our ambushes. We were woefully short of men as the Company was already deployed elsewhere. We gathered at most a platoon and Company HQ so Robert came along too. I dropped them off along a track as best I could until we ran out of manpower. Robert manned a Bren gun beside me at what I hoped was the middle of the cordon. Visibility was about twenty yards to the edge of the estate, and fifty yards to Sgt B Clutton on our left. Gdsm Wood was fifty yards beyond Clutton. For an hour nothing happened. Then between sixteen and twenty men wearing khaki uniforms with packs on their backs were spotted emerging from the estate, heading towards us. Unfortunately Gdsm Wood stood up and shouted to Clutton: "Here come the police". The bandits stopped, chatted among themselves and dived back into cover. Poor Wood, who had killed a bandit himself a week or two previously, had some justification for his action because indeed the police then wore a khaki uniform although not packs.

'Sgt Clutton opened fire on the bandits but sadly our Bren gun could not be brought to bear. Shortly afterwards further bursts of Bren fire were heard from our right, indicating that the bandits were still trying to break out. My dilemma was – what to do? It would be dark shortly. To plunge into the estate to follow up the withdrawing bandits with Clutton, Wood and Robert – and risk a clash with the police beaters with whom we had no radio contact? Or to reinforce to our right in the hope that we might still make further contact? After a short follow-up where I think we found one dead bandit, I moved right to find Gdsm Gatford, a National Serviceman. He had killed with the Bren another bandit who was attempting to break through. The rest must have slipped through the very large gaps. So ended what might have been a really successful ambush where, but for Gdsm Wood's humanitarian instincts, the bandits unwarned would have headed straight for us with Robert raring to have a go! Two bandits eliminated but we might have had a major success.

'On another occasion an informer reported that the leader of the Broga communist group was laid up in a camp above a banana plantation at the edge of the jungle above the village of Broga. He was alleged to have a boil on his bottom and to be immobile. The informer offered to lead us so I assembled a scratch force which included my company 2IC – Capt Lindsey Moorhead. The plan was to move by transport in the dark close to the village. We wanted to arrive at first light within striking distance of the banana plantation where the informer thought he would be able to spot the bandit sentry guarding the camp. It was getting lighter as we climbed up towards the plantation. A running figure suddenly approached us. He turned out to be a Chinese who said that he had escaped from the camp. Through our Chinese interpreter we persuaded him to turn round and lead us back. Reluctantly he did so and took us to a spot where, he said, the sentry was only 100 yards ahead, but he refused to go any further. I split my patrol here into two parties – one under myself and the other under Lindsey Moorhead. We had no means of communication so I gave Lindsey two hours to work his way round to the right flank and to try and establish an ambush or ambushes on any tracks leading out of the banana plantation into the jungle.

'It was a long wait but silent movement was essential. To cut down the risk of my party being heard, I decided to stalk the sentry myself. I left them with orders to follow up as soon as shots were heard. Armed with a No 5 rifle I advanced gingerly, trying to avoid any noise. Suddenly I spotted the sentry. He was wearing a khaki shirt and carried a rifle. He must have been 35 yards away. Taking careful aim, I fired. He disappeared. My party joined me and we advanced cautiously into the clearing on the top of a knoll where he had been standing. No sign of him but I hoped that, if he had run, he would run into Lindsey's ambush. Suddenly at the foot of the knoll a figure in a khaki shirt appeared from behind cover. Thinking it was the sentry, I shot him, only to hear a plaintive cry from Lindsey identifying himself. Indeed Lindsey appeared from behind the same cover. I had shot the Chinese who had said he had escaped from the bandits' camp. A sad mistake but the identity of this Chinese was never discovered. The police thought he was a disillusioned bandit who had decided to surrender. The noise of the shooting caused the other bandits to flee from the camp which, needless to say, was not exactly where the informer had indicated. Subsequently the police confirmed that the sentry had died of wounds – so two bandits were again eliminated, but the leader with his boil had escaped. It must sound very amateurish and almost boy scouting, but in these early days of the Emergency, we had none of the sophisticated equipment such as reliable radios with throat microphones, more suitable weapons and helicopters.'

According to one platoon commander: 'The enemy were better in the jungle than us, and we weren't bad.' The bandits' life was not an easy one. It later transpired that their cooks were usually up at 4.30 am to prepare meals for those such as food collectors who had to leave early. There was often insufficient to eat and everyone slept fully clothed. Unsophisticated military training followed – insufficient ammunition prevented shooting practices: only those who had fought the Japanese were skilled at weapon training. Some form of education such as reading and writing may follow. The bandits' afternoon was spent in repairing their huts, cutting paths, cleaning the areas and listening, perhaps, to a lecture. At about 4 pm they bathed in a stream and mended their clothes. Supper was the big meal of the day, followed by singing for an hour which was regarded as good for morale. At least one sentry was always on duty. Should he open fire, the bandits scattered into the jungle, meeting up at a collecting point.

The Battalion intelligence summaries recorded some successes against them. In the Kajang area alone between 19 May to 19 June 1949 eleven armed bandits were killed, had surrendered or been captured. Nine were wounded and six important Min Yuen organizers captured. Sixty confirmed bandit associates, food suppliers and money collectors were detained. Contacts were usually due to extraordinary luck or excellent intelligence. Many patrols inevitably saw no action. Understandably enough, some police information was faulty. An exhaustive sweep of the Serdang area by three rifle companies resulted in the detention of one rubber tapper who was subsequently released. But the discovery of bandits' documents gradually led to more arrests.

For example in April 1949 2Lt JGC Wilkinson's patrol came under fire near Kachau some twenty miles from Sungei Besi. While the reconnaissance group returned fire he led the other two groups round to a flank. A trail of blood was found not far from the bandits' camp which consisted of native shacks covered by dried leaves in a clearing about fifteen yards square. The principal find was a diary which contained the names of the whole Communist bandit network in the area. Some camps were more extensive. One near Kachau had a messroom, cookhouse and a small armourer's shop.

By early 1949 it seemed that some success was being achieved throughout Malaya in anti-terrorist operations. The number of incidents began to drop. However, it later transpired that the bandits had withdrawn to the deep jungle to regroup and rethink. Realizing that an early mass revolt was no longer possible, their leadership decided upon a more systematic strategy to gain control of selected rural 'liberated' areas by destroying the local government structure, village by village. The North Vietnamese were to adopt the same tactics against the Americans, but the campaigns had little else in common.

It became apparent that as long as the squatters remained outside the effective control of the Government, the bandits would thrive. Unfortunately the question of providing land for squatters who were to be moved was the responsibility of individual State Governments who did not welcome the idea of allotting Malay land to Chinese squatters. Operations had begun in January 1949 which permitted the collective removal and detention of all the inhabitants in a proscribed area. Such operations were, on balance, counter-productive and alienated the entire Chinese population, just at the time when their support was most needed. Lt Col Winnington was horrified at the way the Police treated suspects.

'Initially the operations and general policy were very ham-fisted,' recalls one Grenadier Company Commander. 'It was the fire and sword. We were told that everyone in a Chinese village was suspect. If they weren't in their huts, it meant they had fled and the huts should be burnt as they may be sympathetic to the bandits. There was no 'hearts and minds'. Brig Erskine watched such an operation. On his enquiring why the huts were burnt, he was told that the policy was that unoccupied houses presumably belonged to illegal squatters and so should be destroyed; few clear orders on such matters were received from higher formation: Commanding Officers acted on their own initiative. The Brigadier seemed uneasy at such indiscriminate destruction but did nothing to discourage it, although, with the benefit of hindsight, we know that the Chinese people were being driven into the enemy's hands. The policy was to support Malay supremacy. The Administration was being impelled by planters who thought the British Government was too weak. Another political pressure was from the Malays who believed that any Chinese should be dead.'

But Sir Henry Gurney became determined to resist further police pressure to detain all inhabitants in proscribed areas. The practice had virtually finished by the end of 1949, when the war-winning Briggs Plan for moving the squatters into new villages and giving them the title deeds to land of their own choice was introduced.

In September terrorist action flared up again and continued to escalate throughout the winter. The victory of the communists in China provided a much-needed fillip to the morale of the terrorists in Malaya. But by then the 3rd Battalion had mounted King's Guard at Buckingham Palace for the first time since September 1938. The advance party of the Suffolks had arrived in early July to take over.

'The memory that will stand out most vividly will be that of the fine spirit of comradeship and friendship which existed throughout the Battalion,' wrote Lt Col Winnington. 'We all had a proud feeling that with some 100 bandits accounted for, dead or captured and 54 of their jungle camps destroyed, we had upheld the great tradition of the 3rd Battalion.'

Brig Erskine and Sir Henry Gurney, the High Commissioner for Malaya, were full of praise for the Battalion's successes. Tragically, neither survived the Emergency. On 28 October 1949 the Brigadier's aircraft went missing and was not located until 1954. His body was never found. Perhaps it had been eaten by wild animals.

The High Commissioner had always refused to allow 'a bunch of rebels' to dictate his actions. He insisted on travelling in his official car with the Union Jack flying and only a minimal escort. His aide, as usual, telephoned ahead to make arrangements. On 7 October 1951, as his car came round a bend, a fusillade of shots rang out. His car was disabled and all the accompanying policemen were wounded. Lady Gurney dived to the floor as Sir Henry got out, closed the door and walked to cover, perhaps to draw the bandits' fire away from his wife. He was shot dead at once.

Eventually peace came to Malaya, due largely to the appointment by Sir Winston Churchill of Gen Sir Gerald Templer with the powers of a 'Supremo'. By the powers vested in him he was able to refine the Briggs Plan and ensured that the civil government, the police and the military worked together in close harmony. By his character and leadership he did much to win the hearts and minds – and the confidence – of those living there.

Some Grenadiers returned to Malaya. Among them were Maj AN Breitmeyer who was Military Assistant to Gen Sir Rodney Moore, Chief of the Armed Forces Staff, and Col AG Heywood the Chief of Staff. Breitmeyer and Heywood had been Intelligence Officer and Adjutant respectively when Moore was commanding the 2nd Armoured Battalion in North-West Europe in 1944–1945.

Lt Col PH Haslett returned to Sungei Besi in 1991 and visited one of the small villages where he had operated as a platoon commander in 1949, not far from where a bandit had wounded L/Cpl Smith in the leg before having his head blown off at close range. Lt Col Haslett asked the friendly villagers how they had all got on since the Emergency. They cheerfully replied that they were too young to remember it. That is as it should be.

So at least three members of the 3rd Battalion were happy to return to this beautiful country in which the multiracial society is both cheerful and prosperous – a shining example of the success of the British policy of building a sound infrastructure and political system prior to granting independence.

CHAPTER 5

THE ARMY OF OCCUPATION

AND BRITISH ARMY OF THE RHINE 1945 – 1952

The primary role of the British Army of occupation in Germany in late 1945 was the collection, identification and repatriation of all those – both displaced persons and prisoners of war – who had to be sent back to their countries of origin. The second priority was to avert conditions of famine, disease and civil disturbance which might endanger the army, involve them in police actions or delay the demobilization of those soldiers whose release could now be contemplated.

Much reorganization was necessary. Many Guardsmen had felt deeply about giving up their armoured role and the loss of their tanks. Some kept with pride their Guards Armoured Division black berets. Even so, the date of their release to civilian employment was a major preoccupation. 'Roll on my age and service group' was a common cry. Fortunately the demobilization system, based on points, was very well handled – in contrast to that after the First World War.

*   *   *

One Grenadier who was given an unusual commitment in June 1945 was Maj AJ Shaughnessy who had fought with the 1st Battalion throughout Europe. Before the war he had been a revue producer, writer and composer and so was made the Divisional Entertainments Officer under another Grenadier, Maj N Llewellyn-Davies. 'I resuscitated the German music, opera singers, orchestral players and conductors in our area,' recalls Maj Shaughnessy. 'We put on concerts and operas for the troops and German civilians to share; we staged the first performance in post-war Germany of any work by Wagner. We also put on a big, colourful revue called, after our Guards Divisional sign, *The Eyes Have It*' which I wrote; we found a Gunner jazz band which played in the pit, several professional comedians and a wonderful pre-war stage manager in the Coldstream. The main compere, stand-up comic and star was Brian Johnston MC, later the distinguished broadcaster, who was a Captain in the 2nd Battalion. We found

45

four nightclub dancers in Brussels who joined the cast despite their initially horrified mothers. A left wing poet later criticized us in a book, saying that it was an example of privileged young Guards officers spending money on lavish staging of very ordinary entertainment. It was a clear case of class hatred. We had put it on to cheer up and amuse battle-weary, war-scarred British soldiers; the revue was not intended for itinerant poets following in the wake of a victorious army.'

2ND BATTALION 1945 – 1947

The 2nd Battalion at Siegburg had its three rifle companies split up in neighbouring villages, just as in the war the armoured squadrons had supported units far from Battalion Headquarters. It was felt that a long and slow haul was essential to restore order and discipline within the Battalion. Difficulties were created by the internal separation, no Officers' or Sergeants' Messes, the Battalion's new role, transforming from war to peace, and, above all, major problems of losing highly experienced NCOs to their civilian employment. The 2nd Battalion had always had a high number of senior Sergeant reservists who were very difficult to replace. 3,000 Grenadiers left the three Battalions in Germany during the last months of 1945. The turnover was enormous.

It took many months to get the Battalion feeling like one. A high priority was reintroducing peacetime Battalion routine and concentrating on drill. Bringing men in for Battalion parades was not popular but regarded as essential. Officers lived in flats from which the Germans had been evicted.

In January 1946 the 2nd Battalion moved to a reasonable German barracks in Lübeck which had a large riding school. For the first time for over two years they were together. This became the turning point. Some wings were clipped; strict and harsh discipline proved necessary; spring drills were started. Soldiers were still forbidden, under the fraternization policy, from saying 'Good morning' to Germans. No friendly gesture was permitted. Public opinion had been inflamed by the shocking pictures of naked corpses in the concentration camps, sometimes located not far from complacent well-fed Germans who knew what was going on; they were forced after the war to see the dreadful sights for themselves. Many Germans had been ardent Nazis.

There was very little time for training or exercises due to the important commitment of guarding POW camps. This dominated everything for the first twelve months of peace. The camp at Lübeck contained German generals and admirals who spoke excellent English. They occasionally asked Grenadiers to play bridge with them but this was forbidden.

5. Princess Elizabeth, Colonel of the Regiment 1942-1952, riding Winston prior to the King's Birthday Parade in 1949.

6. Princess Elizabeth inspecting the Corps of Drums 1st Battalion at Prinn Barracks, Tripoli, with, behind her, left to right, Captain M S Bayley, Lieutenant General Sir Frederick 'Boy' Browning and Lieutenant Colonel E J B Nelson.

7. King George VI's coffin, draped with the Royal Standard and surmounted by the Imperial State Crown on a purple velvet cushion, being carried into Westminster Hall, by the Bearer Party of The King's Company. The Bearer Party is led by CSM F J Clutton with Lt A D Y Naylor-Leyland in the rear. Members of the Royal Family watch from the side.

8. The relief changes during the Lying in State at Westminster Hall. The King's Company Colour, the Royal Standard of the Regiment, lay at the foot of the coffin until after the vigil. The Grenadier Officers from left to right were Lt P D Pauncefort-Duncombe, Lt R M O de la Hey, Lt W M Miller, Maj R W Humphreys and Maj A G Heywood.

9. The Queen with the Officers, CSM and CQMS of The Queen's Company after she inspected her Company in 1953.
   Standing: CSM F J Clutton, Second Lieutenant The Viscount Boyne, CQMS C H D Hayes.
   Seated: Lieutenant A D Y Naylor-Leyland, Major A G Heywood, The Queen, Captain P D Pauncefort-Duncombe, Second Lieutenant B H R Hudson-Davies.

10. RQMS G Armstrong DCM wins the Queen's Medal and the Army Championship at Bisley 1955, having won the King's Medal and Army Championship in 1951. The other Grenadiers in the photograph are from left to right: Sergeant G Davies, Major N W Alexander (*at back partly hidden*) and in front CSM R Shepherd.

11. The 1st Battalion undertakes a Commemorative Parade in Bruges, 6th June 1956 to celebrate the 300th anniversary of the formation of the Royal Regiment of Guards.

12. The Queen inspects the 3rd Battalion on the Tercentenary parade at Windsor on 23rd June 1956. Accompanying The Queen are Lieutenant Colonel A M H Gregory-Hood, Colonel Sir Thomas Butler Bt, and General The Lord Jeffreys.

Lt N Hales-Pakenham-Mahon was the camp's Assistant Adjutant: 'It contained General Rancke who had commanded a brigade in Crete. I put him in charge of the fire picquet and he became much less difficult. 300 low level, unimportant people escaped by marching formally up to the main POW gate, saying that they were being transferred. One returned saying that conditions were worse outside than in the camp, so could he be let back in? 'No', he was told. No senior officer escaped. The camp had a very mixed bunch, including Hungarians and Romanians who had fought with the Germans. A Russian liaison officer collected them: they were later liquidated perhaps.

'There was an operation called Fox in which German POWs were to be sent to rebuild London but it was cancelled. I saw displaced persons in the camp wearing only pyjamas. They were very hungry and would have eaten any stray dog straight away. I was later posted to Wuppertal where there was a fire on a hillside one night. I turned out the emergency platoon but a German advised us not to go there as the area was full of unexploded phosphorous bombs dropped the previous year by the RAF. That was why no German fire brigade was on the scene.'

The temptations posed by the black market proved irresistible to some. 'I returned from instructing at the Staff College to the 2nd Battalion at Lübeck in mid 1945,' recalls Maj the Hon Miles Fitzalan Howard MC, later Maj Gen the Duke of Norfolk. 'It is a well worn custom in the Regiment for a new company commander to attend pay parade for several weeks to learn the Guardsmens' names. But CSM White told me, to my surprise, that the men had not been paid for at least six months because, although it was not really allowed, each Guardsman received 50 cigarettes per week which they were selling for Reichsmarks on the black market, and with the Reichsmarks they were easily able to buy their allocation of 150 cigarettes from the NAAFI. The profit the Guardsmen made was put into the Post Office Savings Bank. I understand that the British Government lost about £80 million pounds before this practice was stopped. The officers did not sell their cigarettes on the black market. But it was unofficially agreed that they could use them to tip the German women who cleaned up the houses we lived in, and to purchase fresh vegetables for the mess, and even a haystack or two to feed our horses.

'The Battalion next moved to Neumunster where our main task was to guard about 10,000 German civilian prisoners who were in a brick factory compound. They were being sifted by allied intelligence officers. I often wondered when one saw them in their shabby clothing as to how many were horrible Nazi activists and whether they would ever be discovered.' Also in the camp was the crew of the *Graf Spee* sunk off Montevideo in 1939. Their morale was excellent.

45 years later infantry battalions were apt to remain in the same German garrison for up to six years, whereas in 1945–6 they moved almost every four months to wherever the pressure was greatest.

After Neumunster the 2nd Battalion moved to Berlin, to Wuppertal and then in 1947 to Caterham as related later.

4TH BATTALION 1945 – 1946

The 4th Battalion, commanded initially by Lt Col Lord Tryon DSO, had spent the winter months of 1945 in Euskirchen Barracks, 25 miles south west of Cologne. It was a dreary neighbourhood but there was excellent game shooting for all ranks close by. The biggest shoot ended with the death of a fox and boar shot by Capt JAJ Murray and Maj FE Clifford respectively. As with the other Battalions, nearly all the older men were demobbed, making training and administration very difficult. About twenty officers joined the Battalion from England and so a new 4th Infantry Battalion replaced the old 4th Tank Battalion. The Education Officer ran numerous courses while the Intelligence Officer, Lt PN Railing, was equally busy coordinating searches for wanted Nazis. There was insufficient snow to ski but a few officers ploughed through the slush, including the new Commanding Officer, Lt Col JDC Brownlow, who had forgotten how to stop and ended each run in a cloud of snow and a large ditch. Football matches against German teams were very popular as were the lavish entertainments organized by Captains WT Agar and AJ Davenport. German artists gave shows in the magnificent Battalion theatre. One Sergeants' Mess party ended with a dancing bear.

In March 1946 the 4th Battalion moved to lovely countryside in Schleswig-Holstein to guard and patrol the frontier between the Russian and British zones.

The Red Army soldiers who came from the west of Russia were very friendly – to the extent of everyone firing each others' weapons. The Asiatic Russians, on the other hand, were suspicious and hostile.

Three months later the Battalion moved close to Hamburg to guard some 7,000 Nazis in the ex-concentration camp of Neunengamme. The camp's towers contained searchlights and there were vast coils of barbed wire everywhere to prevent escapes.

Some Grenadiers felt that the relationship with Russians was frosty, while others considered it perfectly amicable. The Adjutant, Capt Murray, and RSM AJ Spratley MM, were arrested by the Russians for being on the wrong side of the street. Language difficulties prevented any interrogation. They ended up being given a very large lunch and plenty to drink.

In the afternoon the RSM gave a demonstration of the slow march which so pleased the Russians that the two were duly released.

The same ambivalent feelings extended to the Germans. Many Grenadiers had seen their closest friends killed in action and were reluctant to forgive, whereas the fighting qualities of the Germans were widely respected: most had fought honourably. 'I did not dislike the Germans in the war,' recalls Maj Frederick, 'and I went all through the Italian campaign – nor can I remember anyone in the line expressing hatred of them. True, the War Office tried to make us hate them – battle schools, visits to slaughter houses, buckets of blood and stirrup pumps. It did not work. My father told me that they tried the same thing in the First War. It didn't work then either. I think the fighting troops often felt more camaraderie with each other than with the base wallahs. I remember a New Zealand officer telling me that they once turned their mortars round and dropped a few shells into their own Divisional Headquarters to liven their ideas up!'

The disbandment of the 4th Battalion – the last non pre-war regular Battalion of the Household Brigade – was always inevitable. The farewell parade was held on 25 October 1946 before moving to Pirbright for final dispersion to other units.

A GUARDSMAN'S REMINISCENCES OF GERMANY

The 1st Battalion moved in 1945 after Victory in Europe Day to Stade, Berlin and Bonn. In February 1946 they moved again – some 350 miles to Kiel where there were ships and displaced person camps to be guarded.

Gdsm F Wright was then serving in the Battalion: 'CSM Pug Cooper warned us of the predatory German women, pointing out the wholesomeness of the Army Territorial Service ladies but we secured ourselves attractive frauleins. The pleasant Kriegsmarine Barracks on the Kiel Canal was almost luxurious after Victoria Barracks, Windsor and Chelsea Barracks. I was detailed to do a 'ship guard' which involved nine of us crossing the Baltic Sea in a beat-up merchant ship to Warnemünde in the Russian Zone to collect German POWs. The town was unbelievably lifeless but our Royal Navy Officer ladled out our 100% proof rum ration in one fell swoop, almost filling our mess tins.

'One night several Russian soldiers quizzed me through an interpreter about the election result when Churchill had been deposed by Attlee. They could not comprehend that this was democracy. One evening they lectured us Guardsmen on the need to use prophylactics. We were a bit nonplussed until we realized they were referring to condoms. Ambition took hold of me back in Kiel and I took the Corporals' course. We practised drilling a squad on a concrete pad which sloped gradually towards the water. Course members were made to stand 400 yards away to drill us. Potential NCOs lacked penetrating voices: consequently, we would march towards the water, ears cocked like rabbits, listening for the about turn which would save us from stomping through the canal. Only the bell-like voice of Drill Sgt 'Dicky' Everett could save us from soggy clothing. He only failed us once.

'We in The King's Company did some field training at Putlos. From there Maj PC Britten, and Lt IHJL Gilmour, later Secretary of State for Defence, took us to Copenhagen to be guests of the Danish Guards Regiment. On the ferry I recall our delight on being served a beautifully prepared giant omelette, and our astonishment on being told it was for only two of us. We thought it would be carved up battalion-style for ten men. We had good beer – it was paradise – The Danes invited us to compete in a rifle shoot so our best marksmen were chosen. The Danes arrived toting enormously long Swedish rifles, like so many Davy Crocketts on a squirrel shoot: as a result of their rifles and all the beer, we lost. On the return journey we sat in the back of a three tonner while Sgt Joe Hickman quoted loudly from *Lady Chatterley's Lover* which he had bought in Copenhagen. We listened in rapt attention.

'Later I was stationed in Neumunster. One of my colleagues was 'Slim' Yates from Manchester who was a messman – a valuable ally for getting extra scoff and other appetizing goodies. At the end of the war Slim was in a battle that captured a German village where he befriended a pretty fraulein who faced raping by soldiers. Slim intervened. At Neumunster he was close to the village where she lived, so I was pressed into carrying a huge wooden suitcase jammed with foodstuffs to a far corner of the barracks where I was to heave it over into the waiting arms of Slim. I lugged the case as instructed, passing the adjutant, Capt AEP Needham, who wished me a pleasant good afternoon. Sweating profusely with exertion and fear, I heaved the bloody case over the wall. As a young Corporal I was jeopardizing my military career, but loyalty meant something along with all that extra grub Slim had fed us. He returned a short time later, very subdued. He later divulged after my constant badgering, that he had located the house of his dream girl. Yes, she had answered the door, so surprised, but not as surprised as the Royal Army Service Corps Sergeant who was staying with her. Slim put the suitcase on the kitchen table, wished them both well and took the lonely road back to barracks.'

GRENADIER WIVES

Within a year of the war ending a few Grenadier wives joined their husbands in Germany. One of them, married to Capt WRCEC Lowe, was hidden in a lavatory while going through the check points on the train to Berlin, being under age. Another wife, Marigold Wedderburn, provides a unique account of those days. Her husband, Capt DMA Wedderburn, was on the staff near Hamburg. 'I had earlier been led to believe by British propaganda that the allies had mainly bombed German communications, marshalling yards and military targets. It was a dreadful shock for which I was completely unprepared in April 1946, on arriving in Hamburg, to find the city a grey, flat wilderness of ruins. It took three hours of shunting before the train could reach the platform. The devastation was far worse than London had suffered. Only the street names in German Gothic lettering were still erect.

'We married families welcomed the opportunity to employ displaced persons such as Latvians, rather than Germans, for they lived in crowded huts, trying desperately to hang on to their pride and identity; the United Nations relief organizations were endeavouring to reunite surviving members of families for eventual repatriation. My two Latvian servants had lost everybody and everything. They had been raped by the Germans going east and Russians coming west. They had been "non-people". One asked me if she could have a table mat which she could cut up to make into a collar for her dress.

'After a week in our large villa which had a grand salon and two Bechstein pianos, I discovered that a German woman was living in a shed at the bottom of the garden; she was the house's evicted owner and had been a concert pianist. So I took her a bar of soap, a much-prized possession then. "I am the wife of the British officer living in your house," I said in my schoolgirl German. The German's resentment boiled over. She spat in my face. Thereafter our only contact was through cigarettes which I put on the back doorstep. They quickly vanished for they were a valuable form of currency.

'Being the oldest Grenadier wife, although still in my twenties, I became the welfare officer for the several dozen others, sorting out their problems. There was only British Forces radio in the way of house entertainment and it was up to the wives to devise their own social life. Some of the senior NCOs wives were very supportive. We had lots of get-togethers and were one big family in a foreign country. One could only shop at the NAAFI, using British forces plastic coinage. Our only connection with the Germans was bartering – giving them a bar of soap in exchange, for example, for having a dress made. To my shame, I did that; it was accepted practice.

51

Even five years later, at Hubbelrath, I gave Herr Weitjens a tin of coffee in return for six dressage lessons; he later trained the German Olympic team.

'The non-fraternization policy was still accepted. I knew that I would be sent home if I told my husband, but when our Grenadier driver took me through Hamburg in the open jeep I often asked him to stop while I nipped down the side streets clutching my well-filled shopping bag, saying that I was going for a walk. I then emptied it, putting tins surreptitiously into half-starved Germans' baskets. I'm quite certain it would have been seen as an act of great disloyalty. It was a dichotomy – a feeling of being so anxious not to be caught and so let the Regiment down. I asked if I could supply the local school with bread and cheese. I was told that I couldn't deliver them to the headmaster because food would be sold by him on the black market, but I could make sandwiches if I watched the children eat them. And so with other Grenadiers we did just that. I had asked the Quartermaster to get extra rations on the 'old boy' net. I expect that he had consulted the Commanding Officer who had agreed. It was all quite extraordinary, being part of a victorious force.

'One of Hamburg's undamaged buildings was the Atlantic Hotel beside a lake. It became one of the many officers' clubs – a little oasis cut off from the city's devastation. It had a dining room covered in pretty brocade and a string band playing operetta music. Our lifestyle, luxurious by post-war British standards, made some of us wives feel uneasy; after all, it was our husbands and brothers who had fought the war, not us.'

Germaine, wife of Capt WGS Tozer, was living in Wuppertal in 1947 in a detached, pleasant house, although its top floor had been demolished after being bombed. Their staff consisted of a German male cook, maid, part-time gardener, plus Gdsm Beresford, the soldier servant. There was no charge for the house or army rations. There was still no NAAFI there. 'One was certainly expected to observe the rules on not participating in the black market,' she remembers. 'But later at Paderborn a few couldn't resist the temptation, for example of giving Germans cigarettes, coffee or chocolate in return for receiving small Meissen china figures of monkeys playing musical instruments, collections of which gradually built up under the envious eyes of those who were following the rules.'

BERLIN 1946

Both the 1st and 2nd Battalions served in Berlin in 1946. Officers shared villas overlooking the lake, looked after by German servants. By then new bridges had been built. All over the city small repairs were being done; the German women were still cleaning up the mess; most of the burned-out tanks had been moved to dumps; the same drab queues were still lining up but now there were far more shops open and brighter displays. Reopening the opera was seen as significant progress, although there was no proper set. The cast was given extra rations to fortify themselves from the cold and hunger.

The Russians in Berlin were very anxious that their men would have no contact with Westerners. They walked in the streets in bodies, rather like school outings. Their officers could be seen in the opera but contact with them was minimal. The Russians were accusing the West of militarizing the Germans. One young German officer POW told Capt JFD Johnston MC: 'Your allies, the Russians, are going to be your enemies. You must rearm us.' 'My reaction was to tell the German to stop talking and that he would be in trouble if he continued,' recalls Capt Johnston, 'but thinking about it later I can see how astute he was.'

Little was seen of the French. The Americans had a good golf course in their zone and so many were met there. Servicemen always wore uniform when walking out although this was not popular. Officers often carried pistols off duty. There was much to enjoy – sailing, riding, all kinds of sport, wildfowling, and shooting where one liked. Eventually one had to have permission to shoot deer, although there were no restrictions on shooting pig. The shooting was free and there was plenty of it because the Germans were not allowed even shotguns. The dead game were sometimes exchanged in the market for chickens and geese.

The 2nd Battalion moved from Neumunster to Berlin in 1946. Maj Fitzalan Howard took trouble to keep his men entertained: 'While in Berlin, I took my company to see round the Potsdam Palaces in the Russian Zone. Senior ranks used to go to the opera in East Berlin. I spoke Yorkshire German which I had learned at Ampleforth. I found it very useful. Although fraternization was still discouraged, both the officers and Guardsmen were beginning to meet Germans and I can remember a small cocktail party in one of the officers' billets, but we were still cold towards them. The Commanding Officer, John Davies, went back to England for Christmas leaving me in command. We were suddenly ordered to leave Berlin and join 2 Division in Wuppertal to take over some very bombed-out barracks from 2nd Battalion South Wales Borderers.'

WUPPERTAL 1946 – 1947

'The barracks, which had earlier housed displaced persons who had stolen everything, still contained German women who had nowhere to go,' continues Maj Fitzalan Howard. 'On one handover between Battalions,

German prostitutes were found in an attic. I nearly persuaded the GOC to let us postpone the move until after Christmas but we had to go on 17 December. It was a great tribute to the Battalion's morale and discipline that all went so well. We had some very experienced company commanders and a magnificent Quartermaster, BH Pratt. The men were wonderful with great inter-company rivalry. Christmas in a battalion was great in those days.

'In February 1947 I chose my company for a commitment to guard the Commander-in-Chief at Ostenwalde. He was Sir Sholto Douglas of the RAF. I discovered that Montgomery had been at Ostenwalde when the war ended. His guard company had been disbanded there. We found armoured cars in the stables, endless mess and a disgraceful muddle which we put right. On our return to Wuppertal I organized a swan to Prague of four officers – myself, Hugo Money-Coutts, Reresby Sitwell and Donald Pearce with four Guardsmen and two jeeps with trailers. This took a lot of planning; we had to get petrol from the Americans as we passed through Frankfurt. We went beyond Prague to the Tatra Mountains and into Poland. We were rather cold-shouldered by the Russians but Czechoslovakia was still friendly. We had sixteen punctures due to the endless horse traffic which dropped nails for our tyres to pick up.'

One highlight of the Battalion's tour at Wuppertal was a battlefield tour of Waterloo. Virtually all the officers and senior NCOs participated. Moreover the Battalion organized a glamorous Waterloo Ball. On other occasions the Battalion's armoured battles of 1945 were studied with interest. Some of the Guards' knocked-out tanks were still there. The Earl of Kimberley enjoyed telling the story of how he 'had been left with Peter Carrington in Brussels to collect the remnants of the 2nd Armoured Battalion's tanks in 1944. We stayed in a very comfortable hotel which was full of every sort of luxury and all for free. After a while, we realized that it was a very high class brothel!' Another officer, Maj T Tufnell MC, recalls his amazement at the large amounts of sugar, food and champagne found in Brussels on its liberation whereas thousands were dying of malnutrition in Holland.

Lt Col GC Gordon Lennox DSO took command of 2nd Battalion in Wuppertal in 1947.

One of the few operational commitments was the rounding up of Yugoslavs and separating them from their German 'wives'. This was done with fixed bayonets in the middle of the night and proved heart-rending. Another time-consuming job was listing all the war material, such as knocked out tanks and guns. Companies were given large areas in which to work.

'By now the Guardsmen were drilling well,' recalls Capt DW Fraser, then the 2nd Battalion's Adjutant. 'Musketry was good; sport was going well.

Geordie Gordon Lennox said, "What about tactical exercises and battalion training?" which we had not had since the war. We therefore started to concentrate on that.

'As in 1918, there was still no general deployment plan or area to rush to in the event of hostilities. There were, instead, quite a lot of tactical exercises without troops – deploying, from the back of jeeps, companies which existed only on paper. We would go on living on Second World War equipment, particularly vehicles and wireless sets, until the late 1950s. The tactical pattern was related to the last war: there was no nuclear, biological and chemical training to impinge on the tactics. The Commanding Officer taught us the night attack; it thereafter remained in my mind as a model on how to do it.'

Sport was to play a notable part in the Regiment's history over the generations. In 1946 the 1st Battalion won the British Army of the Rhine Athletics by 200 points over their nearest rivals. The variety of sport included speedway, riding, rowing and yacht racing. The tug-of-war team, coached by Capt PGA Prescott MC, won the Rhine Army competition and pulled at the Royal Tournament. In England the 1st Battalion had several boxing Army champions and Maj AR Taylor MBE won the Army racquets. The 2nd Battalion athletics team won both the 2nd Division and BAOR championships in 1950, 1951 and 1952. They also won the Division Hunter Trials and BAOR bayonet fencing competitions. Sport was not just for the gladiators. Everyone was encouraged to participate.

In December 1947 the 2nd Battalion moved to Caterham Barracks, Surrey, for only six months.

2ND BATTALION SENNELAGER AND KREFELD
1948 – 1951

In May 1948 the 2nd Battalion was back in Germany with the most unusual role – that of training squads of National Service Junior NCOs from units throughout Germany.

Normally units are responsible for training their own NCOs but the reduction in the length of National Service from two years to 18 months made training very difficult – hence the Battalion's commitment. It was a difficult one because those being trained ranged from Lance Corporal up to Sergeant with varying degrees of knowledge. The short courses covered drill, weapon training and low-level tactics, rather similar to the training undertaken at Pirbright. The Battalion was based at Sennelager, adjacent to the training area – then regarded as the best in the world. The courses were hard work; it was difficult finding sufficient instructors because the establishment was reduced to 319 all ranks, despite having the grand title of 'Leader Training Battalion for Rhine Army'! It was a great compliment having such a role. 1200 NCOs were trained in all. A Coldstream battalion had a similar commitment in England.

It has been suggested that giving these two Guards Battalions such a role enabled them to survive the drastic reduction in the number of infantry battalions from 113 to 72 of which only 24 were to be fully operational. The strength of the army was to be reduced from 527,000 to 339,000 between 1 April 1948 and 1 April 1949.

Money was so scarce that army uniform was difficult to obtain. Some NCOs arrived on the 2nd Battalion's courses with almost none. 'Lt Col Gordon Lennox rang the Army Commander whom he knew,' recalls the then RQMS EC Weaver. 'As a result I was given his authority in writing to get absolutely everything I wanted from the Ordnance Depot. The Brigadier there was horrified but couldn't argue. Every NCO was therefore given a new battle dress suit and a second tailored one for the passing out parade.'

After ten months the Battalion's role reverted to that of a conventional infantry battalion, now under command of Lt Col TP Butler DSO. Battalion exercises were resumed. In April they moved to Krefeld in the Ruhr. Sennelager's open countryside was forsaken for the popular, bright lights of Düsseldorf. The Germans there were friendly, while the shops and restaurants revealed the country's growing prosperity. After two years as the only Guards Battalion in Germany, the First Battalions of the Welsh and Irish Guards arrived, enabling 4th Guards Brigade to be reformed.

On 7 June 1950 the King's Birthday Parade was held in Düsseldorf. The Guards marched through the city centre with fixed bayonets and Colours flying. Thousands watched the parade and were much impressed. It was a beautiful day and perfect setting. The Battalion was regarded as being exceptionally professional.

Germany was a good posting. The status of the soldier was higher there than in Britain. 'In this we inherited something of the traditional status of the German Army,' wrote Maj RH Whitworth MBE in 1950. He was serving in the 2nd Battalion and provides an interesting insight into soldiering in Germany then. 'We knew that, however much attention may be drawn to the Far East by the encroachments of the communist Empire, this is the only place in which the Cold War can be lost in a day. We have ceased to be members of a victorious army occupying a totally defeated and disintegrated Reich, and are now outposts in the defence of Western Europe. No longer are we the guardians of law and order: we are now the defenders of the defenceless against the common enemy – the threat of Russian attack, direct or by proxy through its puppet army, the East

German *"Bereitschaften"*. Eight million refugees from the East now live in the West.

'Western Germany has made prodigious strides towards the re-establishment of her immense industrial and economic growth. People coming from England stagger at the sight of heaped counters. Butter and sugar can be bought amidst the ruins of Düsseldorf: there is no rationing, unlike in England.'

Maj PC Britten had an unusual responsibility. He was serving on the staff of 4th Guards Brigade and had the authority to marry in his office British soldiers to German girls who could not get married in their local church if they were divorcees. His wife did the flowers on his desk to add to the occasion.

Lt Col Gordon Lennox strongly discouraged his men from marrying Germans; every obstacle was put in their way. It was no different for the officers. When Capt WGS Tozer had applied to marry Mlle G Gaysor he received a grudging letter from the Regimental Adjutant which said that the Lieutenant Colonel, Col RBR Colvin DSO, 'views with disfavour regular officers marrying foreigners. . . . After a war of this nature it is even more frowned upon for very obvious reasons. It is undesirable for officers of the professional army to have closer connections with foreigners who might, when war takes place, be enemies of the British nation.' The future Mrs Tozer came from Luxembourg and her father had fought for the allies!

In any event, the German Army was reforming. Drill Sergeant ST Felton was sent with Sgt T Pugh to Osnabruck to drill Germans for their new army. 'They were a very odd selection,' he recalls. 'Ex Air Force and Naval officers, several with Iron Crosses including an ex-submarine commander. We drilled them in English with Tom Pugh translating if necessary. They learnt fast.'

The married warrant officers regarded the quality of life as being very good. The requisitioned German houses were comfortable; rations, augmented by the NAAFI, were delivered by a Grenadier 3 ton truck. They included tinned pineapple, tinned milk and, for some, a chunk of meat still with the sawdust on it. 'If brother officers came to dine,' recalls the then Mrs Marigold Wedderburn, ' a formal four-course dinner was mandatory. The German cook and I were inexperienced and dinner was usually filthy. A brown Windsor tinned soup was followed by meat, perhaps a roast if one was rash enough to try and cook it that way, but usually done as a casserole. Then the tinned pineapple and mostly cheddar cheese with horse biscuits – served by the Grenadier soldier servant with great pomp. All helped by a great deal of German wines and very strong Gordons gin which cost six shillings and eight pence a bottle. Germany was now much more fun.'

The Germans were seen by all to be working hard. Each house had its tiny vegetable patch. Buildings were constructed round the clock – under floodlight at night. Similarly, farmers had lights on their tractors and worked at night too.

The British Army was gradually doing more for its families. Children went to Army schools in Germany and there were good all ranks wives' clubs. The senior ranks did a great deal to keep the younger wives happy. There were coffee mornings and parties in the NAAFI. Almost all Guardsmen were single and so occasionally were invited out by the married. Few could go far afield; hardly anyone had cars which had not been the case immediately after the war when many looted ones were available. Lt LAD Harrod gave an American a Luger pistol in return for a jeep, but he was not allowed to keep it long.

The 2nd Battalion was kept busy with drill almost daily, inspections, and training at all levels involving lots of trench digging. There was sufficient transport and so no excessive marching. Some training was unusual – firing rifles on skis, and crossing rivers in canvas assault boats. The exercises were taken more seriously as the international situation steadily deteriorated.

THE FORMATION OF NATO

In 1946 there had been no clear conception of British strategy in a major war. There was no strong Western bloc which would provide protection from an invasion from the East. There was an urgent need for a political, economic and military union of the Western European and Mediterranean countries. France and the Benelux countries had little military value. In 1948 Russia took over Czechoslovakia. On 24 June they began the blockade of West Berlin. The British Government, now thoroughly alarmed, agreed on 9 July to the need to establish a Western Union defence organization to prepare plans for combined action in case of attack.

The United Nations Organization could not guarantee peace; their proceedings had been constantly blocked by Russia's abuse of her powers of veto. On 4 April 1949 the North Atlantic Teaty was signed. Twelve nations had joined together in a defensive alliance to maintain international peace and security and to promote stability and well-being in the North Atlantic area. The long struggle began to get Western Germany integrated, politically and militarily, into the Western camp. The pace quickened in June 1950 when the Korean War began.

The following month the War Office announced that all voluntary retirement and discharge by purchase were suspended. The period of National Service was increased from 18 months to two years. More battalions in Germany were brought up to strength and the training of the Rhine Army was greatly intensified. Even so, the Officers' Mess always accompanied Battalion Headquarters in the field. It was not unusual seeing the priceless George II silver being tossed by Sgt T Yardley into trucks at the end of the exercise.

In 1951 special orders were received that the 2nd Battalion, still at Krefeld, should be capable of deploying from barracks very quickly. Some vehicles were kept loaded up and quick deployments were tested, although designated points to move to were not reconnoitred.

Also in 1951 the Corps Commander ordered that everyone had to sleep out for three nights each month. The Battalion therefore spent 12–14 December in the woods and snow above the Mohne See. Companies kept warm with route marches around the lake.

On Christmas Eve the Officers met the Sergeants to the accompaniment of thunderflashes and smoke grenades for the annual, traditional football match. The previous year Maj EBM Vaughan hit the ball with both barrels of his shotgun: the game had continued with a rugby ball.

Life in Germany was neither dull nor drab, but some missed the glamour of Public Duties which will be covered in the next chapter. Others, more operationally minded, would have preferred active service in Malaya (Chapter 4), or Libya (Chapter 7). The Battalion had now been in Germany almost continuously since 1945. Few therefore were disappointed when they all moved to Chelsea Barracks in 1952 for several very eventful years.

CHAPTER 6

EVENTS AT HOME

1945 – 1956

The return of Grenadiers from abroad provided splendid excuses for parties. The 3rd Battalion at Hawick near Edinburgh on their return from Austria were honoured by the presence of Princess Elizabeth, the Colonel of the Regiment. 'She loved the Grenadiers because she had been made their Colonel when the Duke of Connaught died when she was only 16 years old,' wrote Maj Gen Sir Julian Gascoigne who later commanded the Household Division.

All Grenadiers over the generations have rejoiced in having such a gracious Colonel and then, when she became Queen, Colonel-in-Chief. Her interest in Grenadier activities, her encouragement and her very presence at great occasions have ensured that Grenadiers have special admiration and affection for her. 'She has an eye like a hawk,' continues Sir Julian. 'She once asked me why the Scots Guards after an Opening of Parliament were wearing their second greatcoats. At the Royal Tournament she asked why the RSM had drawn his sword. On both occasions I hadn't noticed and had to find out.'

It could not have been easy for Princess Elizabeth, at Hawick in 1945, attending a Regimental All Ranks dance which was held in the local drill hall. DSgt RE Butler visited all the breweries and distilleries in Scotland and spun them a long story that whisky had to be provided due to the presence of the Princess. They were all most generous and crates appeared on the big night. The evening was a resounding success, except for the caretaker who tried to close the hall at 10 pm. He was thrown into the nearby river.

* * *

At the war's end the Grenadiers had their own Training Battalion at Windsor and a Holding Battalion at Stobs Camp near Hawick. Both

Battalions were over 1000 strong. The Holding Battalion, commanded by Lt Col HRH Davies and then Lord Ardee (later the Earl of Meath) looked after the flood of Grenadiers home from the war.

Recruits started their training at the Guards Depot, which was then at Caterham. During 18 weeks they learnt drill, musketry, physical training and received some education. They then spent eight weeks at Windsor in the Training Battalion, followed by a month of field training at Bridestowe in Devon. Regimental Training Battalions were replaced in 1946 by Nos 1 and 2 Guards Training Battalions at Windsor and Pirbright respectively. Both had a Grenadier Company. Recruits continued to be trained at the Depot as before but then spent 12 weeks field training at No 2 Battalion at Pirbright which included a camp in Wales. They were then posted to No 1 Battalion at Windsor before being drafted to a service battalion or the Westminster Garrison Battalion.

Maj AR Taylor MBE, who had served as the Adjutant of Grenadier Battalions in North Africa and Italy during the war recalls: 'In 1946 I was giving a lecture on military law at Sandhurst when I was suddenly ordered, half way through, to abandon it and report to Wellington Barracks at once. Apparently the Major General had inspected the previous day three Public Duties teams of composite companies drawn from each Regiment of Foot Guards which formed the Westminster Garrison Battalion. He was so unimpressed he had sacked the Commanding Officer, Adjutant, several Company Commanders and Warrant Officers. A small frog-like creature, dressed as a Lieutenant Colonel in the Grenadiers, met me and said, "I'm Guy Drury. I suppose you realize that we will be the most unpopular men in London District within 48 hours?" Guy and I were appointed to sort out the mess.

'Next morning the Grenadier Company mounted guard at Buckingham Palace. I inspected them. A number of men were marched off in close arrest because their turnout was not up to scratch. The Guard Mounting was very poorly performed. I had no sooner got back to the Orderly Room when the telephone rang: "This is the Lady-in-Waiting to Her Royal Highness The Colonel (Princess Elizabeth). She was displeased with the showing of her Grenadiers in the forecourt." I replied, "Please convey my humble duty to Her Royal Highness, the Colonel and assure her that nothing like that will happen again!"

'There followed a three-week reign of terror. Most officers found themselves confined to barracks until 10 pm. Captains and Subalterns had drill parades at 7 am and 7 pm. Guy Drury then said, "That's enough". We attended the Sergeants' Mess that night and Guy led a conga line round the square. Shortly after, I was best man at his wedding!'

Both the Westminster Garrison Battalion and Windsor's No 1 Battalion were disbanded at the end of 1946, leaving only the Guards Depot, and one training battalion at Pirbright for the whole Brigade. The Holding Battalion, redesignated the Guards Composite Battalion, spent a year on garrison duties in Norway, commanded by Lt Col TP Butler DSO, before also being disbanded.

Grenadiers resumed Public Duties in January 1947 in battledress. The 1st Battalion was stationed in Chelsea Barracks. Public Duties involved guarding Buckingham Palace, St James's Palace, the Tower, the Bank of England at night, and the Central London Recruiting Depot. On moving to Windsor in September the Battalion was joined by HH The Kabaka of Buganda, Mutesa, who had been granted an honorary commission in the Regiment by King George VI. He was nicknamed 'King Freddy' and subsequently became his country's first President after the abolition of the monarchy. He was probably the only coloured person in those years to wear a Grenadier uniform. The very tough reputation of the Guards Depot may have, understandably, deterred some. However, today, with an expanding ethnic population, more have joined and others are expected to do so in the future.

On 20 November 1947 Princess Elizabeth married Prince Philip. The Regimental wedding present, subscribed to by all ranks, consisted of a table-canteen of silver, ivory-handled knives, and also two silver five-branched candelabra, four silver statuettes of Grenadiers, six silver menu holders and two silver grenade table-lighters. The 1st Battalion lined part of the processional route.

The 3rd Battalion replaced the 1st in 1948 in London but moved to Malaya in September (Chapter 4). A year later it was back on Public Duties again, but now in scarlet tunics and bearskins rather than Malaya's jungle green or London's post-war battle dress. Officer's Home Service Dress was at last provided at public expense, rather than having to be paid for by the officer.

To save manpower the Guards Training Battalion at Pirbright between 1948 and 1951 was run largely by a Scots Guards Battalion, rather than by a staff drawn from all Regiments.

By now recruits were spending thirteen weeks training at the Depot at Caterham, and ten weeks at Pirbright on range firing and elementary tactics in the field. (The number of weeks training has continued to change over the years.)

National Service officers formed the overwhelming majority of the platoon commanders throughout the Army. Those seeking commissions in the Brigade of Guards by 1953 served eleven weeks at the Depot and three at Pirbright in a special Brigade Squad representing seven regiments. They were treated the same as recruits, being hustled and harried everywhere.

Earlier service in their school Combined Cadet Forces did little to prepare them for the very tough regime at Caterham. After Pirbright the potential officers went to the War Office Selection Board at Barton Stacey. About three-quarters of the Brigade Squad candidates usually passed. A high proportion of the under-officers at Eaton Hall Officer Cadet School were destined for the Brigade of Guards. During their sixteen weeks at Eaton Hall, National Servicemen were taught to be efficient platoon commanders in battle. After six months training in the ranks the potential regular officers went to the Royal Military Academy Sandhurst for 18 months.

In addition to the frequent guards and fatigues, battalions spent three weeks training at Pirbright or elsewhere in the UK. The fatigues ranged from officers escorting packing-cases full of pound notes to Holland to Guardsmen taking the part of archers in a musical comedy.

By 1950 so many Guards Battalions were overseas that there were insufficient men to line the Mall during the King's Birthday Parade. The Mounted Squadrons of the Household Cavalry resumed participating in the Parade after a gap of fourteen years, to the delight of King George VI. He had earlier told Sir Julian Gascoigne, now commanding the Household Division and London District, that 'I cannot possibly have the Household Cavalry coming under command of the Inspector General of Cavalry: bring them under your command in the same way as the Brigade of Guards.' Sir Julian successfully accomplished this.

Sadly, for the second time in succession, King George VI's health prevented him mounting a horse, and so he attended the Parade in an open semi-state landau pulled by a pair of greys.

On the previous day the 3rd Battalion found the Guard of Honour at Field Marshal Wavell's funeral. It was so hot that the streets' tarmac melted.

Due to the King's ill-health, Princess Elizabeth presented new Colours to the 3rd Battalion at Buckingham Palace in May 1951. The old Colours were laid up in Manchester Cathedral.

The 1st Battalion returned from Tripoli (Chapter 7) in September. After five months at Warley Barracks in Essex they moved to Windsor where they were unpacking and settling in when they heard on Wednesday 6 February 1952 of The King's death at Sandringham. The Union Jack Flag on the Round Tower was immediately hauled down to half-mast. All those on leave were recalled. His death caused the greatest sorrow throughout the country.

The fifteen years of King George's reign were years of strife and stress. War, the threat of war, and the 'cold war', intermingled with economic and social problems, harassed the country and her rulers. Yet during that time The King, together with his Queen, built up throughout the

Commonwealth immense good will and affection. King George always regarded the Guards as his own. They were his Regiments and he enjoyed being with them. During operations he insisted on receiving the casualty lists of the Brigade at the earliest possible moment. When, after the Second World War, it was suggested that the 3rd Battalions of the Grenadier and Coldstream Guards should be disbanded, thus reducing the Brigade of Guards to eight battalions, the King intervened and the battalions were reprieved.

The proclamation of HM Queen Elizabeth II was read by Garter King of Arms in Friary Court on 8 February. 350 men under command of Lt Col EJB Nelson DSO, OBE, MC, lined the route. That same day Lt ADY Naylor-Leyland, CSM FJ Clutton MM, L/Sgt D McMahon with twenty-one men of The King's Company went to Sandringham for duty as bearers and escort. The Queen commanded that The King's Company should continue to be designated as such until further notice.

CSM Clutton remembered all the events which followed with great clarity: 'I was taken to The King's bedroom to see my dead Sovereign in the semi-darkness. Maj Sir Arthur Penn KCVO, MC, a former Grenadier, came along later and told me that Prince Charles and Princess Anne wanted to meet the Guardsmen. I put a tunic on him, although it trailed on the ground. Walking behind him, holding on to the bearskin on his head, we went out to Princess Anne in her pushchair.

'Disaster was averted on a number of occasions. On the rehearsal of carrying the coffin at Sandringham, the studs of our boots made a dreadful screech noise on the marble floor, so we pulled out all the studs. On approaching a low arch all the six-foot-four-inch Grenadiers had to bend their knees low, otherwise the arch would have swept everything off the top of the coffin.

'On 11 February I woke the Warrant Officer of the King's Troop at 5 am so that the horses would be fully exercised. Later, watched by all the Royal Family, we were about to rest the coffin on the gun carriage when a voice in my head said, "Wait, wait." We stood still – just as well – as the horses lurched off prematurely. They were pulled back and all was fine. Perhaps the horses were nervous due to the lament played by the bagpipes, or else they hadn't been exercised enough.'

Meanwhile Maj FA Magnay MVO, the Captain of The King's Company in 1936 and Capt FJ Carver, the CSM in that year, gave advice and produced photographs of the previous King's State Funeral.

The King's body was brought on 11 February by train to King's Cross and placed in Westminster Hall for the Lying-in-State until 15 February. The coffin was guarded by officers of the Household Troops, and members of the King's Bodyguard of the Gentlemen-at-Arms. 300,000 people filed reverently and silently by the catafalque during three cold, wet winter days and nights.

The King's Company Colour, the Royal Standard of the Regiment, lay at the foot of the coffin until after the vigil. It was then taken by the Second Captain, Capt TH Marshall, assisted by CQMS A Chatfield, to Windsor. Meanwhile the Grenadier bearer party accompanied the coffin in the formal procession to Paddington Station.

The 1st Battalion found two Guards of Honour at Windsor – one at the station (Maj RW Humphreys) and the other at St George's Chapel (Maj AG Way MC). The King's Colour was carried by 2Lt BHR Hudson-Davies.

The Escort marched to St George's Chapel in two files, on each side of the Royal Navy drawing the Gun Carriage. Lt WM Miller carried The King's Company Colour between The Captain (Maj AG Heywood MC) and Lt PD Pauncefort-Duncombe who led the files.

The marching Servicemen passed the street-lining party which included three Grenadier half companies commanded by Maj the Lord Worsley and Captains J Wendell and AH Gray respectively.

Maj Heywood recalls that, on arrival in the quadrangle of Windsor Castle from the Long Walk, the troops traditionally lining the route were drawn from Eton College Combined Cadet Force. He had been a member of the Eton College Officer Training Corps, as it was then known, in 1936, and remembers being forced back on the grass by The King's Company Escort marching alongside the Gun Carriage at the funeral of King George V.

There was nearly a crisis over the availability of the small Company Colour of The King's Company which is always buried with the Sovereign. The previous Captain, Maj HC Hanbury MC, had taken the Colour as was customary on leaving the Company. He had had it framed in a fire screen. A new Colour had been ordered but was not ready. The old one was hastily removed from the screen for the funeral.

Inside St George's Chapel Col GC Gordon Lennox DSO, the Lieutenant Colonel Commanding the Regiment, handed the Company Colour to The Queen who placed it upon the coffin immediately before the committal.

On 21 February Queen Elizabeth the Queen Mother received the bearer party at Buckingham Palace to thank them for the exemplary manner in which they had carried out their duties. She presented each man with a signed photograph of them carrying the coffin. On the same day The Queen's command was received that The King's Company was to be redesignated The Queen's Company.

In April The Queen held an inspection of the whole Regiment at Windsor Castle, her 26th Birthday and tenth anniversary of her appointment as Colonel. Despite the heavy rain it was a very memorable day.

The parade was followed by lunch in the Officers' Mess, Victoria Barracks, and attended by The Queen, The Duke of Edinburgh and Princess Margaret. The Captain of The Queen's Company was put in charge of the five-course meal and rather rashly chose salmon for the second course; fortunately with the help of his orderly, Gdsm Carter, Maj Heywood had managed to catch two from the River Wye the previous Sunday. In a letter afterwards Maj Sir Arthur Penn GCVO MC wrote: 'You would I think have felt that your catering triumphs were fully rewarded if you had heard the eulogies by which The Queen and her sister made their mother's mouth water on their return.'

By now the Queen had become Colonel-in-Chief of the Regiment. General Sir George Jeffreys KCB, KCVO, CMG, had succeeded her as Colonel. He later became a baron. He was held in enormous respect and awe by all who knew him.

Maintaining the Public Duties commitment in 1952 proved particularly difficult. There were only three Guards battalions in England, all of which were much under strength. The 1st Battalion had to provide guard mounting at Horse Guards during May which necessitated leaving Windsor at 5 am daily. Guards for Windsor Castle and other state ceremonial were also required. Only five Guards could be provided for the Queen's Birthday parade in June instead of the usual eight.

The international situation was so uncertain that some former National Servicemen were recalled for a fortnight's training, and they had to remain on the Reserve for 3½ years after completing two years with the Colours.

In August the Royal Assent was given to the affiliation of the 1st Battalion Royal Australian Regiment with the Grenadier Guards. The Australian Battalion was then serving with the Commonwealth Division in Korea.

In November the 2nd Battalion returned to Chelsea Barracks from Germany. There were then two Grenadier Battalions in London because the 1st had moved to Wellington Barracks. They were the only Guards Battalions in London – an unusual situation. The 2nd Battalion mounted its first Queen's Guard since 1939 on 30 December 1952.

Both Battalions provided detachments for the Funeral Procession, street lining, Guard of Honour and bearer party for Queen Mary who died in March 1953. The following month The Queen presented a new Queen's Company Colour to her Company in the Quadrangle of Windsor Castle. The parade was very special because only once in the Sovereign's reign does The Queen or King present a new Colour to her or his Company. Five officers and 117 men were on parade on 14 April to receive the Colour. They included 39 National Servicemen. The average height was six feet three inches and that of 1 Platoon, the senior platoon in the British Army, was six feet five inches.

The old Colour was marched off parade. The Colonel, General Jeffreys, then escorted The Queen to the saluting dais where she was received with a Royal Salute. Her Majesty then inspected her Company and presented the new Colour.

She then addressed the parade declaring, 'I now commit this new Colour to your charge, confident you will guard and honour it as the symbol of your loyalty and devotion, and my personal trust in each one of you as a member of the Company which bears my name.'

On 7 May 1953 the 1st and 2nd Battalions received new Colours from The Queen in the garden of Buckingham Palace. Just before 11.30 am they were called to attention by the Lieutenant Colonel to give The Duke of Edinburgh, wearing the uniform of a Field Marshal, the appropriate compliment. Soon afterwards The Queen took her place at the top of the terrace steps. After the inspection, the old Colours of the 1st and 2nd Battalions (carried by 2Lts MKB Colvin, RA Lindsay, BC Beaumont-Nesbitt and MGC Jeffreys) were trooped before being marched off parade in slow time to the hushed strains of Auld Lang Syne.

Then, to herald a new chapter in an unfinished and undimmed history, the consecration of the new Colours followed. The Queen presented them to two subalterns from each Battalion (Lt ADY Naylor-Leyland MVO, 2Lt CHD Everett, Lts JPB Agate and EHL Aubrey-Fletcher).

These two Battalions had not stood side by side since 1944 when they had landed together on the Normandy Beaches and fought their way with the Guards Armoured Division to Germany and victory.

The Queen in her address recalled that she particularly remembered inspecting the 1st and 2nd Battalions in Brighton just before D Day nearly nine years previously. After an Advance in Review Order and three cheers for Her Majesty, the Battalions marched past The Queen and away.

It had been another great parade. The Queen congratulated both Commanding Officers (Lt Cols PW Marsham MBE and C Earle DSO, OBE) and RSMs (LE Burrell and AG Everett) on the standards achieved.

'Some of those present may have recalled that Her Majesty whose grace lent such an unsurpassed distinction to the parade, had worn a grenade for more years than most of whom she addressed,' wrote Sir Arthur Penn. 'And on this occasion, on the eve of her Coronation, there can have been few Grenadiers, past or present, who did not feel towards her a sense of personal devotion which transcended even that which the First Regiment of Foot Guards has always been proud to offer to their Colonel-in-Chief.' The tens of thousands of those who followed in these Grenadiers' footsteps had equal loyalty and affection for their Queen.

THE CORONATION 2 JUNE 1953

The Coronation will be remembered by all those who participated for the rest of their lives. It is unlikely that such a ceremony, involving so many troops from so many different countries, will ever be held in London again.

8,000 men and women marched in the procession. The Regiment played a very prominent part. The 1st Battalion provided a marching party in the procession (commanded by Maj WGS Tozer), while the rest of the Battalion, less The Queen's Company, lined the streets near the Victoria Memorial under command of Maj ECWM Penn. The 2nd Battalion also found a marching party, and lined the streets in the Mall under the command of Maj the Hon GNC Wigram. The 3rd Battalion in the Middle East was represented by a Company under Maj PGA Prescott MC.

The Captain (still Maj Heywood) and the Ensign (2Lt the Viscount Boyne), escorted by CQMS CHD Hayes and Sgt J Walmsley, provided the Colour Party. At 8.30 am on 2 June 1953 they proceeded in slow time up the aisle of Westminster Abbey until they reached the Great Screen under which the Captain and the Ensign, on either side of the aisle, took post facing one another.

Capt Pauncefort-Duncombe with CSM Clutton and fifty men of The Queen's Company were on duty lining the walls of the Annexe which had been specially erected outside the west door of the Abbey. Space in the Annexe was almost insufficient to allow room for the procession. The Guardsmen found it impossible to slope arms because they had to stand so close to the fabric-lined wall. 'Present arms' was therefore, uniquely, carried out from the order. The Guardsmen worked on signals given by the Second Captain with his sword; verbal orders were considered inappropriate. During the service five sentries were posted on the inside of the doors of the vestibule, with two at the Royal entrance and four Lance-Sergeants mounting guards over the Regalia.

Serving and former officers of the Regiment were literally everywhere. Col Winnington, as Field Officer in Brigade Waiting, and the Regimental Adjutant, Maj FJC Bowes-Lyon MC, both rode behind the State Coach. Within the Abbey the Sword of State was carried by the Marquess of Salisbury. Two standards of Quarterings of the Royal Arms were carried by Lord De L'Isle VC and the Earl of Derby respectively. Also present were Brig The Lord Tryon (Keeper of the Privy Purse), Capt the Rt Hon Harry Crookshank MP (Lord Privy Seal), Lt Gen Sir Frederick Browning (Treasurer to the Duke of Edinburgh), Sir Arthur Penn (Treasurer to Queen Elizabeth the Queen Mother), and Maj Gen Sir Allan Adair (Lieutenant of the Yeoman of the Guard). Capt the Earl of Harewood accompanied other members of the Royal Family while the Master of Forbes and Capt Alexander Ramsay bore the coronets of the Princess Royal and Lady Patricia Ramsay. Others carried out the duties of Gold Staff Officers within the Abbey. Such a large involvement is indicative of the service given by serving and retired Grenadiers to their Sovereign throughout the twentieth century.

Lt Col EJB Nelson, serving in HQ London District, had been responsible for organizing much of the procession. After the fireworks display, which had lit up the whole city, and dinner, he walked through St James's Park amidst the cheerful crowds towards the Palace. 'The heroine of the occasion was still apparently utterly unfatigued, appearing from time to time on the balcony of Buckingham Palace, acknowledging the cheers from the enthusiastic and vast assembly,' he wrote later.

The rain earlier in the day had little effect on the brilliance and dignity of the proceedings or on the spirits of the great crowds of thrilled spectators. After many post-war years of rationing and austerity, there was a new atmosphere of optimism and good cheer.

THE TERCENTENARY OF THE GRENADIER GUARDS

In 1656 a body of loyal Cavaliers, who had followed King Charles II abroad after the battle of Worcester, were formed by him at Bruges into his Royal Regiment of Guards, later known as the First Guards and, from 1815, the First or Grenadier Regiment of Foot Guards.

Since 1656 there have been many changes; changes in weapons, in methods of warfare, in dress – to mention only a few. 'But some things remain unchanged,' wrote Gen Lord Jeffreys, KCB, KCVO, CMG in 1956. 'The same loyalty, the same high sense of duty and discipline, the same reliability, have characterized the Regiment – and, indeed, the Brigade of Guards, of which they are so proud to form a part – throughout the three centuries which have followed. In these three centuries, many and varied have been the duties which have fallen to the Regiment, though always in peace-time that of guarding the Person and Palaces of the Sovereign has been its special duty.'

In 1956 the Regiment celebrated its tercentenary. All the main festivities, with the exception of the Review of the Regiment at Windsor, took place in London between 29 May to 3 June. The Exhibition at St James's Palace required the longest planning. Nearly a year was needed to collect the exhibits, commission pictures to be painted, dioramas to be modelled and valuable and historic relics, many no longer in Great Britain, to be traced. 'Maj Hew Hamilton-Dalrymple, the Exhibition Secretary, had a room the size of a monk's cell at Regimental Headquarters and was told to get on with it,' wrote Col Sir Thomas Butler Bt DSO OBE. 'The Queen and Duke of Edinburgh attended the special view of the Exhibition on 28 May. Earlier that same day carpenters were still hammering nails into the scaffolding and half the dummy figures were still half-naked except for their boots and head-dress. However, we opened on time, thanks to Hew's dynamic energy.' A Private View was held two days later attended by 2,000 guests before the Exhibition was open to the public.

The following day the 3rd Battalion provided the Escort for the Colour on the Queen's Birthday Parade while The Queen's Company of the 1st Battalion was similarly employed at the Birthday Parade in Germany. The Colonel of the Regiment, Gen Lord Jeffreys, riding in the Royal Procession, had the almost unique experience of watching his grandson, Mark, on parade as Subaltern of the Escort. (Lord Jeffreys' only son, Capt CJD Jeffreys, had been killed in action while serving in the Regiment in the Second World War. Lord Jeffreys' successor as Colonel was Sir Allan Adair. He, too, lost his only son, Capt DAS Adair, in the war.)

On the evening of the Birthday Parade the serving officers of the Regiment gave a dance at 23 Knightsbridge in the presence of The Queen, The Duke of Edinburgh and Princess Margaret. As the strains of the dance band of the Regiment floated across the ballroom on that warm spring night, the slight figure of our Queen, wearing an exquisite dress and superb jewels, could be seen among the throng of dancers.

June 1 had dawned before the last guests had departed. Only a short while after the sun set on that same day, 400 members of the First Guards Club assembled for their Annual Dinner at the Dorchester Hotel. None of those who were privileged to be present will ever forget the charm, the grace, and the wit with which The Queen proposed the toast of the Regiment. Two guests present (in principle, no guests, apart from the Major General, are ever invited to this dinner) were Field Marshal Earl Alexander and Gen Sir Charles Loyd.

Possibly Saturday 2 June was the busiest day. As part of the celebrations, the Band of the Regiment staged a Musical Pageant at the Festival Hall before a Grenadier audience of 3,000. Maj DW Fraser, the Regimental Adjutant, and Capt AJ Shaughnessy produced the pageant which was entitled 'Once a Grenadier . . .' The background and framework was of course the Band under their Director of Music, Maj FJ Harris MBE, who was widely known for his outstanding contribution to Army music. Other Grenadier Directors of Music have been equally distinguished, in particular Lt Col RB Bashford OBE.

After the glorious and moving pageantry of colour and music, 1,300 serving and past Grenadiers sat down to dinner for one of the biggest banquets given in London since the war.

On the following day, Sunday 3 June, 700 members of the Regiment marched to Horse Guards Parade on Regimental Remembrance Day.

The echoes of these events taking place in England had scarcely faded when the eyes of the Regiment turned to Bruges because the 1st Battalion, based in Germany, paid a two-day visit to the city en route for England. This visit became a magnificent example of international hospitality at its most inspiring.

6 June was a lovely day. The 1st Battalion marched along Langstrasse, past the house where Charles II stayed, and into the vast market place where the Burgomaster, Mr Van Damme, inspected the Battalion. At a luncheon after the parade he spoke most graciously in honour of the Regiment, describing the close ties which have linked Grenadiers to his city since 1656. That afternoon The Queen's Company gave a drill display to music, provided by the Regimental Band. The next morning the Battalion embarked early at Ostend, sailing for Dover, as had Charles II and the Royal Regiment of Guards before them.

For some Grenadiers Ascot week was not quite as carefree as usual. The final rehearsal of the Review by The Queen at Windsor Castle took place on the morning of Gold Cup Day – enabling some to lose their names in the morning and their money in the afternoon.

On Saturday 23 June the entire Regiment marched from Victoria Barracks Windsor by Battalions up the hill towards the East Terrace. No sooner had the 3rd Battalion taken its place on the smooth expanse of green turf when in the distance there appeared a long line of Comrades marching six abreast behind the Band.

Shortly before 11.30 am The Queen, accompanied by Gen Lord Jeffreys, and followed by The Duke of Edinburgh, Queen Elizabeth The Queen Mother and many other members of the Royal Family, appeared on the steps of the East Terrace. The scene which greeted them was impressive. The Regiment was formed up by Battalions in mass, the 1st Battalion on the right in battledress with The Queen's Company Colour, the Royal Standard of the Regiment, in their midst. In the centre was the 2nd Battalion and on the left the 3rd Battalion, both in tunics together with their Colours. Behind all three, the Band, together with the massed Corps of Drums, were six very proud Companies of Comrades (as former members of the Regiment were then called). Each Company of Comrades numbered over 200. They were distinguished by the medals of many campaigns from Omdurman to the final victory medal of the last war.

After the inspection The Queen returned to the saluting base, and the Regiment and Comrades marched past before the Advance in Review Order. Before the 6,000 spectators, 2,300 Grenadiers gave three earth-shaking cheers for Her Majesty the Colonel-in-Chief. So ended a memorable parade the like of which nobody would ever see again.

That same evening on the lawn in front of the Officers' Mess at Victoria Barracks, the Warrant Officers and Sergeants gave a wonderful cocktail party which was honoured by the presence of Queen Elizabeth The Queen Mother, and followed by a brilliant ball attended by 1,000 guests. It had been a wonderful day and was a fitting climax to the Tercentenary of the Grenadier Guards.

CHAPTER 7

MIDDLE EAST KALEIDOSCOPE 1948 – 1956

LIBYA, EGYPT, MALTA AND SUEZ

The retention of base installations in Egypt was regarded since the Second World War as being of critical importance. Without the Suez Canal in peace it would be difficult to maintain communications with India, Australasia and the Far East. In the event of war the bases would be required to maintain the integrity of Egypt and the Arab states: post-war European recovery depended upon the Middle East oil.

Field Marshal Montgomery, about to become Chief of the Imperial General Staff, visited Egypt in June 1946. 'It was necessary to have the use of the Suez Canal and other facilities we needed,' he wrote in his *Memoirs*. 'It followed that there must be maintained in peacetime adequate base facilities for the British forces in Egypt.' He came to the conclusion that, without occupying much of Egypt, the British could still dominate the eastern Mediterranean and be in a position to protect vital interests in war provided that Britain could station forces in Libya and Malta. He also saw the need to keep advanced forces, air and ground, in Cyprus and air forces in Trans-Jordan with full military rights in Palestine. The eventual withdrawal from Palestine increased the importance of Libya and Cyprus, which will be covered in the next chapter.

The 1st Battalion, as recorded in Chapter 3, having withdrawn from Palestine in June 1948, leaving the civil war raging between the Jews and Arabs, arrived at Gialo Barracks, on the outskirts of Tripoli. Gaddafi, Libya's President in the 1980s and 1990s, was bombed by the Americans in the same barracks many years later. The advance party, commanded by Maj PAS Robertson, had arrived a month earlier, spending most of one week quelling Jewish–Arab riots in Tripoli. But no further trouble was expected until the future of the Italian colonies, including Libya, was decided by the United Nations. Meanwhile the British continued to administer the country with King Idris as the nominal ruler.

The 1st Battalion was part of 1 Guards Brigade. 6 Field Regiment, Royal Artillery, and 4/7 Dragoon Guards were in neighbouring towns. The other Guards Battalions were 1st Coldstream and 1st Irish Guards. They were all under command of 1 Division; other brigades were distributed all the way to Egypt. The town of Tripoli occupied the centre of a well-irrigated five-mile semi-circle on the Mediterranean sea, otherwise surrounded by desert. Sixty miles south is the steep Gharyan escarpment marking the northern range of hills up to a thousand feet high. There were, therefore, ideal unrestricted areas for training and plenty of scope for enterprising young officers to disappear with their platoons into the hills and desert.

After settling in and a 'Brigade Flag March' through the centre of Tripoli, the tempo of training increased. All the Companies ran their own training exercises, the Company Commanders being Maj AMH Gregory-Hood MC (The King's Company), Maj the Lord Worsley (No 2), Maj DMA Wedderburn (No 3), Capt WA Spowers (Support) and Capt DAC Rasch (HQ). In due course Maj HC Hanbury MC took over The King's Company, Maj RW Humphreys No 2, Maj AG Way MC No 3 and Maj PR Freyberg MC Support Company.

In February 1949 the weather deteriorated. Snow fell in Tripoli for the first time in living memory. No 2 Company, on a march in the hills, was cut off for 48 hours. Drifts of snow up to 15 feet were recorded. Gunners doing a survey went missing. They were found by the Grenadiers and carried on their shoulders to safety. Some Arabs, who had never seen snow before, curled up and died, claiming that it was the will of Allah.

After the snow had melted the Battalion moved to a camp at Tarhuna, a village 1,500 feet up in the hills for a fortnight's battalion training. The temperature and the flowers were both lovely. The surrounding mountains reverberated to the sound of rifle, gun and mortar fire. The training concluded with an attack on Castel Benito airport which was occupied after an approach march of 50 miles on foot in two days. 'On one exercise,' recalls Maj HW Freeman-Attwood, 'our supporting troop of 6th Field Regiment RA were firing over our heads while we took up a defensive position. One 25 pounder round fell short and hit our position. Geoffrey De Bellaigue, Capt TH Marshall and I were standing together when this round exploded 15 yards from us. Luckily it detonated under a large rock so the blast went sideways. We were all thrown violently to the ground and Geoffrey was hit in the chest by shrapnel, fortunately not seriously.' Geoffrey's nickname was 'Double Egg.' On hearing of the accident, one Guardsman retorted: 'Mr Double Egg has had his chips now.' There was certainly egg on the face of the Gunners!

The operational role of the Battalion was to keep the peace between Arabs and Jews. But there was no fresh disturbance to justify operational deployments.

The Battalion largely consisted of National Servicemen. Rather to their surprise, most settled in cheerfully enough initially, believing that their tour in North Africa would not be a long one. A few extended their engagements, but it was unusual for a National Serviceman to do so.

Battalion routine consisted of a very early drill parade after the 'passion truck' had collected the married men from afar. Breakfast was followed by training, lunch, a siesta and sport. The Commanding Officer, Lt Col CMF Deakin, and the Adjutant, Capt JFD Johnston MC, usually met the RSM, AJ Spratley MM, at 5 pm in the Orderly Room, but life was relaxed, and, as is often the case, enjoyed particularly by the officers and their wives who had their cars, their dinner parties and could afford the leading hotels and casino. Two officers lived with their wives on a yacht in Tripoli harbour. About twenty wives of Non-Commissioned Officers and Guardsmen were with the Battalion, partly in married quarters and less satisfactory hirings.

Marigold Wedderburn, whose recollections of Germany were recorded in Chapter 5, lived in an Italian villa on the outskirts of Tripoli. She decided that she would like to meet Arab women and so was taken to a harem owned by a respectable local potentate. 'The conversation was almost entirely by sign language.' she recalls, 'but women with a common interest have no difficulty in understanding each other. The children were dressed in white satin and had fat stomachs, while their mothers were pretty, plump, heavily veiled and wore silks and large gold jewellery. I suggested that it must be terribly boring – sitting around all day long nibbling sweet-meats. They screamed with laughter, indicating that men were dull creatures, mere stallions, whereas women had such fun chatting together with no responsibility. It was clear that they were all great friends.

'The rooms had marvellous intricately carved screens with brightly coloured mosaics in geometric designs. Later on, there was a terrible famine in North Africa caused by crop failure. Thalia Gordon-Watson, the wife of the Irish Guards Commanding Officer, a few others and I went to the Quartermaster to get tins of milk, biscuits and other rations to feed the Arab nomad mothers and their children who were crowding into the city in search of food. We got together what clothing we could all spare and had extra ones sent out from England. Six Guardsmen accompanied us to a distribution point where we were immediately and alarmingly over-whelmed by massive crowds. Arab mothers had stripped their children, we were later told, and hidden their clothes beneath the rocks to obtain ours.

"What did we tell you?" joked our husbands. It was a terrible shambles, but we believed that there was still scope for individual initiative. It was the tail end of colonial rule, and we thought we could help and make a difference. We tried and failed.

'The wife of the Commanding Officer, Evelyn Deakin, believed that standards should be kept up. Even in Tripoli, when it could be 90° in the shade, officers' wives had to set an example by wearing gloves and hats in church, which was, of course, obligatory. There was still a strong feeling of pride in the Regiment which we regarded as a family we were privileged to be part of. We had a wonderful choir, coffee mornings for all wives and a children's clinic. As wives and mothers in a fairly primitive foreign station we were all great friends, only observing the niceties of status when the occasion demanded.'

In the summer of 1949 the Battalion moved to Prinn, to the east of Tripoli. 'Air training' followed. Not many Guardsmen had ever flown, and were unlikely to do so in the C-47 aircraft which was parked in the barracks: it had no engine. Although there were a few scattered, burnt out tanks left amidst decaying trenches of the Second World War, no real tank was seen until 4 Royal Tank Regiment sent a squadron to exercise with companies for two days. All-arms training, with artillery, engineers and armour, continued to be almost unknown outside Germany at that time.

The Battalion's second year was similar to the first, although the exercises were now at Brigade and Divisional level. Princess Elizabeth, the Colonel, again sent out Christmas puddings. The Battalion donkey race opened the Christmas 1949 festivities, followed by an evening of tombola. On Christmas Eve a great victory was won at soccer over the Coldstream. The ceremony of 'the brick' had been completed as usual, in the Sergeants' Mess. The Officers' and Sergeants' Mess football match had ended in a draw despite the Sergeants' use of barbed wire and smoke grenades.

Sport continued to be important as Gdsm Joe ('Tiny') Harding, now living in Ontario, recalls: 'My days of real soldiering ended in April 1945 when I was wounded liberating Sandbostel concentration camp and Fred Clutton won his Military Medal. Every Guardsman always found another Guardsman in those hospitals and convalescent depots; we stuck together. I was posted to the 4th Battalion where Fred was my CSM. He told me to teach yachting. I didn't know the front end of a boat from the rear. We were located on the Ratzeburger See in Germany; the days were not long enough for me. I rejoined The King's Company when the Battalion had challenged the Irish Guards to a boxing match and lacked a heavyweight. I was persuaded to "have a go" and was then struck off, the rest of my service being devoted to sport. On returning to England we probably had the strongest boxing team in the country. The 1st Battalion had three of us who were invited to train with the Olympic team: Sgt Jack Daniels, Gdsm Tony Lord and myself. Later Sgt Jack Gardner represented Britain in the 1948 Olympics.

'The athletes in the Battalion in Tripoli were fantastic. Sgt A Furneaux was the best sprinter in North Africa: our soccer team took all the honours. I was told to form a boxing team although some of the men hadn't laced on a pair of gloves before, but the team won every championship in North Africa.

'We were invited to fight a Combined Services team in Malta who could choose from the Navy, Army and Air Force. The day of the boxing match happened to be the birthday of Princess Elizabeth, our Colonel. Our boxers chipped in and we raised enough to buy a bouquet of flowers for our future Queen. I was delegated to present the flowers. That was the easy part. I was also expected to make a short speech. I wrote one which Maj Wedderburn approved and I practised it faithfully until I had it word perfect.

'The presentation was at a polo ground in Valetta where Prince Philip was playing polo. I had spent a lot of time in speech therapy after being wounded, but still stuttered a bit. When the moment came, I was introduced to Princess Elizabeth and, instead of the long speech I had practised, all I could manage was; On behalf, behalf, behalf, – Happy Birthday!' The funny thing was that I was not booked for being idle.

'I played rugby for the Battalion with Capt Michael Bayley and Lt PNS Frazer on a pitch converted out of an olive grove. Despite the casualties, all the Companies had their teams. A new speedway track led to the Battalion's team, the Hellcats, having a successful season; the cross-country team had decisive victories.

'As you see, my stories relate to sporting events: if you did well in any sporting event, the Regiment looked after you. But perhaps we didn't have everything our own way in Tripoli. One morning I erected a huge marquee, for a cricket match. We took an hour off for "Tiffin" and when we got back it had been stolen: even the footprints in the sand had been brushed out.' Basketball, water polo, motor cycling, cricket, sailing and riding also proved popular sports.

Tripoli even had its own Hunt. Lt MGP Stourton was the Master. The hounds had come from Palestine where they had hunted jackal. They did not seem to mind chasing foxes instead – perhaps because they rarely caught either. Jumping cactus fences was fun and the Arab ponies loved it. Not only did the Guardsmen enjoy hunting but they also shot sand grouse. Dinghy sailing was also popular, while bathing in the sea and sun off the rocks and beaches appealed to many. Nearly everyone of all ages acquired snorkels and discovered a new world inhabited by highly coloured fish.

There were other opportunities for those who did not enjoy sport. Many Guardsmen went to Tunis for ten days on a leave scheme run with the generous assistance of the French Army. Numerous Grenadiers visiting the Second World War battlefields in North Africa of the 3rd, 5th and 6th Battalions, laying wreaths on their fallen comrades' graves. The Commanding Officer drove across the desert in a Rolls Royce to visit the battlefields, with the car door open so that he would be blown out, rather than killed, should they go over a mine.

On Col Goulburn's visit from Regimental Headquarters, a convoy set off for Mareth where the 6th Battalion had so greatly distinguished itself. He was amazed, in the midst of the desert, to find tents awaiting him containing tables covered by white table cloths, flowers and the mess silver – all flown up by Air observers in Auster aircraft. The tents' walls were hung with Arab rugs.

On a later occasion Grenadier Hill at Medjez-El-Bab was visited. Although it was seven years after the battle, those who had fought there were fascinated to see the tracks of three German tanks which had approached their position at Grenadier Hill. The soil had been compressed by the tanks' tracks; as a result the texture of the crops was different and photographs revealed the route which the tanks had taken.

There were many other expeditions. Maj Freeman-Attwood planned a desert rescue operation with a group of Grenadiers: 'The Battalion's role was to be able to rescue a possible air crash along the British Overseas Airways Corporation route from Tripoli to Cairo or Tripoli to Khartoum in an emergency. We were therefore training for deep desert penetration which involved advanced desert navigation by both sun compass and theodolite, and gaining experience of driving over the various types of sand which exist between the coast and the Great Sand Sea. On approaching Gialo we were actually shot at from a distance; the offender thought that we were Germans! We were taken to the tent of the Sheik who spoke some English.'

In August 1950 the Battalion visited Malta to take part with 1st Bn Coldstream in a large-scale exercise. They had many friends among the ships' companies of the Mediterranean Fleet, and a break from Tripoli was most welcome. The advance party, which included twenty families, sailed on 5 August but the weather was bad and the voyage dreadful. The main body left five days later. The sea was now calm, and the men slept on deck and fed better than they had ever done in the Army. On arrival at Valetta everyone enjoyed themselves: the shops never seemed to close; every other building was a bar; the Military Police were indulgent and contained a high proportion of ex-Guardsmen among their ranks.

Preparations for the high-level exercise had begun well before. The

13/14. In 1956 the Regiment celebrated its Tercentenary. After a glorious and moving Pageant of colour and music, 1,300 serving and past Grenadiers sat down to dinner on 2nd June for one of the biggest banquets given in London since the 1939–45 War.

15. The 1st Battalion march through the City after returning from Hubblelrath in October 1957. The Adjutant, Captain B C Gordon Lennox, is followed be RSM L C Drouet, Major M S Bayley, and left to right, Second Lieutenant D O J S Lort-Phillips, Lieutenant R M Micklethwait, Lieutenant J P Smiley, Major L M Moorehead. The CSM on the left is C H D Hayes.

16. Major P J C Ratcliffe with Guardsmen and a donkey-mounted 62 set during Operation Springtime with the 3rd Battalion in Cyprus 1958.

Guards were to 'defend' the Island against an 'invasion' by the United States Marines transported by the American Sixth Fleet. On 17 August the Battalion dug in on a ridge above St Paul's Bay. There was tremendous 'enemy' air activity and innumerable small Royal Marine craft were placed under the Battalion's command. At dawn on the third day the long-expected assault began. The ships of a prodigious task force thronged the shores. Reports from observation posts began to arrive. The directing staff had removed all defensive positions from the northern bays. News came from there of landings by an enemy in a strange uniform. As dawn broke a rather exhausted group of them climbed the rocks. Alas! Instead of being the Americans, they were friends from the Royal Malta Artillery who had been taken from their guns, searchlights and radar to become the enemy. The American Fleet and Marines had been ordered away at the last moment to support the integrity of Jordan, currently under pressure from various neighbours.

The Battalion, now under command of Lt Col EJB Nelson DSO MC, returned to barracks for a last glorious weekend. Nelson was a great trainer according to CSM FJ Clutton MM: 'The umpires on exercises dreaded him because he would cheat to win, in the nicest possible way. "If you could get over that hill by 5 am tomorrow," he would tell us, "you can cut off the Coldstream enemy and finish the exercise a day early." The hill would be a distant spot on the horizon. Getting to it entailed a forced night march carrying full kit, but we were invariably successful.

'No wonder the umpires always wanted the Commanding Officer to be on their team! After drill and breakfast the companies marched down to the beaches to play cricket or shoot at balloons with the Bren guns. On exercises we built sangars and stuck figure targets among them. (Sangars are shallow trenches protected by rocks.) Then after 'tiffin' of sardines or cheese sandwiches and a swim, we would fire at the sangars and drop two-inch mortar bombs on them, before inspecting the damage. We then marched back to barracks, with the CSM shouting arms drill movements on the march to sharpen men up. The large Corps of Drums met us at an RV and played us in.

'There were plenty of section and platoon attacks, live firing, long forced marches against the clock and company camps by the sea. Support Company got some .22 rifles to shoot birds and hit instead a popular Arab light-house keeper who was having a siesta. Four Grenadiers attended the funeral, carrying concealed weapons in case they were attacked. I saw half a dozen wailing wives and the dead Arab's body being tossed into the grave; the way the body ended up, we were told, indicated whether the dead man was going to Allah or elsewhere.'

One of the happiest days occurred when the Colonel, Princess Elizabeth, flew over from Malta to inspect them. It was a great honour to be visited abroad. She was accompanied by Princess Margaret and Lt Gen Sir Frederick Browning, a very distinguished Grenadier. She had succeeded the late Duke of Connaught in 1942. All Grenadiers over the generations have been very proud of the genuine interest she always showed in all the members of the Regiment. Princess Elizabeth inspected the Battalion on parade and visited the Sergeants' Mess and had lunch with the officers. Both Princesses also met Grenadier families in the YMCA, before driving round barracks. In 1951 Prince Philip was commanding the frigate HMS *Magpie* and sailed into Tripoli harbour on a courtesy visit giving impeccable hospitality on board to many of the British garrison.

A visit by the then Vice Admiral Lord Mountbatten was memorable for different reasons. He was visiting Tripoli in October 1950 and had Grenadiers on board his Cruiser Squadron when a small tornado hit the city, causing the ships to sail with the unfortunate soldiers still on board.

The Chief of Police in Zavia was a former Lance-Sergeant of the 3rd Battalion from Palestine. He told the Commanding Officer that if he wanted any labour he had but to ask. The Battalion wanted a sports ground and so the former Grenadier brought Arabs by the truckload to build the sports centre. Other Arabs were quite happy to work for the food and no pay.

By now the Battalion had been abroad for more than three years including active service in Palestine. They had been led to believe that they would only have to serve a few months in Tripoli and were wondering how long it would be before the promises made to them would be fulfilled. 'Time and again hopes had been dashed with a consequent lowering of morale,' wrote Lt Col Nelson later. 'We were to be relieved by the 3rd Battalion Grenadiers in June 1951 and although it appeared on time, its arrival coincided with a crisis in Iran caused by a threatened nationalization of the Abadan oil installations.

'1 Guards Brigade were alerted to fly in to prevent this occurrence. As that area was one of the hottest in the world, the 3rd Battalion were to be allowed three months of acclimatization.' The 1st Battalion was therefore ordered to remain alert and in Libya.

'In those days, after three years of foreign service, any British soldier was entitled to return to the United Kingdom. Their anxiety for further service in North Africa naturally increased. By the time the expedition to Iran was called off and Mossadeq left in charge of the Anglo-Iranian property, the 1st Battalion was disenchanted with waiting to return to England.

'When we heard that the Commander-in-Chief, Gen Sir Brian Robertson, was coming to say goodbye to us on parade, and still no date fixed for our departure, we found this a little hard to comprehend,' Nelson wrote. Three men purposely dropped their rifles when ordered to slope arms prior to the Adjutant's inspection. RSM LE Burrell had them marched off parade and in front of the Commanding Officer.

'I discovered,' Lt Col Nelson continues, 'that quite a few Guardsmen had agreed, as a form of protest, all to drop their rifles. I subsequently talked to each company separately and encouraged everyone to speak up and grumble for a good half hour while I was with them. This seemed to have the desired effect and when Gen Robertson's visit came the parade was as smart and impressive as it should have been.

'I asked the General beforehand not to say goodbye but just to explain the background for our delayed departure. He did so to such effect that we all felt, quite unjustifiably, very pleased with ourselves. I was very fortunate in having Leslie Burrell as my RSM. He was outstanding even among the very high calibre Warrant Officers which the Household Division has always produced. Bill Nash was the senior Drill Sergeant. He had had an exceptionally gallant war record and was a wonderful friend and adviser.

'I was also lucky to have Roger Humphreys as a Company Commander who took endless trouble to keep in close sympathy with every man under his command. The summer of 1951 dragged on. The dreaded southerly winds, which blow for five days at a time, came and went, bringing with them sand storms and frayed tempers. Indigenous Tripolitanians say that murdering a wife does not count during the winds because no one can be expected to be reasonable under stress while the hot, sandy, relentless wind evaporates all self-control.'

The 1st Battalion left Tripoli in September 1951, moving to Warley Barracks in Essex.

3RD BATTALION IN TRIPOLI AND EGYPT

The 3rd Battalion, commanded by Lt Col PAS Robertson, arrived in Tripoli on 30 July 1951. 'The Company Commanders were Bobby Steele (No1), Tim Bradley (No2), me (No3), Jock Rowan (Support) and Charlie Frederick (HQ Company),' wrote Maj PGA Prescott MC. 'This command structure was a very strong and experienced team. The Battalion quickly settled down and became a very effective, if unconventional force, with the Company Commanders showing strong leadership in their individual styles; yet when required the Battalion operated together with enormous joint enterprise and elan. Having served in all three Battalions, I felt that the 1st was the smart "cream", the 2nd was the solidly professional, appropriately nicknamed "the Models" and the 3rd was just a bit mad, but efficient with the flare of individuality.

'We were stationed in a barracks built by the Italians in Zavia near Tripoli. It had been unoccupied for a while and had drifts of sand in passages and barrack rooms when we arrived. However, it was a good compact place with modern barrack blocks. Zavia itself offered little in the way of recreation. Fortunately, the numerous beaches on the coast not far away offered wonderful swimming for all ranks, and Tripoli was sufficiently close to provide a good day's leave in recreational transport. We also made good friends of the 4/7 Dragoon Guards further west up the coast from us. Sport didn't play a great part in our life, but we were rather pleased to beat the 1st Battalion at cricket who considered themselves the champions. (I recall breaking a window in their Orderly Room block with a six clean over the bowler's head!).

'We trained very hard for our role for the protection of Abadan. Dehydration (called "sun-stroke") was our particular enemy, and eventually we carried two water bottles until we got acclimatized to the needs of severe water discipline in the desert. We, also, had *our* way of defeating umpires; during a Brigade exercise in which the 3rd Battalion was to attack Castel Benito airport, held by the rest of the Brigade, a left flanking attack by one rifle company, having cut its way through the very permanent perimeter wire, caused total confusion and was said to be "unfair" by the 1st Battalion Highland Light Infantry who got caught looking the other way. We also did much training with the RAF Transport Command at Castel Benito airport.

'The US Air Force Base at Wheelus Field was a typical overseas American enclave, but it offered numerous delights to gamblers with poker schools, one-armed bandit machines, craps and many others. We, impecunious Brits, although experienced gamblers, were reluctant to sit down at the poker tables with "eye-shaded" USAAF officers with dollars galore. We used to take over two one-armed bandits instead and feed them until they spewed out the jackpot. Julio's Restaurant in Tripoli became famous to the Guards Brigade as the best there – Julio himself became honorary cook to the Officers' Mess of the Coldstream Battalion, and used to accompany them on exercises dressed as a Guardsman. We were warned never to go into the kitchen of his restaurant or we'd be put off going there for ever by the cockroaches. The food was, nevertheless, always superb so we never bothered with the kitchen!

'Soon after the troubles began in Cairo the 3rd Battalion with the Highland Light Infantry embarked in HMS *Sheffield* and HMS *Manxman* to meet up with 13/18 Hussars who were already based at El Adem.

'When we were in Tobruk movement in the desert off the roads and tracks could be hazardous because of the minefields still extant since the war. It was not unusual to see Arabs minus a hand or a large part of an arm, and to hear the thud of an explosion in the desert: the results of scavengers trying to boil the explosive out of a mine or shell, or remove the copper driving bands from shells which could earn them good money. The loss of a hand, arm or foot was often greeted with a shrug of the shoulders and "Allah's will".

'We were tasked to drive as fast as possible to Alexandria and Cairo via the coast road if need be,' continues Maj Prescott. 'The role of No 3 Company with a troop of Hussars was to capture the mole of Alexandria harbour. Air photos provided by RAF El Adem showed a number of gun emplacements along the mole. Neither the Troop Leader nor I were much comforted to hear that whereas they were 14 inch guns, they only pointed out to sea. I have always been somewhat relieved that the operation never took place, however much fun it would have been to follow in the steps of Rommel.

'We flew from El Adem to Fayid in the Canal Zone, eventually to be billeted by Companies in whatever accommodation could be found for us adjacent to Tel-el-Kebir. No 3 Company got the best, a disused NAAFI shop. This didn't last long, however, and we were soon concentrated in a tented camp within the perimeter of the vast Ordnance Depot of Tel-el-Kebir. It was very basic with no electric light until our Senior Major, Charles Earle, after a protracted and alcoholic lunch with the Depot Commander, got him to sign a release note for a vast generator which lit the whole camp.'

*   *   *

Had Tel-el-Kebir not been a Battle Honour, few would have heard of it. The sprawling, impoverished town is adjacent to the Cairo-Ismailia railway and Sweet Water Canal. It has nothing of interest apart from a small, pleasant, Commonwealth Cemetery. A Grenadier, visiting Tel-el-Kebir precisely fifty years later, found the Egyptians there were still not the friendliest of people.

Egypt in the 1930s and 1940s has been described as a three-legged stool – the King, the Wafd political party and the British. The three legs might be uneven, the stool might rock and tilt, but somehow it stayed up. Year by year, after the Egyptian army brooded over its ignominious defeats by the Israelis, the King, whose reputation and authority were at a low ebb, became the scapegoat; there was certainly corruption in high places. By 1951 the Wafd had been re-elected with a large majority. The party was at the height of its power, although the economic situation was fast deteriorating and the Government could not cope. They chose the remedy of history – foreign quarrels.

In October 1951 the Egyptian Prime Minister denounced the 1936

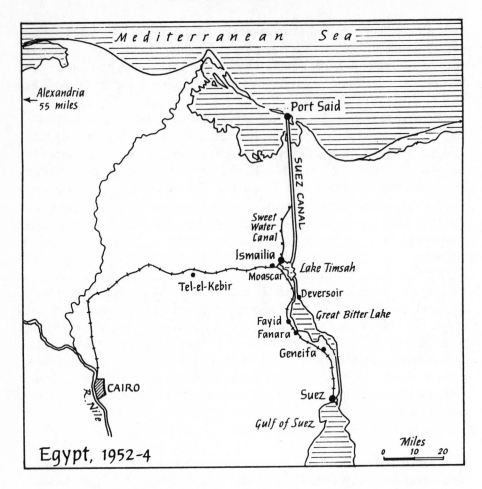

Egypt, 1952-4

Anglo-Egyptian treaty which still had five years to run. Even before this, terrorist activity was growing, particularly in the Canal Zone. It was encouraged by a violently extremist press. 'Liberation squads' extorted subscriptions and forced thousands to stop working for the British forces although there was little work elsewhere. The Middle East base installations were still regarded as a vital element in Imperial strategy. They stretched from Port Said on the Mediterranean coast to south of Suez, and inland to Tel-el-Kebir.

The bases depended on Egyptian labour to keep them functioning, on the Egyptian port facilities and railways for transportation, on the Sweet Water Canal for their water supply and on the Egyptian police for the maintenance of law and order. The British forces in the Canal Zone, which then amounted to little over two infantry Brigades, were therefore lodgers in a foreign and unfriendly land.

There was no time for the 3rd Battalion to enjoy themselves. Three days after their arrival they were given 48 hours in which to draw up and issue all their equipment and move out to a Divisional five-day exercise south of Geneifa. On their return to Tel-el-Kebir they had to take over twenty-eight different guards and duties for ten days, the principal one being the huge ordnance depot containing the major part of the ammunition and equipment for the Middle East forces. Its perimeter was 25 miles in length, necessitating 280 sentries day and night on patrol and in watch towers. Despite searchlights, dogs and anti-personnel mines, equipment was still stolen.

At the Abu Sultan camp every evening the outer perimeter track was swept by a vehicle pulling a mat so that a patrol within the wire could see any footprints. Egyptian thieves covered themselves in grease and rolled in the sand to camouflage themselves. East Africans from Uganda and Kenya also guarded the camp, as did smart Sudanese, who were in the Egyptian army. The contents of the depots were fairly chaotic judging by one officer's attempts to find mortar ammunition amid the thousands of huts. The designated bunker was found to be full of saddlery and bridles.

On completing the ten days of guards 300 men were sent to Port Said to unload ships. The Adjutant, Capt HF Hamilton-Dalrymple, and RSM RE Butler chose a ship, working together, influenced by the type of lunch they would be given. However, Maj Prescott told them to work on a refrigerated ship moving kippers which was not so popular. As a result of Grenadiers working in the docks the average daily tonnage unloaded immediately shot up from one ton to four tons per man. The ships' captains were delighted with the speed of 'turnround', saving enormous sums in harbour dues. On 4 December a convoy under Capt WRH Brooks, returning from the docks at Suez, was fired upon by Egyptians, wounding Gdsm Plowman.

The second period of guard duties was notable for Egyptians shooting at the searchlight posts, a bomb being planted in the military hospital which Grenadiers were guarding, and a section of the railway close by being blown up. On 21 December the filtration plant was attacked, but the enemy was driven off by Sgt Miles.

The situation in Tel-el-Kebir deteriorated rapidly. A patrol of the Queen's Own Cameron Highlanders was ambushed. No 1 Company was deployed to support them and shot, amid a hail of fire, an armed thug who approached the Brigade Commander. Two days later the Battalion put a cordon at dawn around the village of El-Hamada. All 1 Guards Brigade was involved.

Launched at dawn on 16 January across the Sweet Water Canal from Tel-el-Kebir, the operation took the Egyptians by surprise, but it was thought that some seventy terrorists had left the hamlet beforehand. Armoured cars of the Royal Dragoons and four tanks took up positions supported by 25-pounder field guns. At 7 am six Meteor jet fighters flew overhead, reappearing over a period of several hours. The Grenadiers were fired upon when they established the cordon. Two Egyptians were shot dead trying to escape. The 3rd Battalion Coldstream then entered the village with Bren and armoured troop carriers and began a house-to-house search. Both Battalions took prisoner 170 Egyptian armed police. 162 rifles, sten guns and considerable quantities of ammunition were found at the police post which normally numbered only twenty policemen. It was assumed that the weapons were for the terrorists who called themselves the Liberation Army. The principal suspect was a police Major General. He was promptly arrested.

The operation was completed by a successful cordon and search by the Battalion of Tel-el-Kebir. This was the biggest operation launched by the British Army in Egypt until then. One Grenadier recalls that: 'My recollection of the operation was that, having traversed the reed-strewn roofs of the shanty town in our search for weapons, we returned to camp bitten to death by fleas. Gaiters at the ankle were useless and we soon reverted to the puttee at the ankle and a pair of ladies' stockings under our khaki drill trousers.' Within a week the terrorists had lost twelve killed and fifteen wounded.

On 26 January 1952 an emergency platoon was urgently required to rewire a section of the Tel-el-Kebir perimeter. Few men were available in barracks and so the nominal role of storemen, messmen, cooks and others was called, going down to those whose surname began with 'S'. Two Guardsmen, FL Smith and AN Smith, were killed on British mines during the rewiring. They were buried at Moascar Military Cemetery.

By now the situation had deteriorated further. A bloody affray in the Canal Zone between British forces and Egyptian police posts resulted in heavy Egyptian casualties. Next day a number of British-owned buildings were set on fire amid anti-British violence. At the Turf Club the Canadian Trade Commissioner and nine British subjects were murdered by a mob. The country was teetering on the edge of anarchy. The British Government ordered reinforcements to the Canal Zone.

At 10 pm on 2 February 1952 the Battalion was ordered to move by 8.15 the following morning. They left Tel-el-Kebir, hoping never to return. After further moves, they ended up in a camp at Deversoir where the Suez Canal joins the north end of the Bitter Lake.

Training facilities in the Jebel country were superb. The Battalion's operational role was to make a broad sweep with the Royal Dragoons towards the south-west and to enter Cairo, should it become necessary, from the south. The group was known as Desforce named after the commander of the Royals who became Gen Sir Desmond Fitzpatrick, the Colonel of The Blues and Royals.

Sailing on the lake could be dangerous due to violent, unpredictable storms. On 8 March two officers, Lt DL Gregson and the Hon ET Fitzherbert, were drowned when their small sailing boat capsized in such a storm. Their bodies were later recovered.

French ships, taking their servicemen to the war in Indo-China, were shadowed along the adjacent road by British military vehicles to pick up the French deserters who jumped overboard. Later one Guardsman decided to join the French Army and got aboard, but he was eventually detected and returned to his company.

Much of the remainder of the year was taken up with guards. The Egyptian thieves were both notorious and everywhere. When the Grenadier jeeps drove through crowded villages pulling trailers, it was necessary to look back because the thieves cut the trailers' canopies to steal the contents. A Royal Military Police vehicle, sent to guard an aircraft, had two of its wheels stolen.

Classifying and company training in the desert provided a welcome relief from the monotony of camp life. Cairo was out of bounds. Some questioned the need for so many British troops in Egypt. Although street maps were held of Cairo and plans existed to move into the Delta and elsewhere, a scenario necessitating such an invasion was difficult to imagine. The Russian threat, on the other hand, was taken fairly seriously. An attack through Iran or the Syrian desert may have seemed far-fetched but not impossible. The Suez Canal continued to be strategically important; the Middle East's oil was worth fighting for if necessary, and so troops, which now amounted to three Divisions, continued to be required to guard both themselves and the massive stock-piles of equipment in Egypt.

The training continued to be first class – better than in Germany. The free-running, realistic exercises increased in complexity, and were undertaken anywhere – down to the Red Sea, over hills and through ravines and wadis. Moreover, battalions trained alongside artillery, armour and engineers who were seldom, if ever, seen in England. Officers learnt how to handle all Arms.

By contrast, life in camp was dreary, although sport and swimming kept everyone fit. The Durham Light Infantry joined the Brigade from Korea. Some Grenadiers were disappointed that they had not fought there, but Guards battalions were fully committed elsewhere. Fortunately, companies were detached in turn to Aqaba. The more enterprising visited Petra, Amman and Jerusalem. Another change, but less welcome, was the fortnight of spring drills which, traditionally, are carried out almost regardless of where the Battalions are serving.

By now a few families had reached the Canal Zone. Some lived in houses resembling Nissen huts at El Ballah near Moascar.

Families' morale was fragile for they lived in a barbed wire, heavily guarded compound. The families' bungalows in Moascar were preferable. The atmosphere there was comparable to Aldershot. The staple food was a skinny chicken from the market. Everything was prepared by the Arab cooks on small calor gas stoves. It was impossible to tell when the stoves were running out of gas, which they occasionally did before a dinner party. When this occurred some Arab cooks preferred praying frantically for more gas rather than alerting the hostess to the probability of cold, uncooked food.

Although Cairo was still out of bounds several enterprising young officers managed to visit the town concealed in the boot of the Rolls Royce representative's car. He had served in the Coldstream whom he visited in Fayid occasionally.

Due to the uneasy internal security situation, the Queen's Coronation celebrations in Egypt in June 1953 were very modest. A two-day holiday started with a swimming gala in the Great Bitter Lake. The Battalion sent a large detachment, under command of Maj RH Heywood-Lonsdale, to a Divisional parade on the RAF aerodrome at Kasfareet on Coronation Day. That evening the Warrant Officers and Sergeants waited on the Guardsmen at a special Coronation dinner. This was followed by an open-air concert, The Queen's speech being relayed over loudspeakers and a grand draw, the first three prizes being free trips to England. The occasion was also celebrated by the families in Tripoli where Majors CB Frederick, HWO Bradley and Capt RAG Courage were responsible in turn for commanding the rear party.

Maj Prescott and a detachment of Guard size flew back to London to represent the Battalion at the Coronation. 'I took with me Robin Brooks, Tom Richardson, Crispin Gascoigne, Chris Keeling, CSM Graham and CQMS Staniland,' wrote Maj Prescott. 'We felt enormously privileged and honoured to be given this role.'

At last the Battalion's tour came to an end. They returned to their families and friends in England in February 1954. Chelsea Barracks proved a welcome change from the trials, excitements, frustrations and boredom which they had endured in the Canal Zone for over two years.

2ND BATTALION IN EGYPT

In 1952 the 2nd Battalion had returned to England after seven almost unbroken years in Germany. They were expected to remain there, but on 27 March 1954 they arrived at Port Said, to be welcomed by Brigadiers GC Gordon Lennox and CMF Deakin. Their destination was Fanara where they marched behind the Corps of Drums of the Irish Guards into St Pierre Camp. The Battalion was commanded by Lt Col C Earle DSO OBE, with Capt RMO de la Hey as Adjutant and AG Everett as RSM. After innumerable duties, guards, fatigues and participating in the Queen's Birthday Parade on the polo pitch at Moascar, the Battalion settled down to a similar way of life in Egypt which the 3rd Battalion had experienced.

The Middle East was still bedevilled by the Arab/Israel dispute, by Russian infiltration and by the jealousy and lust for power of some Arab states. It was a dangerous situation in an area where Britain still had important interests and commitments such as the treaty with Jordan, the Baghdad Pact, the tripartite declaration guaranteeing the armistice line in Palestine, and arrangements for the Canal Zone base. The Saudis were trying to gain mastery over the sheikdoms of the Persian Gulf, where Britain had a major interest in oil supplies. There were also commitments in Aden, threatened by Yemeni subversion, and in Cyprus torn by 'the sad sight of friend attacking friend for an illusion'.

During 1953 and 1954 both 1 Guards Brigade and 32 Guards Brigade were in Egypt. The latter occupied the northern half of the Canal zone from Port Said to Ismailia. Although 32 Guards Brigade had no Grenadier battalion in it, the Brigade staff was wholly Grenadier 'with Brig Algy Heber-Percy, Ronnie Taylor and me,' wrote Maj Freeman-Attwood. 'Sadly Brig Algy had a stroke and Ronnie a duodenal ulcer. They were replaced by Brig Peter Deakin and David Wedderburn.'

In December 1954 the 2nd Battalion moved to Golf Course Camp at Port Said where Lt Col the Hon MF Fitzalan Howard MC, later the Duke of Norfolk, took over from Lt Col Earle. The Senior Major was Maj AG Way MC.

According to one young officer, it was a dreadful camp 'between the station and the jail. I could hear the screams of the prisoners who were chained to heavy balls. The bed bugs and lice were awful.' Although the camp was dreary, the urban environment was a change and some families were there. It even boasted the odd night club. The Commanding Officer took the Quartermaster, Capt GC Hackett, to see a belly dancer on the dubious grounds that they should know what the young officers were getting up to. The English community in Port Said were not particularly friendly and some chose not to fraternize with the Army in order to demonstrate their independence and non-involvement in the political situation. So Grenadiers joined the French Club where the food was excellent.

The Battalion was made responsible for trials on the new FN Rifle which gave the Commanding Officer every opportunity to send officers all over the Middle East. Maj GW Lamb took a team to Khartoum, and 2Lt ATW Duncan to Tripoli. Lt Col Fitzalan Howard even succeeded in sending officers for training in jungle warfare and to fight the Mau Mau in Kenya. (Chapter 20).

Lt HA Clive, Sgts R Sharratt and J Ridd trained forty Iraqi officers in the hills of Kurdistan. They were rather cautious about writing course reports since, on a previous course, run by Iraqis, the only subaltern to fail slipped a grenade into his instructor's pocket which blew them both up. The Iraqi officers were initially gravely affronted at the prospect of being instructed by NCOs, but quickly overcame their prejudices on seeing the calibre of Grenadier NCOs.

In 1955 the Regimental Colour was trooped on the Queen's Birthday Parade which necessitated leaving Port Said for Moascar at 3.30 am in the morning for rehearsals. The Commanding Officer commanded the parade with Capt PJC Ratcliffe as Adjutant. Officers of the Escort were Maj MS Bayley, Capt GC Anderson and 2Lt ATW Duncan. No 2 Guard comprised Maj DW Hargreaves, Lt MGC Jeffreys and 2Lt DV Fanshawe. The Army Commander, Lt Gen RA Hull, considered the parade to be excellent. With two Guards Brigades in the Canal Zone during 1953–54, there were more guards on parade in 1954 than could be found on the Horse Guards that year.

Battalion life was dominated by the increasing number of outside guards. It was not unusual for the rifle companies to find themselves without anyone left.

The Battalion borrowed a twenty-ton ketch, *Meander*, from the Royal Engineers. An ambitious expedition to Cyprus was frustrated when she sprung a leak after four hours' sailing. An attempt to sail to Alexandria ended aground for a week near Damietta. It was decided that *Meander* should thereafter be true to her name and potter happily about the outer basin of Port Said.

Cairo had been out of bounds for several years but Maj GW Lamb, commanding Support Company, arranged through an Egyptian tourist agent to take his Company there as the ban seemed to have been lifted.

'Went to Consulate on 11 March 1955 to report our arrival,' he wrote in his diary. 'Saw a Maj Green who went right off the handle: apparently other ranks can not stop in Cairo and Green said we must leave. Went to the Guardsmens' very nice hotel on the Nile with the ghastly prospect of telling them, but the Egyptian Officer in charge of the party refused to allow anyone to leave. 12 Mar. Saw David Saunders, Welsh Guards, at the Officers' Club who said all hell is let loose.'

The Company visited the museum and other tourist sights before retiring gracefully back to Port Said. But on 21 May many of them were off again to Aqaba. 'The party with me consisted of Robert Wolridge-Gordon, Monty Cholmeley, Peter Petrie and 80 men,' continues Maj Lamb: 'Arrived on 23rd to be met by Jack Wendell and the Band of the Bays. Went for a drive round the tiny village on the edge of the Red Sea where Israel, Saudi Arabia and Jordan all meet. There is just ourselves, the Garrison, the Bays and the Arab Legion with which Nigel Bromage is serving. It is very, very hot and the flies are perfectly appalling.' Their role was to defend the area from Israeli incursions. There was no relationship with the Jews on the other side of the bay, but everything passed off peaceably. Maj Bayley eventually took over with another Company.

When Lt BC Gordon Lennox was the Assistant Adjutant 'I was made responsible for moving ammunition and high explosives on a barge between Cairo and Ismailia,' he recalls. 'We were not allowed to tie the barge up for safety reasons, but I did so, by a quay near the French Canal Zone offices. The Commanding Officer got a bill for two million pounds and was most upset. However, it was later cancelled. Nor did I make myself popular with the Commander-in-Chief of the Middle East, Gen Dick Hull, who had expressed the view that nuclear weapons would not change warfare much: tanks would hold their own. Fresh from Germany, I said nuclear weapons would change everything, dominating the battlefield. My views were not well received!'

Fortunately the days of the British Army in Egypt were numbered. Early in 1953 Anthony Eden, then Foreign Secretary, had decided on a phased withdrawal of all British troops from Egypt. While the Suez Canal continued to be of supreme importance, together with the freedom to transit through it, the base was clearly less so. By now the cumbersome tangled mass of workshops and railways occupied an area the size of Wales and was still dependent on Egyptian labour. It was considered improbable that, in the nuclear age, a base of that scale was required or would survive an attack. Moreover, service in the Canal Zone was a poor recruiting agent. Terrorism decreased when it became clear that the British were finally leaving.

The Commanding Officer persuaded the senior military authorities that the 2nd Battalion should be the last to leave Egypt, on the dubious grounds that the Coldstream and Scots Guards had played a greater part in the 1882 Battle of Tel-el-Kebir. The final withdrawal was scheduled for 1 April but the Foreign Office said they could not leave on April Fool's Day. 'So we departed on 2 April 1956,' recalls Lt Col Fitzalan Howard. 'All our camp was perfectly clean when we left in secrecy during the night, but, as so often happens, the Officers' Mess was in bad order. Peter Thwaites of No 1 Company and I cleaned it up before we left.'

The Battalion collected the families at Tripoli, en route for Liverpool and Pirbright. Grenadiers thought that they had seen the last of Suez. Most had, but there were a few notable exceptions.

MALTA AND SUEZ

After the last troops had left, the administration of the waterway remained the responsibility of the Anglo-French Suez Canal Company which had originally constructed the Canal. On 23 June Colonel Nasser was elected President of Egypt and nationalized the Suez Canal Company on 26 July. The next day the Prime Minister, Anthony Eden, established a 'Suez Committee' of senior Ministers. It did not include the Minister of Defence nor any of the three Service Ministers.

At that time, before the birth of the giant tankers, the Canal was the essential means by which Western Europe could be supplied. Seventy million tons of oil a year passed through the canal, representing at least half the oil supplies of Western Europe. Nasser was boasting of his intention to create an Arab Empire from the Atlantic to the Persian Gulf.

The British Cabinet proceeded to make a number of miscalculations, the gravest of which was to misjudge the likely reaction in Washington to the Franco-British intervention. Eisenhower seemed to regard the action of Britain and France as a personal affront. Eden decided that force should be used in the last resort to bring Nasser to his senses. The Services were geared either for all-out war with Russia or for counter-insurgency operations in the colonies; little attention had been paid to the requirements for a limited conventional war against a third party. The nearest effective base from which a seaborne attack could be launched was Malta.

The 3rd Battalion was mobilized on 3 August and sailed from Southampton with 400 reservists on 14 August. The Battalion expected to be committed to battle within days. There was no question of taking civilian clothes, only battle dress. On arrival, however, the Commanding Officer, Lt Col AMH Gregory-Hood MC, found that he needed a white dinner jacket to play bridge with the Governor, Maj Gen Sir Robert Laycock. Nevertheless, initially, there was a great sense of urgency: wills were made out and mail was censored.

The Battalion's tented camp in Malta was close to a disused air-strip. Conditions were appalling. Even the straw for mattresses had not arrived.

There were insufficient boxes for many beds to be off the ground, so when it rained everything got wet because the camp was on a slope. Red dust was everywhere and the hurricane lamps were proving unsatisfactory.

Gradually the likelihood of operations seemed to recede. The repeated postponements were creating practical problems which worried the Army Board in London. The recalled reservists, throughout the Army, were increasingly anxious about their families and jobs. Wives were writing from England saying they were not getting their allowances. Reservists' morale in Malta was declining fast due to the uncertainty and the squalid camps.

The British press was divided between support and opposition to the Suez venture. It has always been recognized that international boundaries should not be overturned by force. This was the argument to rally opinion against Argentina when she invaded the Falklands in 1982 and when Iraq attacked Kuwait. Yet here was Britain planning to invade Egypt.

To sustain morale, attempts were made to keep the Guardsmen occupied – physical training, drill, low-level exercises, a novices boxing competition and route marches. But it was not enough. A group of forty Grenadier Reservists with twenty inquisitive onlookers went to the Officers' Mess to protest. They were followed by Lt WLA Nash, the Quartermaster, and Drill Sgt FJ Clutton MM. They were all met by the Senior Major, DMA Wedderburn, who listened to their complaints. They had come to Malta, they said, to fight, not drill; kit checks were preferable to formal inspections with a Guards Depot layout which RSM C White had apparently ordered.

Maj Wedderburn told the men that the complaints and misunderstandings would be sorted out. Drill Sgt Clutton walked back with them to their tents: the atmosphere was friendly. 'The following morning RSM White said that the men probably wouldn't go on parade,' recalls the Commanding Officer. 'I saw the men sitting on the grass before the parade and walked down the line chatting to them. When the moment came, they all came on the square. The most telling point I made to them was that there was a battalion of the Parachute Regiment in Malta: how delighted they would be that a battalion of the Household Division had behaved in such a manner. I talked to all the Sergeants' Mess members afterwards. The problems had arisen over discomfort and worry over jobs. One Guardsman had a relation on the *Daily Mail* and so the news broke. A few realized at the time the political risk involved of going to war: if Suez was taken by force – what was next? The Occupation of the whole of Egypt? The Egyptian Army could be defeated, but to what purpose? The British Army had just left Egypt. Now were they really going back for a further ten years?

We were told that the Egyptian ships' pilots could not run the canal – but they did.'

On 29 October the Israelis launched their attack on Egypt. The British bombing began two days later. The paratroops dropped successfully at Port Said on 5 November. Among them were Capt MP de Klee, Scots Guards, with eight Guardsmen of the Guards Independent Parachute Company. The rest of the Company landed by sea. Among the Grenadiers were CQMS R Maxfield, Sgt Coleman, L/Sgt Keech and L/Sgt Allcock. 'Although an Irish Guards piper piped us ashore, much to the delight of an American cameraman,' recalls Maxfield, 'we were under effective enemy fire from a light machine gun. I stood behind a Centurion tank of 6 Royal Tank Regiment. We headed south to meet up with the airborne group, having a few skirmishes on the way. We were well down the canal when the ceasefire came, ruining our chances of reaching Ismailia by breakfast. The French parachuted in supplies and petrol. Having loaded some on my Austin vehicle, I was shattered when a French officer at pistol point took them off me. We got our own back later, relieving them of reserve parachutes which became very welcome up Mount Olympus. We could clearly see the friendly tank-borne Israeli army about 2000 yards away.'

The 3rd Battalion, meanwhile, were still in Malta. Their vehicles and stores were loaded into ships. Their state of exhilaration and anticipation slowly turned to anti-climax. Only the machine-gun platoon, commanded by Capt DW Martin, reached Egypt. (He had transferred into the Regiment after being decorated for gallantry during the Korean War.) The platoon went into Port Said on minesweepers to provide close protection from shore-based fire.

L/Sgt J Tidridge was there: 'We were called on deck and told by the skipper that we were at war with Egypt. We set sail: the ship had not been to sea for well over a year and everyone was sick for 36 hours. The sea was extremely rough. During the voyage we practised with the machine guns. The sailors swept for mines as we drew near to Port Said. British aircraft were still firing rockets into the area as we entered harbour. There was no response from the enemy. We ran aground on one of the obstructions placed by the Egyptians at the canal's mouth. After disembarking we spent 18 hours guarding a deserted village: our only food was very green grapefruit thrown to us by soldiers in a Jeep. On the return journey our minesweeper caught fire and we were towed back to Malta's Valetta harbour. We didn't enjoy acting as Royal Marines at all. While our contribution to the Suez campaign was obvious to us, it had not been registered with the rest of the Battalion. In our absence they had moved to Cyprus.'

Brig CMF Deakin commanded 29 Infantry Brigade at Dover which was committed to the Suez operation. They reached the Sweet Water Canal. He knew the area well for he had commanded 32 Guards Brigade there in 1953–55. At one stage it was necessary to resume bombing and shelling the Egyptian Army in front of them: 'I was amused to see the Egyptian officers coming out of their trenches, carrying suitcases, scurrying away,' he recalled later. 'The Egyptians started to blow up the Sweet Water Canal which contained fresh water. I told them to stop destroying it, otherwise they would have nothing to drink. They stopped. I stayed on to fix the arrangements with the United Nations representatives. They eventually deployed between us and the Egyptians who were reluctant to cooperate with the UN. In view of the loss of the bases in Egypt, Cyprus was now regarded as all-important, but a terrorist campaign was fast developing there.'

CHAPTER 8

THE CYPRUS EMERGENCY

1956 – 1959

The British Government was determined to have a military base in the Middle East to safeguard their interests in that volatile region. We had a treaty of alliance with Jordan and were members of the Baghdad Pact. Troops, aircraft and stores had to be close by. If they could no longer be in Egypt, Cyprus was an attractive alternative with an excellent military airfield and staging facilities. It suffered only from having no deep-water port.

As the military build-up associated with the siting of Middle East Headquarters progressed in 1954, rioting broke out in Nicosia and Limassol. Five servicemen were murdered in one week – two in an ambush led by the self-styled General George Grivas, a retired Greek Army Colonel.

The Greek Cypriots, numbering 419,000, looked to Archbishop Makarios who initially advocated union with Greece (*Enosis*), although Cyprus had never belonged to Greece, having been ceded to Britain in 1878 by Turkey. The 105,000 Turkish Cypriots were never going to agree to any form of *Enosis*.

Field Marshal Sir John Harding, the Governor of Cyprus, offered the Cypriots self-government as soon as the security situation allowed. Talks about eventual self-determination continued, although the importance of Cyprus as a British base was always emphasized. Both Turkey and Greece were members of NATO: the quarrel was weakening the alliance.

Throughout 1956 Grivas's terrorists were increasingly active. A Hermes troop transport aircraft at Nicosia airport was blown up by a time bomb. Had it not been for a delay in the aircraft's departure, sixty-eight passengers, all British soldiers and their families would have been murdered.

Makarios was deported to the Seychelles, although it was Grivas who controlled the EOKA movement (Ethniki Organosis Kyprianou

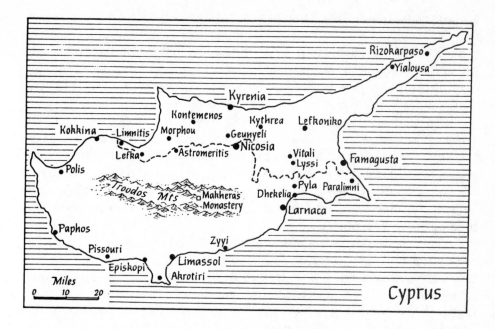

Agonistou). During the next three months an average of two members of the security forces were killed each week. The Army was forced on to the defensive with the safeguarding of life and property taking priority. The difficulties were exacerbated by inter-communal violence with Greek and Turk Cypriots fighting each other. EOKA regarded Turkish policemen as legitimate targets.

Nasser's seizure of the Suez Canal necessitated two Parachute battalions and 3 Commando Brigade being removed from Cyprus for the abortive Suez operation. Such was the shortage of infantry battalions in Cyprus that two Gunner regiments assumed that role. During November 1956 casualties inflicted by EOKA amounted to thirty-three – the highest figure for any comparable period during the entire emergency. Not surprisingly, 3rd Bn Grenadier Guards, with little to do in Malta, was ordered to move immediately to Cyprus. On 2 December 1956 they disembarked at Limassol and arrived at a dreadful, water-logged camp near Nicosia. Fortunately a Parachute battalion left their camp, called Tunisia, near Nicosia's airport. Maj DMA Wedderburn, the Senior Major commanding the advance party, quickly moved the Battalion there. The new camp was on a plateau of solid and uneven rock. Everyone slept in tents but messrooms, stores and some offices were later in corrugated iron huts.

The Battalion first had to release many of the experienced reservists and National Servicemen whose service had expired: secondly to learn internal security drills, and finally to make the camp comfortable for a cold Cypriot winter.

Lt Col AMH Gregory-Hood MC, who had steered the Battalion through public duties, training for war in Malta and the opening period of the Cyprus campaign, was posted to the Northern Army Group in February 1957, and was succeeded by Lt Col PC Britten.

The Battalion's overseas tour which followed can be divided into two phases. The first, as part of 50 Brigade, was concerned with operations in the area of Nicosia and the surrounding 140 square miles. The second phase, a year later, was much more exciting: the Battalion was employed all over Cyprus as the Island Reserve, some 50,000 square miles, in 3 Brigade, forming a striking force, with the emphasis on mobile operations making use of helicopters.

Initially the Battalion undertook such varied tasks as a static guard on the central prison, Nicosia, including the roof of the condemned cell, to burying executed terrorists at dead of night; to cordon and search some remote village at 3 am, to trying to disperse hostile crowds in towns. 'The bomb throwers still continue their malicious pastime,' reported the Commanding Officer to Regimental Headquarters. 'The new draft from the Depot appear keen and interested. We gave them a week of lectures and demonstrations on internal security operations. These included a spirited performance by Nick Hales Pakenham Mahon's Company on how not to disperse a riot which was worthy of the highest traditions of the Aldershot Tattoo. Also a dramatic demonstration of searching personnel and buildings laid on by Peter Ratcliffe.' The tactics of cordon and search in Cyprus were little different from the campaign in Egypt and were discussed in that chapter.

Dispersing hostile crowds in built-up areas was a difficult operation which also required sound training. If possible the platoon commander tried to persuade the crowd, through an interpreter using a loudhailer, to go home non-violently. If this failed, a magistrate read the Riot Act. Soldiers would then advance in a line with fixed bayonets as long as there was no danger of close contact with the crowd, but hand-to-hand fighting had to be avoided. Ringleaders were photographed and, if necessary, tear gas or a riot control agent were fired into the crowd. If there was no alternative, for example if the platoon was about to be overrun, the commander, and he only, could order fire to be opened. A bugle call would first draw the attention to a banner which proclaimed that the crowd must disperse or fire would be opened. If this was ineffective, the commander indicated to a marksman the ringleader to be shot and the number of rounds to be fired. Fire had to be effective. It was a case of shoot to kill. In the gravest emergency, rapid fire from rifles or bursts from machine guns were acceptable. There were variations on the above, but such tactics, learnt by all British infantry officers, were the basis for keeping the peace in the numerous pre-independence crises overseas. It may all sound a trifle simple, but dispersing a noisy and a bitterly hostile, projectile-throwing, rioting crowd who were infiltrating round an isolated platoon was far from easy.

Early in the Battalion's tour an important operation took place near the Makheras Monastery on the eastern end of the Troodos Mountains. The mission was to kill or capture Gregoris Afxentiou, Grivas's second in command, an officer in the Greek National Army and one of the surviving leaders of the Troodos gangs. Maj JD Makgill Crichton Maitland's No 3 Company was attached to 3 Brigade for the operation, with No 1 Company holding the cordon. Four Grenadier patrols were also involved. The operation was helped by searchlights provided by the Royal Engineers. They illuminated the area by reflecting the light off low cloud.

It was a cold, wet morning when Lt Col Britten's party which included Maj PH Haslett, Lt WLA Nash, and RSM C White, set out in two landrovers. They found No 3 Company busy searching the sides of a valley for terrorists' underground 'hides'. Extended lines of Guardsmen were wading through undergrowth. On driving on, they were surprised to hear the sound of small arms fire and an explosion at midday, close to the monastery.

The 1st Battalion the Duke of Wellington's Regiment (DWR), assisted by an informer, had found at first light half a mile down the valley a very cleverly concealed oil drum containing a brand new 2 inch mortar. This led to a closer search. Suddenly the DWR patrol discovered a hole, covered with corrugated iron, which was skilfully hidden by boulders. A Greek-speaking officer called on the terrorists to come out. Four did, but Afxentiou was not among them. The last one to emerge was pushed back inside and told to bring him out. A shout was heard in English saying, 'Come on in and get us out.' Afxentiou suddenly appeared firing a sten gun and killed a DWR corporal. A grenade was tossed out. He bolted inside again and a seven-hour siege began, watched by the Grenadiers and a large section of the press. Petrol cans were rolled down the hill to the terrorists' hide and ignited by Very lights, but this only set fire to the undergrowth.

A Royal Engineer officer was sent up by helicopter to explode charges, while a DWR assault section formed up, fixing bayonets. Finally the combination of grenades, explosives, tear gas and petrol together with the DWR charge into the hide settled the matter. Beneath Afxentiou's charred body was the Abbot of Makheras's book, *Christ Recrucified*.

The remaining terrorist, who was found crawling away in the confusion, was captured. 'I shall certainly remember it as an unforgettable afternoon packed with intense excitement,' wrote Lt Col Britten. 'The arrival of the press turned this unique and privileged occasion into the resemblance of a fantastic fox hunt, with the fox run to earth, watched by the field as he fought back bravely, determined to sell his life dearly. That night a No 3 Company ambush shot at an individual in the hills but he escaped. We have now established a platoon permanently in the monastery since it has been conclusively proved that EOKA have been using it.' Abbot Irineos of Makheras Monastery had served in the RAF Regiment during the war. He suggested to Lt Col Britten the day after Afxentiou's death that he had not been protecting him. However, the Abbot later told a Greek journalist that he usually gave up his own room to Afxentiou; they took it in turns using the Abbot's bed, while the other stayed awake. Other Abbots were also supporting EOKA, giving the terrorists food and clothing while some monks carried messages.

A Grenadier who visited the area 35 years later found that a shrine had been erected to Afxentiou and courteous priests occupied the attractive monastery.

The operation described above had culminated in several months of exceptional success for the Army. Sixteen terrorist gangs had been reduced to five; sixty-nine EOKA had been killed and rioting and bombing had virtually stopped.

As the Battalion's tour progressed, there were the inevitable changes to the officers' order of battle. A brief analysis of those there in May 1957 may be of interest. The Battalion contained fifteen regular officers, eight national service officers and two on short service commissions. The Senior Major, DMA Wedderburn, was the only substantive major. Of the five company commanders, two – Majors N Hales Pakenham Mahon and PH Haslett – were to command Grenadier battalions while Maj PJC Ratcliffe was to command the Guards Depot. The other two Majors, EBM Vaughan and JD Makgill Crichton Maitland, were to retire early. Capt GEV Rochfort-Rae had handed over as Adjutant to Capt SJ Loder. The remaining eight regular and short service commissioned officers – Captains TN Bromage, DW Martin, MKB Colvin, Lieutenants E Luddington, CJ Airy and 2nd Lieutenants DHC Gordon Lennox, CTF Fagan, and FWS Hopton Scott filled such appointments as Intelligence Officer, Machine Gun platoon commander, Transport Officer, Assistant Adjutant, Signals Officer and platoon commanders. Lt WLA Nash, the Quartermaster, was the only officer commissioned from the ranks. All the eight National Service platoon commanders – 2nd Lt RHN Wills, GTG Kenyon Slaney, AG Neville, NP Nicholson, JGR Williams, DH Miller, CJ Shaw and DRA Hardy – had joined the Battalion during the last sixteen months: none of them converted to a regular commission, although the Battalion was short of officers. The three attached officers were the Padre, Paymaster and Medical Officer.

The above can be compared to the 2nd Battalion serving in Cyprus with the United Nations twenty-five years later. The differences are not very discernible but they consist of the 2nd Battalion having officers of longer service – the two-year National Servicemen having been replaced by officers on three-year commissions, more officers overall, significantly more commissioned from the ranks, and a smaller Battalion, overall.

*   *   *

The Army was still largely made up of National Servicemen. Everything inside the walls of Nicosia was out of bounds, and sometimes the surrounding suburbs as well. There was little the ordinary soldier could do when off duty apart from sit in his small tent plagued in the summer by flies and dust. The monotony was broken by patrols, operations and exercises. The British soldier never met a Cypriot socially and never a girl, except the kind that threw stones. If the soldiers were allowed out they had to move in groups of four, armed and in uniform.

However, the tension unexpectedly eased. On 14 March 1957, encouraged by the Greek Government, Grivas, very reluctantly, agreed to cease hostilities if the British Government would open talks with Makarios. Harding acquiesced and Makarios was flown in triumph to Athens. Lord Salisbury KG, then Lord President of the Council, was horrified and resigned. (He had served in the Regiment in the First World War.) Makarios' release caused a wave of rejoicing among the Greeks but damaged British relations with Turkey. Dr Kutchuk, the Turk Cypriot leader, alarmed at any drift towards self-determination, announced that partition was the only answer. Despite this, Greek Cypriot celebrations and remembrance services to the EOKA fallen continued. The problem of enforcing the curfew appeared insurmountable until a thunderstorm of galeforce strength hit Nicosia dispersing the crowds. The Grenadiers' camp was flattened; the tent pegs snapped and the canvas tents were ripped. Such was the desolation that a rum issue was authorized.

By the summer Cyprus had lost much of its colourful charm. Everywhere was hot and dusty. During the ceasefire EOKA rearmed and recruited but refrained from active operations. Grivas was relying increasingly on women couriers; he found them less talkative and more trustworthy. Hundreds of them worked for EOKA and none let him down.

The Battalion continued to find a large number of static guards. There was little to relieve the boredom of guarding a mine to prevent explosives being stolen or sitting alongside the Cyprus police in the village police stations. The Corps of Drums beat 'Retreat' in one rural area and were greeted with apathetic amazement. Although everyone was kept busy, there were few operations against EOKA. Company camps were established by the sea, swimming replaced forced marches and rigorous physical training.

By the autumn of 1957 there was no complacency that EOKA's cease-fire would last for ever. The 3rd Battalion carried out night patrols and road blocks. Every week hundreds of civilians and vehicles were searched in the rural area where each Company had its own defined parish.

The British Government adopted a conciliatory approach: Harding retired in November 1957. He was replaced by Sir Hugh Foot who had plans for Cyprus's self-government.

In January 1958 the 3rd Battalion came under command of 3 Brigade to be deployed wherever the threat was greatest throughout the Island. All Companies practised training from helicopters found by the Joint Experimental Helicopter Unit. Tracker dog teams and a recce troop mounted in Ferret armoured cars were formed. The emphasis was at last on mobility rather than static guards. This proved to be a much more exciting and worthwhile role.

There were numerous occasions when the Companies were deployed to cordon and search villages, and maintain order against aggressive crowds in Nicosia and elsewhere. On one exercise in the Troodos Mountains the whole concept of a quick anti-terrorist operation was practised. Helicopters dropped observation patrols on prominent high features before the main operation commenced, thereby hindering movement in daylight, and resupply by air was tried out. Rigorous patrolling, searching and ambushing took place.

In March 1958 the murders resumed. The most shocking inter-communal atrocity occurred on 12 June at Geunyeli, a small Turkish Cypriot village on the main road leading north-west from Nicosia to Kyrenia. Thirty Greek Cypriots had been arrested by the Blues who took them towards the central police station in Nicosia. But a Turkish demonstration gathered ahead of them and the cells were already full. The Blues were therefore told to take them out of Nicosia and dispatch them home on foot to their village of Kontemenos. A Platoon of No 1 Company, under command of the Blues, was following in 3 ton trucks. 2Lt RS Corkran told the Greeks to walk home and watched them leave across country, before patrolling elsewhere. Driving home to Kyrenia at 2 pm, Maj PJC Ratcliffe, accompanied by his driver and wireless operator, came across fields of corn on fire close to Geunyeli. He found bodies. Some of their heads had been cut off. The Guardsmen collected the dead – a harrowing experience.

The Turks were said to have attacked the Greeks with axes. Lt Col AG Heywood MVO MC, Senior Major of the Battalion, heard Maj Ratcliffe's radio report and sounded the general alarm in Tunisia camp. Within an hour the whole Battalion had deployed, cordoned off the area and patrolled, while a helicopter hovered overhead. A Squadron of the Blues, commanded by Maj RMF Redgrave, was already there. The Turkish massacre of eight Greek Cypriots had occurred less than three miles north west of Geunyeli.

Lt Col Britten, accompanied by a policewoman, came across an English girl in great distress in the village. She had just married a Turk Cypriot, having arrived from Birmingham a week earlier. She insisted that her new life was destined to be there. Arrests were made while a cordon was placed round the village. A thorough search was conducted at dawn the following day. Although more arrests were made, nobody was convicted of the atrocious murders. An official report on the massacre was later published. It said that the security forces had been 'indiscreet' in leaving the Greek victims in an area where they could be ambushed by Turks and considered the attack itself to be of 'a most savage nature indicating an extraordinary bloodlust.' Such was the speed of modern communications that the incident was being controlled by London before nightfall.

A few weeks after the incident an officer, driving through Geunyeli, had a puncture. He got out of the vehicle with considerable trepidation, aware of the recent murders. But the Turks produced chairs and coffee while the tyre was changed. 'What a strange lot they all were,' recalls one Grenadier.

The terrorists hid themselves, their weapons and ammunition in the most ingenious places. It was not surprising that so little, overall, was found, even if, due to some informer or luck, the patrols and search parties were looking in the right places. The police had sniffer dogs which were said to be trained to locate breathing pipes connected to underground hiding places. Prodders ripped walls apart while water was thrown over flagstones to see where the water drained away – possibly into the hide. If Grivas is to be believed, British security was poor. One high-level intelligence conference was tape-recorded by one of his supporters. Three Cypriot informers were named during the conference. They were all quickly murdered.

Lt ATW Duncan, the Battalion's new Intelligence Officer, on hearing that a cordon and search was imminent, would order air photographs by despatch rider or wireless. An aircraft would then drop very detailed photographs over the Battalion's camp in time for them to reach the Company Commanders who were planning the operation. 'Some recces in the early days were very amateurish,' recalled Duncan. 'David German and I set off surreptitiously to a suspect village in a private car. We drove through the village and found a road blocked on the far side so came back the same way and found a wedding in progress. All the Greek Cypriots surrounded our car and persuaded us to join the wedding party. We danced with the bride, participated in the feast and were sprinkled with very smelly water which horrified my orderly the following morning. "Where have you been, Sir?" he enquired nonplussed. I was quite pleased because that particular operation may have been cancelled.'

Occasionally the terrorists were less fortunate, blowing themselves up, just as British troops shot each other by mistake.

Hard work paid dividends. At 2 am 2Lt JGR Williams's patrol saw a man with a hurricane lamp in a narrow village lane. He gave chase and found twelve big pipe bombs. The terrorist was arrested and the police arrived, but the culprit escaped by jumping out of the vehicle. He later blew himself up with one of his own bombs.

One incident of note occurred when Maj WRH Brooks took out an ambush patrol. At 3.30 am the CSM saw a figure crouching in front of the ambush position. He challenged. There was no reply so he fired a burst from his Sterling sub-machine gun. Maj Brooks fired two more bursts. The figure turned out to be Gdsm Williams, the company clerk, who was hit only in the ankle. Williams had been under the delusion that the CSM had been out in front, beckoning him forward. Fortunately he was back at duty within a few days. Williams was heard to remark that, despite the close range, neither the Company Commander nor the CSM could hit him! The Royal West Kents were less fortunate, losing two men that week in a similar muddle.

In May the Battalion was put on 24 hours' notice to move to the Lebanon with the task of capturing the airport. Conventional training for war was resumed. Companies practised with Beverley aircraft and daily briefings added to the excitement. However, by June the situation in Cyprus had deteriorated so badly that the 3rd Battalion reverted to internal security operations.

In August they were operating against a group of villages on the Famagusta area. They arrested twenty-six men at Vitali and found a large quantity of bombs, wires and mortars. A further terrorist bomb hide was detonated by a lucky 2 inch mortar illuminating bomb that happened by chance to light the straw covering the hide.

Most of the Battalion was unexpectedly moved on 11 August to the neighbouring village of Lyssi, which lay under a ten-day curfew. Everyone was told to smarten themselves up as an unnamed VIP was coming. A helicopter arrived on the football pitch. Mr Harold Macmillan, the Prime Minister, emerged with the Governor. 'The Battalion was commanded by Lt Col PC Britten, son of Charles Britten who was in the 4th Battalion with me at Loos in 1915,' wrote Mr Macmillan in his memoirs. 'The officers, NCOs and men looked in splendid shape. I spoke a few words to them. I found all this very moving, and I almost broke down in speaking, for it all recalled so many memories.' The Prime Minister visited the operations room which was in the school's infants' classroom before walking down to No 1 Company chatting to the Guardsmen, telling them amusing stories about Grenadiers in the First World War. He said they looked just like those whom he had known forty-six years earlier when he joined the Regiment. He added that he was mystified by modern weapons and modern arms drill. Mr Macmillan's visit gave a great boost to morale. He was totally unpompous, and had clearly come in his capacity as a former Grenadier, rather than as Prime Minister. 'I ask you to assure your readers,' he told the Press afterwards, 'that the Brigade of Guards is not in decline, particularly the finest Regiment of the Brigade.'

*   *   *

The 2nd Battalion, meanwhile, had been stationed outside Lydd on the cold and bleak shores of Dungeness. It was therefore with tremendous enthusiasm and a spirit of adventure that on 18 June the Battalion, with 1 Guards Brigade and 16 Parachute Brigade, were ordered to Cyprus for operations in the Lebanon. Some drummers were told that they were too young for overseas service: they were so upset that they burnt down the gymnasium. The Commanding Officer sympathized with them and they remained unpunished.

Forty-eight hours after being warned to go, the Battalion flew from Abingdon and Lyneham in Comets, Hastings, Beverleys and Shackletons. 1 Guards Brigade had become part of the strategic reserve. The Battalion's camp at Limassol was on a bare patch of rock and sand. No sooner had the first deep trench latrine been dug than they were ordered to the airfield to stand by for an immediate move to Jordan. They remained there for a fortnight, packed up and ready to go. One crisis was averted when the American Sixth Fleet reached the Lebanon. This most unsatisfactory period was followed by very intensive training to test the Battalion's skill and endurance to fight on an air-portable basis. Two full-scale brigade group exercises were set by the Divisional Commander, Maj Gen GC Gordon Lennox. The Fleet Air Arm pressed home their 'attacks' with such vigour that two pilots collided in midair, luckily without loss of life.

On 18 July it became evident that King Hussein of Jordan was threatened by an insurrection planned and fomented by the United Arab Republic to overthrow the regime. 16 Parachute Brigade was sent to protect the King and his Government. The Grenadiers were earmarked to be despatched in support. Fortunately the Parachute battalions were adequate to preserve the independence of Jordan. The value of Cyprus for launching this operation was significant – it could not have been done from

Germany or England. This strengthened the Government's resolve to keep at least part of Cyprus.

At one stage the 2nd Battalion was warned that it might have to go to Khartoum to capture and hold the airfield there, but nothing came of this.

Two large-scale internal security operations were launched. Both were successful. In the Yialousa area nine wanted Cypriots were arrested. At Kythrea a large amount of arms and ammunition were found.

On one occasion elsewhere RSM ST Felton saw the Cypriot police running a 'pretty rough' interrogation centre which resulted in switching a search to some nearby fields. Sergeant Papworth, the Pioneer Sergeant, had his mine detectors deployed and a bulldozer was digging up the ground. RSM Felton suddenly saw wooden boxes emerging from the newly ploughed earth. They contained rifles, pistols and bayonets.

In December 1958 the 2nd Battalion returned to England bound for Hubbelrath in Germany.

Greek and Turk savagery against each other was appalling. Men, women and children had their throats cut while they slept. A Greek shepherd boy aged eleven was knifed while he watched his sheep. A Mother Superior was shot dead at her convent gates.

From the British point of view, the worst atrocity occurred on 3 October 1959 when Mrs Catherine Cutliffe, the wife of a Royal Artillery Sergeant, was fired on from behind and killed while shopping in Famagusta. Another wife, walking with her, who was badly wounded, described the murderer as 'light-complexioned and fair-haired'.

The 3rd Battalion had some wives in Cyprus. They lived in Nicosia and Kyrenia in civilian married quarters under military guard. Shutters were meant to be kept closed for security, despite the stifling heat. Terrorists tossed one grenade into a flat in Kyrenia but it was occupied only by bachelor Grenadier officers who put the fire out without difficulty, using the soda syphons from the drinks tray.

Mrs Peter Ratcliffe lived in Kyrenia: 'We went out to Cyprus by troop-ship,' she recalls. 'Kyrenia was lovely although it was twenty miles from the Battalion's camp. There was no telephone but we had a radio. I was worried in case my husband was shot, so at night we put stones in tin cans around the outside of the house. The terrorists might knock them over and alert us. Peter sometimes left his orderly with me; he was armed with a Sterling sub-machine gun. I was always concerned that my children might pick up the weapon. When the orderly had his hair cut in Kyrenia, I stood guard outside to warn him if a terrorist approached. The Cypriot bread was always covered in flies so we baked our own. It was difficult getting water in the summer. The Wiltshire Regiment guarded the area and was always very kind to us. We were very friendly with the Greek Cypriots who proved to be pleasant neighbours and gave me oranges. Many years later it was disclosed that the neighbouring building was an EOKA "safehouse". The school had wire netting across the windows and an armed guard. One shopped by oneself: once everyone vanished from the streets and it went unusually quiet. This frightened me'.

Mrs David Wedderburn lived in the Turkish Quarter of Kyrenia: 'We were in a wonderful house. There was still the feel of the colonial atmosphere. Being the wife of the Senior Major, I somehow found myself rather responsible for the local Turkish Cypriot families who were all extraordinarily honest; it was then the Turkish custom to cut off the hand of anyone who stole. Nobody locked their doors. We didn't either; one just tied a string across the front door if one went out. Domestic Cypriot dramas occurred regularly for me to deal with, such as a wife beaten up by her husband or an accident with a child, but I drew the line with curing sick donkeys. The Cypriots expected a British lady to cope and so I gave them stuff to put on their babies' burns or whatever. We were still pioneering really. There were great 3rd Battalion expeditions to beaches. I never felt there was the remotest threat to us from the EOKA. Even so, we weren't encouraged to go into the hills.'

'We only once had a Greek guest to a dinner party throughout our tour in Cyprus,' remembers Maj WGS Tozer. 'He suddenly vanished when a bomb went off nearby. He was found hiding beneath the dining room table.'

The Harbour Club in Kyrenia was run by Judy and Roy Findlay MC. It was a haven of peace with delicious food and a beautiful balcony overlooking the harbour. The EOKA never attacked the club. Some wondered if Roy Findlay had done Grivas a good turn during the Second World War when he had served in Greece with the partisans. Grenadiers serving with the United Nations in 1965 were still enjoying the Findlays' hospitality.

Maj Gen DA Kendrew, the first Director of Operations in Cyprus, had initially ruled that no Grenadier family was to be as far away as Kyrenia because he felt that their husbands could not react, operationally, quickly enough from there with few telephones. He said that they should be in Nicosia instead. Suddenly the Battalion was warned at 11 pm that they must all deploy at 4 am next morning. Most of the officers were at the Harbour Club that night, so there was no difficulty in contacting them and the operation was launched on time. Maj Gen Kendrew never discovered the reason for the quick reaction and the rule of not living in Kyrenia was changed.

The restrictions imposed on families and the bachelors varied, depending on the terrorists' activities. Sometimes the families were allowed out to dinner with friends in their homes while the single men were confined in after 7 pm, after which all restaurants and hotels were out of bounds. The curfew was a most unpopular weapon for everyone, including service personnel who had to suffer from it as well.

When officers did go out to restaurants in Nicosia, or elsewhere, they carried pistols and some, such as 2Lt HML Smith, always sat with their backs to the wall, facing the door. It was a court-martial offence to lose the Army-issued pistol, and so a number of officers purchased their own. It was not unusual for pistols to be dropped into the deep-trench latrine; fishing them out was an unpleasant task. One officer handed his pistol to his orderly and told him to return it to the armoury. The orderly instead took it to his tent and shot himself in the thigh. Lt Nash was first on the scene and had a tourniquet on within seconds. The army doctor who arrived later was most upset: 'All this time I have been waiting to deal with a gunshot wound,' he said, 'and the only opportunity that I get is taken away from me!'

On 19/20 October 1958 a most successful operation was carried out at short notice in support of the Royal Welsh Fusiliers at Astromeritis, near Morphou. Reliable information had been received that EOKA killers were staying that night between villages. The cover plan was that the sirens would go off in Morphou, proclaiming a curfew there. Maj Haslett's Support Company was deployed in Astromeritis. After dawn when the night cordon had been thinned out to daylight positions, the interpreter entered a room and saw a terrorist. He slammed the door and told Lt Fagan. Four terrorists were behind the door and jumped out of a rear window. It could almost have been a scene of 'cops and robbers' were lives not at stake. A terrorist fired at L/Cpl Harlock who was armed with a 'greener gun', similar to a shot gun which fired twelve ball bearings – the maximum allowed under the Geneva Convention. The Lance Corporal returned fire, mortally wounding the Cypriot who was running through a farmyard, armed with two pistols. Everyone fought hard to save his life in hospital. The Grenadier sentries on roof tops kept a sharp look out as three remaining terrorists were hunted down. The cordon was drawn closer. Lt Col Britten joined the CSM and the Turkish police. 'I was poking around in a house,' he remembers, 'when I saw a foot sticking out from under a couch. A policeman pulled out the terrorist who was clutching a grenade which didn't go off.'

The remaining two terrorists were found – one in a cupboard and another under a bed. Their weapons consisted of two .45 sub-machine guns, one .45 pistol, two bombs and a shotgun.

As the cordon was lifted, the women were permitted to leave their houses. Within minutes they had covered the bloody spot where the terrorist had been shot with flowers and black crepe. Almost twenty-five years later Maj Ratcliffe visited his son's Grenadier Company then serving with the United Nations in the same area. A statue in white marble had been erected in the village to remember EOKA's fallen.

The terrorists began ambushing vehicles and mining dirt tracks and little-used roads. Strict Battalion anti-ambush and mine drills were effective in preventing casualties. All Battalion vehicles were sandbagged.

The 'Panhandle', the long strip of land in the Island's north-east, was a terrorists' paradise as it had only one main road up the centre and was flanked on either side by the sea. An efficient information system gave ample warning of the security forces' approach. On 4 November 1958 the 3rd Battalion deployed there. 2Lt MRN Moore was dropped off, under-cover, by boat. It was thought that the terrorists might have a hide on a rock face by a cliff. A Royal Marine Commando flew in some outer cordon stops; their Commanding Officer arrived by helicopter. While climbing out he fell, badly fracturing his leg. One stop was put in the wrong place and was nearly shot by another. The Grenadiers were in a horse-shoe formation around the cliff near Rizokarpaso. One thought he saw someone being lowered over the cliff face, so opened fire. The bullets hit the rock and ricocheted with a flash towards others who assumed they were under fire. Soon everyone was shooting. Battalion Headquarters, having their evening meal, feared the worst, but the end result was only one dead goat. After the Battalion's withdrawal, small parties of Grenadiers were landed by ships off the Cyprus coast to patrol in clandestine operations. They gained useful information but it led to no terrorist arrest. Capt DJK German was often in demand as an interpreter as he spoke good Greek.

On 10 December a successful operation near Famagusta resulted in the find of a small arms and bomb cache, a duplicator and leaflets, all in an olive grove. More information necessitated the cordon and search of Lefkoniko nearby. In one such operation Gdsm NR Townsend recalls deploying with the Corps of Drums at dawn by trucks. 'Drum Major Cornell bashed down a heavy door. We were accompanied by a Special Branch English police officer in civilian clothes. We found only one very elderly Cypriot, but there was a warm mattress indicating that someone had fled. In an outhouse there was a tiled floor which we flooded. The water quickly drained away into a cavity. Four large tiles were prized up to reveal a hole. The Special Branch man shouted in Greek: 'come out with your hands up'. Nothing happened so he fired his pistol into the hole. Still nothing happened so he got a torch and explored the three-foot shaft which branched off into a large cavity. He brought up automatic pistols, and some explosives which were in a highly dangerous state. It's lucky his shots missed them, otherwise we would have all been blown sky high.'

Operations intensified in early 1959 but Prime Minister Macmillan's tireless efforts to find a solution were proving successful. On 11 February a Turk and Greek communiqué stated that the two Governments had reached a compromise. Everyone except Grivas agreed that Cyprus should be an independent republic subject to the retention of British sovereignty over the base areas together with such rights as were necessary to ensure that they could be used effectively as military bases. The claims to Enosis – union with Greece – and partition had been abandoned by the Greeks and Turkish sides respectively. Britain had abandoned sovereignty except over the bases. Some referred to it as a sacrifice. If that was the right description, then it was a sacrifice all round.

Grivas emerged from his well-concealed underground bunker in Limassol. Peace descended on the Island. Gone were arms and escorts, the searches and much of the barbed wire. Ledra Street, known as the infamous 'murder mile', became a pleasant, busy shopping centre once more. The new Adjutant, Capt JP Smiley, held more drill parades in a fortnight than EOKAs activities had allowed during the previous year.

Soon after Easter the Companies started training for conventional war. In April they moved to Tarhuna Barracks, seventy miles south-east of Tripoli where they joined C Squadron of 3 Royal Tank Regiment. After plenty of excellent training and a most interesting and moving excursion into Tunisia to visit the scene of the 6th Battalion's battle at the Mareth Line, they returned to Cyprus to pack up. The Battalion was sad when the Blues (The Royal Horse Guards) left Cyprus; the two Regiments had established a very special relationship both operationally and socially. The Grenadiers embarked on the *Dilwara* on 27 July for England, having been in the Mediterranean area for two years and eleven months. Thus ended a momentous tour which was to be the last for the 3rd Battalion before being placed in suspended animation.

The Battalion had enjoyed its full share of operational successes and much useful experience was gained. However, there were fatal accidents due to instances of poor weapon handling. Traffic accidents accounted for several other deaths due to the dangerous state of the roads.

Both Grenadier Battalions shared one notable achievement; neither lost a man in action against the terrorists, although ninety-nine British troops were killed by EOKA. This reflected the professionalism of all ranks and the sound leadership of officers and NCOs alike.

On 16 August 1960 the Republic of Cyprus came into being after eighty-two years of British rule. Sadly, the British military involvement in Cypriot affairs was still required as the Turk and Greek communities proved unable to live together. The Grenadiers' two tours with the United Nations will be described in subsequent chapters.

General Sir David Fraser

Major General Sir Allan Adair, Bt

General Sir Rodney Moore

Major General Sir James Bowes-Lyon

Major General Sir John Nelson

Major General Sir Julian Gascoigne

Major General C M F Deakin

Lieutenant General Sir George Gordon Lennox

Major General M F Hobbs

Major General B C Gordon Lennox

. Some Grenadier Generals

18. Lieutenant P A J Wright's expedition in British Guiana in 1963 during which ten days were spent on a raft travelling 110 miles down the Demerara River. From left to right, Guardsmen Concill, Smith, Lance Sergeant Perkins, Lieutenant Wright, Guardsmen Harrison and Miller.

19. One of the most outstanding sporting achievements of a serving Grenadier was undoubtedly that of Captain The Hon T R V Dixon (right) who, in company with Tony Nash, won Britain's first Winter Olympic Gold Medal for 12 years in the two-man bobsleigh event on 2nd February 1964.

CHAPTER 9

PUBLIC DUTIES, OFFICERS' CAREERS AND THE 3RD BATTALION

1958 – 1961

This Chapter starts with an account of the 1st Battalion on Public Duties in 1958. It then considers officers' careers and their social life, and the move to Tidworth in 1959. Finally it summarizes the last three years of the 3rd Battalion's long, proud history.

Due to Public Duties the 1st Battalion, commanded by Lt Col R Steele MBE, had very little opportunity for training in 1958, apart from a month on Salisbury Plain and a week on the ranges. There was little else to keep the Battalion properly trained for active service. Public Duties alone have often been a full time commitment if there have been insufficient Guards Battalions in London to share the roster.

In due course, when National Service came to an end, there was an acute manpower shortage which placed an extra burden on those finding Public Duties. At one time thirty Guardsmen of a Scots Guards Battalion left their barracks at Pirbright, established themselves in a local public house and telephoned the *Daily Mirror*, complaining that the manpower shortages, ceremonial duties at Windsor and field training left them little time for relaxation. The papers had a field day.

The 1st Battalion was in Chelsea Barracks which was very old. It was said to have been condemned in Queen Victoria's reign as scarcely suitable for soldiers. The gaunt, spartan barrack blocks were far from comfortable. It was not surprising that many National Servicemen in London counted the days until they would return to civilian life.

Discipline was tight. It needed to be. High standards of smartness were expected. Every sporting competition was entered, but the lack of facilities in London led to little success. There were few opportunities to get soldiers away. Even so, Lt Col Steele accompanied a party to the Lisbon Trade Fair for three weeks which involved a four-day trip with the Royal Navy. The Corps of Drums had a successful visit to the International Horse Show in Paris, and a party went on a cruise with the Navy to Nantes. Maj JFD Johnston MC commanded a British detachment at the 150th anniversary of the Battle of Corunna. By such means 85 members of the Battalion reached the Continent.

During the Battalion's 19 months of Public Duties it was difficult for platoon commanders to get to know their men adequately. The day before the Guard, the Guardsmen were cleaning their kit; the officers had virtually no opportunity of talking informally to them during the Guard, and the following day, after dismounting Guard, the Guardsmen were off duty.

Public Duties still consisted of Queen's Guard – guarding Buckingham Palace and St James's Palace. Guards were also found at the Tower of London and the Bank of England.

When The Queen was not actually at Buckingham Palace, her Guard mounted in the Friary Court of St James's Palace and could only be witnessed by a small number of the public. By the early 1960s the daily ceremony took place at Buckingham Palace where thousands could watch. To save manpower 48-hour Guards were later introduced and an increasing number of other Regiments and Corps were committed to Public Duties when a large number of Guards battalions were away on operations or training.

At that time Queen's Guard was almost entirely ceremonial, and scarcely tactical at night. Later it became more operational with sentries patrolling in rubber-soled boots and using small pocket radios in the grounds of Buckingham Palace. Until 1959 the sentries were posted by day on the pavement outside the railings in front of the Palace. But the behaviour of the sightseers detracted from the dignity of the guard and so the sentry posts were moved to within the forecourt.

In the 1960s the number of sentry posts on the various guards was reduced. However, the ceremony of the Changing of the Guard at Buckingham Palace is regarded as one of the sights of London and there was an obvious danger of reducing the number taking part too much. To get round this, the Tower Guard joined the Queen's Guard for the Buckingham Palace ceremonies. After this the Tower of London Guard was superimposed upon the 'Beefeaters'. They give a nightly performance of the ceremony of the Keys. They also have a major role of protecting the Crown Jewels, and so this Guard has not changed significantly over the generations.

The Bank Picquet began in 1780 when a regular military guard was provided for the Bank of England, following the Gordon Riots. The Picquet marched to the Bank in scarlet tunic, bearskin and carrying rifles. The march coincided with the evening rush hour, causing much dislocation of traffic. Red traffic lights were ignored, and during the Blitz the daily march reassured Londoners that 'life went on'. Two elderly ladies, nicknamed Fortnum and Mason, sometimes ambled alongside the Picquet holding hands (with each other rather than with the Guardsmen).

The Picquet was led by a drummer. Tunic and bearskin did not seem to be the ideal uniform in which to repel marauders intent upon stealing the gold from the Bank's vaults. Moreover the sentry post outside the Governor's dining room was unlikely to be a serious deterrent to a would-be bank robber. So, from 1963, the Bank Picquet travelled by vehicle, clad in battle dress and armed with automatic weapons, and alterations in its deployment were made. The Governor of the Bank of England, Lord Cromer, a former Grenadier, proved most helpful during these innovations. At last, with the arming of the police and faced with military manpower cuts, it was reluctantly decided that there was no longer a justification for this duty to continue, and so since July 1973, the tramp of Guardsmen's feet has no longer echoed round Threadneedle Street.

The fourth regular Royal Guard was at Windsor Castle being provided daily by the Foot Guards battalion quartered at Victoria Barracks, Windsor.

GRENADIER OFFICERS' CAREERS

How were Grenadier Officers recruited for the Regiment? What was life like for an officer commissioned into the 1st Battalion at Chelsea Barracks in 1959? How was his career likely to evolve thereafter?

Regimental Headquarters kept files on potential officers, some of whom were put down for the Regiment by their fathers at birth. Since everyone had to do National Service, there were plenty of applications to join the Regiment. Those doing so could be confident that they would make some life-long friends. The Lieutenant Colonel Commanding the Regiment, Col Sir Thomas Butler Bt OBE DSO (to be succeeded in 1959 by AMH Gregory-Hood OBE MC) examined the candidates' housemasters' reports and interviewed them at schools such as Eton. Some candidates invariably fell at the various hurdles such as the extremely tough Brigade Squad at Caterham, the War Officer Selection Board and Eaton Hall Officer Cadet School. It was not easy to get into the Regiment: some candidates had to go elsewhere.

The Brigade of Guards, like the Cavalry, the Rifle Brigade and some others have been accused of choosing their officers from a narrow background, and it is true that the officers were not a cross-section of society. In 1960 well over half the members of The Queen's Company Dining Club were Old Etonians. The criteria for selecting officers over the generations has been – will they make good ones? Will they put the interests of their men before their own? And – will they 'fit in'? Any regiment, like any business, which selects people who will not 'fit in' is asking for trouble. As one Colonel of a Regiment put it: 'If you can draw people from the same background without lowering standards, it strengthens the family bond. And that sense of family in turn makes for reliability when you're facing bricks and bullets.' However, in fact the Grenadiers have welcomed promising potential officers from all backgrounds, including, for example, the son of a Polish immigrant, who proudly announced his background. He went on to win the Sword of Honour at Sandhurst. A good number of Grenadier officer cadets have won it too, but he, GF Lesinski, also won the Queen's Medal coming first in the order of merit, and he was later to be a fine Commanding Officer of the 2nd Battalion.

The RMAS course was increased in the late 1950s from 18 months to two years. Half the syllabus was academic which was frustrating for those who wanted to get on with proper soldiering. A few Sandhurst cadets applied to join the Regiment 'out of the blue' while there. Some did so with success. If the Regiment had two candidates of equal calibre, the one with a family connection would be chosen, while nevertheless some sons of distinguished Grenadiers were rejected as being unsuitable.

On joining the 1st Battalion at Chelsea Barracks in 1959 the newly commissioned officers were largely confined to barracks for the first few weeks since they were obliged to understudy in turn the barrack duties of the Corporal-in-Waiting, the Sergeant-in-Waiting, and the Picquet Officer. Their only escape was visiting the Regimental Tailor, then Meyer and Mortimer.

In due course they mastered Colour and sword drill and learnt the traditions of the Regiment and Household Division. For example senior officers were not called 'Sir' or saluted apart from the Commanding Officer, unless on a parade; that the Quartermaster, A Dickinson MBE, despite his two rows of medals, rank of Captain and immense experience was, seemingly, junior to them. For the newly joined young officer it took a while to understand all this, but it worked.

The Adjutant controlled the lives of the Ensigns and Subaltern officers. He 'went ballistic' with anger for the most trivial offences. RSM LC Drouet formed up the Adjutant's memoranda some distance from the orderly room. He ensured that there was much stamping of feet (not in itself an unusual occurrence) to drown the noise of young officers being shouted at. The Adjutant was seeking maximum efficiency and was, of course, quite oblivious to personal popularity. The Commanding Officer, seen only rarely, nevertheless exuded to the newly joined an intellect, a professionalism and experience, military achievement and charm, whatever the shortcomings of the young officer.

After several weeks new officers were examined by the Adjutant on their knowledge of the scale of punishment to be awarded to Guardsmen. Dirty Flesh, for example, warranted three drills rather than 'Here's a bar of soap, go and give yourself a good scrub.' It was all quite similar to the frightening test which new boys received, on house colours or whatever, on joining their Public Schools aged 13.

Meanwhile most young officers had received a flurry of notes from the Adjutant, left in the Officers' Mess 'pigeon holes' which read: 'Subject Discipline. You are to attend the Adjutant's Memorandum on . . .' A loose sash on dismounting Guard, for example, could lead to three extra picquets.

Some commitments were rather unusual. Near the Tower of London was a large warehouse run by the Customs and Excise. It was filled with contraband cigarettes. The officer was required to see that the cigarettes were present before destruction. Similarly, stores at an Ordnance Depot, earmarked for write-off, had to be checked. One wondered what corruption had led to Guards officers being required to undertake such irrelevant duties. And who, if anybody, ultimately destroyed 'the goodies' remained a mystery.

The newly commissioned officer often then had to attend a Corporals' Course run for Guardsmen seeking promotion. The emphasis was on the drill square but with tactical and weapon handling instruction too.

Within six months of being commissioned, regular and most short service officers then usually attended their special-to-arm courses at the Small Arms School at Hythe, and the School of Infantry, Warminster, learning how to command a platoon in war.

After a further two years the regular officers took up such appointments within the Battalion as the Regimental Signals Officer or commanding the Support Platoon which was equipped with 3" Mortars and the Mobat anti-tank gun. (These weapons were replaced by the 81 mm Mortar and the Wombat recoilless anti tank gun.)

Alternatively they may have instructed at the Guards Depot, Pirbright, or become an ADC whose responsibilities are described in Chapter 20. Thereafter, they could hope to be the Adjutant of a Battalion. Two Grenadier officers, both future Divisional Commanders in Germany, (MF Hobbs and AA Denison-Smith), were chosen instead to be Adjutants of the

Guards Depot. The latter was married and, traditionally, married officers were not then selected to be Battalion Adjutants, although there were exceptions to this rule.

Since so many officers left the Regiment so young, promotion was rapid for the remainder. Responsibility, such as commanding a company of 100 men, therefore came very early, perhaps at the age of 28, as opposed to mid-thirties for those in many other Regiments or Corps.

Passing the Staff College exam, a hurdle faced at about 30, was very important, not only because of what officers learnt on the one-year course and the useful contacts made, but also because the officer's first staff appointment thereafter was at Grade Two level, rather than Grade Three – and so he was two years ahead of those who failed. The exam then consisted of eight papers, all of which had to be passed, with an overall pass mark of 50%. A pass did not lead to automatic selection. The exam was not easy; of the Grenadiers, at least two future Major Generals failed at their first attempt; but those who worked really hard for up to two years preparing for it usually passed. The exam was not necessarily fair because those on active service beforehand had little opportunity to study for it.

To be promoted to Lieutenant Colonel and command a battalion, officers had to excel both on the staff and, in particular, when commanding a rifle company. The staff-trained officer did not necessarily have a successful career thereafter. As usual, there was considerable luck in being at the right operational place at the right time.

Between 1956 and 1995 there have been thirty-seven commanders of Grenadier battalions, of whom five have not attended a Staff College course. (All the five have received decorations or awards, two of them having the Military Cross.)

As officers moved up the military ladder, they served increasingly frequently away from their Battalion. Chapter 20 indicates the sort of appointments they took upon the staff and in command elsewhere.

There was no typical career to which retiring Grenadiers turned. Of the 1st Battalion's officers, photographed together in 1960, only three served a full career in the Army. They reached the ranks of (full) General, Major General and Colonel. And so what happened to the rest? Two became stockbrokers, and the remainder a judge, a university administrator in Australia, the director of a charity, a member of the Royal Household, a farmer, an overseas British Council official, a well known photographer, the manager of a chain of bookshops, the Household Division Public Relations officer, the representative of De La Rue in India, of Cathay Pacific in the Far East, the Treasurer to Household Division Funds, leaving only one who did not choose to work for long.

THE SOCIAL LIFE IN 1959 – 1960

The Regiment was pleasantly free of unnecessary customs in the Officers' Mess. The Queen's health is seldom drunk because several centuries ago only Regiments whose loyalty was in doubt were ordered to drink the Sovereign's health. True, hats may be worn in the Mess, perhaps because at one time they had to be pinned to the wearer's wig. Officers are expected to treat the Mess as their club or home. Before the First World War the Brigade of Guards had no Mess unless overseas – officers lived at home.

Some other Regiments dine their officers 'in' or 'out'. The Grenadiers do neither. A newly joined Ensign (a Second Lieutenant) was sometimes not talked to in the Officers' Mess, on principle, for the first few months. This may be due to him having been a prefect, or the equivalent, at school, followed by being a privileged Under Officer during his potential officer training. His seniors in the Regiment therefore occasionally ostracized him to ensure that he was not bumptious. He quickly discovered that he was on the bottom rung of a ladder which hopefully led somewhere.

Officers on Queen's Guard at St James's Palace could invite guests. They ranged from those of great distinction, to say, the French actress Brigitte Bardot, from the pretty deb to the impressionable father.

Some years before, BC Gordon Lennox had been asked by Harold Macmillan, then Prime Minister, while on Guard: 'How do you think the Government's doing?' Gordon Lennox, aware that Macmillan had just appointed the Duke of Devonshire to be Minister of State in the Commonwealth Relations Office, replied: 'I think the Government's doing very well, Sir, but some say you shouldn't have made your cousin a Minister.' The Prime Minister, without turning a hair, replied 'Unless you've known a chap all your life, you can't trust a fella.' Since he and his Government were subsequently let down so badly by his Secretary of State at the War Office, the remark has some significance.

Mess bills on Queen's Guard were large, making life difficult for those who had no private income. Some drank water on Guard: the pay was very inadequate. Fathers sometimes sought the guidance of the Lieutenant Colonel on what allowance they should give their son. He would reply, 'Give him a car and enough money to run it.' Not all fathers asked for or took this advice: four officers were without cars. Several had not even passed their driving tests; they were to be seen in Land Rovers with large L plates, driving under the instruction of a transport platoon NCO 'bunny hopping' around Sloane Square. They were given a wide berth by other motorists.

There was no ostentatious wealth; some officers' first cars were little, convertible Morris Minors, the forerunner of the Mini. There was also no

117

gambling in the Officers' Mess, partly because there was nobody there in the evening apart from the Picquet Officer. Since the Mess had few bedrooms most lived at home or with friends, although they could draw no allowance for doing so. (Guards Officers, traditionally, never buy each other drinks so the expense of 'buying rounds' is unnecessary, and excessive drinking is avoided.)

In some Regiments young officers' free time was strictly supervised, whereas in the Brigade of Guards officers did whatever they wished. And, inevitably, officers found that they had ample spare time in the 1st Battalion on Public Duties in 1959.

Some have never enjoyed such freedom before or since. Many chose the novelty of a very social life, as had, for some 300 years, their predecessors in London.

For a young officer who had been at school or Sandhurst before being commissioned the avalanche of invitations to the deb dances in the best hotels came as rather a surprise. However, young officers had the reputation of being fun and polite. Their availability depended upon the Adjutant's ruthlessness in awarding extra picquets (duties), but every hostess was only too happy to accept a substitute at the eleventh hour; for she did not necessarily know her guests. A very prompt and excessively warm thankyou letter was sufficient to ensure that the young officer was added to the list of eligible young men without which some hostesses could not make up the numbers. One hostess noted the name of a butler she was recommending to a friend. His name was absent-mindedly added to a list. He was amazed to receive a flood of invitations thereafter.

House parties at weekends, and lavish dinners, seldom seen since, added a new dimension to life.

It was difficult combining a hectic social life with some military duties. One officer's abiding memory is attending a ball at the Ritz, taking the 'milk train' to Folkestone, sitting upstairs on the open, double-decker bus to Hythe, arriving just in time to change from white tie and tails into battledress before breakfast, setting out for the ranges, and paying a Gurkha officer to mark up his target in the butts while he slept, despite the noise, before that night's party.

Gradually the whole scene changed. Soldiering became more serious and, to the regret of some, more competitive. Short overseas tours, for example in Northern Ireland, Cyprus, Belize or the Falklands, removed the young officers from the social scene. Moreover the grand 'deb' dances became considerably fewer due to the vast expense. The girls became more career conscious. By the mid 1990s those aged in their early twenties would not be surprised to receive an invitation saying P.B.A.B., standing for Please

Bring A Bottle! Many hostesses were quite happy for their guests to bring sleeping bags due to the drink-drive laws.

The Household Division point-to-point near Aldershot was an annual event. In 1959 Captains FJ Abel Smith, RM Micklethwait and 2Lt CR Glyn rode in numerous points-to-point. Everyone lived life to the full and enjoyed it immensely.

The Guards Boat Club at Maidenhead, meanwhile, was decreasing in popularity and later closed. Not many young officers regularly used the Guards Club in Charles Street although membership became compulsory. Free meals and drinks were available in abundance at so many parties that there was little need for a club.

Quite a few officers married the girl friends whom they met during the deb season. Most wives thereafter, but certainly not all, 'followed the drum' loyally and unselfishly, forsaking permanent lived-in homes or careers, in order to accompany their husbands in difficult, and occasionally dangerous, postings all over the world. When accompanying the Battalion overseas, as this History relates in other chapters, they provided invaluable help to the younger Guardsmens' wives whose husbands were being bombed or shot at in Northern Ireland or elsewhere. There were notable exceptions. Before the Second World War Lady Browning (Daphne du Maurier) 'was never happy in the role of being the wife of the (Grenadier) Commanding Officer. She didn't try to adapt and wasn't interested in foreign postings, regardless of her husband's career.'

When one Grenadier took command in Germany in the 1960s he was astonished to discover that the wives of all the Company Commanders lived in England. He refused the married officers' applications to go home at weekends, asking them what they were doing for their Guardsmen each weekend.

TIDWORTH

At last, after 19 very long months on Public Duties, the 1st Battalion moved in 1959 to Kandahar Barracks, Tidworth, joining 1 Guards Brigade in Hampshire. The barracks was magnificent compared to the old Chelsea buildings. The well-furnished, centrally heated barrack room was a novelty to some. This had a noticeable effect on morale and, it was hoped, on recruiting.

The Battalion was now equipped with the new self-loading rifle and the '1958 pattern' equipment. A much publicized Strategic Reserve or 'Fire-Brigade' was established in England. As a result the force level in Germany was reduced by 9,000 to 55,000. The 1st Battalion formed part of this reserve which gave extra motivation to the training. Everyone became fitter. The Battalion marched the length of Salisbury Plain in four exercises in short succession, culminating in the divisional exercise when, as enemy, the Grenadiers showed the supposedly fast-moving Parachute Brigade that they still had something to learn. The highlight of the exercise, in more ways than one, was the burning of the command vehicle in the middle of the night. It was a fine sight to see the bearded Pioneer Sergeant, LSgt Creswell, felling trees that were in danger of catching fire, as the petrol tanks exploded.

The Battalion entered two teams for a difficult 1st Guards Brigade patrol competition. It was won by 2Lt DH Morland's Queen's Company patrol and the signal platoon.

A very successful rifle meeting was held in September 1959. The Queen's Company, commanded by Maj GW Lamb and then PH Haslett, won every event most convincingly, and L/Sgt Bailey of that Company won the beautiful miniature Gold Gun which had been competed for most years since being given to the Battalion by the Hon James Lindsay when the Lieutenant Colonel in 1860. (By 1960 the other Company Commanders were Majors CCW Hammick, RAG Courage and MH Wise.) Two weeks later the Battalion won the 3 Division shooting competition. Maj Gen GC Gordon Lennox DSO who commanded the Division presented the cup. It seemed surprising how quickly a former Public Duties Battalion regained their professional military skills.

The Queen's Company also won the platoon competitions on patrolling and platoon attacks. The Company has always considered itself elite. The Company Commander, CSM and Platoon Sergeants were specially selected for the appointments, while the tallest of the newly joined officers often went to the Company.

The minimum height of the Guardsmen was six feet two inches while No 1 section of 1 Platoon, the senior platoon in the British Army, was six feet five.

The Queen's Company Dining Club dined annually at Boodle's Club in St James's Street. The Duke of Windsor, who had served in the Company, was always invited but no longer came.

The high-sounding title of 'Strategic Reserve' was something of a misnomer as the Battalion discovered when they participated in Exercise Starlight. 1 Guards Brigade deployed by air to North Africa. The Battalion was concentrated by the sea at Tmimi, 80 miles west of Tobruk. Old soldiers said how like/unlike it all was to Transjordan/Sinai/Egypt and Tripoli. Young soldiers thought how the desert was just as they expected. The Battalion marched all night, every night for a week covering 70 miles on foot. By day some slept, some planned the next move, while others stood by to meet the coveys of senior visitors. The Prince of Wales's Company, Welsh Guards, was attached to the Battalion.

The exercise culminated in an ambitious Brigade attack on foot at dawn. The Battalion's objective was the furthest of three enemy positions. With growing amazement the umpires watched the Battalion attack the right hill at the right time. Grenadiers routed and pursued the enemy, found by the remainder of the Welsh Guards. Obscene Welsh curses faded into the distance as the order came 'Exercise Starlight ends'. The exercise was most enjoyable but, with the benefit of hindsight, we saw that all three Services were not then equipped for such operations.

Back at Tidworth, renewed efforts were made to make soldiering fun for the Guardsmen. Lt Col Steele instigated a Go-Kart Club which was very popular. The three Karts were to be seen many evenings hurtling round the square, sometimes spinning in complete military abandon.

Tidworth was far from the main Grenadier recruiting areas such as Lincoln, Manchester, Nottingham and Derby and no Guardsman had a car then. Buses were therefore hired to take them home for the occasional long weekends.

Much sport was played and everyone was encouraged to lead a full life, getting on to the adjacent training areas when fatigues and barrack guards permitted. The Battalion was well understrength as National Service gradually finished.

The Regiment, and indeed the British Army, owes a great debt to the very willing National Servicemen, almost all of whom contributed much. They accepted discipline and hardship with good humour. The most able reached the rank of Lance Sergeant (ie a full Corporal) within two years. Many thereafter joined the police. The Chief Constables of half a dozen counties regularly advertised in *The Guards Magazine* for men from the Guards, such was the Guardsmen's excellent reputation.

Inevitably there were a few criminals in every regiment. Young officers gave evidence as to their character in court. Wearing uniform and carrying a sword, they were despatched at short notice to distant prisons to identify the miscreants, and speak highly of them, if possible. Defending soldiers at courts martial provided a new dimension to the young officer's career.

Fewer then than now of the Battalion were married and so there was no Married Families Officer. Even so, the number of married quarters was quite insufficient. Some quarters at Tidworth were tiny and most uncomfortable. In desperation several caravans were obtained and parked in a disused quarry, but they were no better. In those days the Army scarcely catered for young married soldiers. It will be seen, as this History relates, that much has changed for the better over the last 35 years.

By the summer of 1960 the Battalion, now commanded by Lt Col DW Fraser, had attacked or defended nearly every feature of the adjacent Salisbury training area at least once, and dug and then filled in their slit trenches on most reverse slopes.

After a week of Christmas celebrations culminating in a Battalion route march and leave, there followed a month's training with platoons dispersed from Devon to Staffordshire. When the Battalion reassembled, the Commanding Officer told them of the impending move. Rumour suggested Laos or Canada: money changed hands when we learnt it was the Cameroons – one of the most exciting overseas tours many experienced. Chapter 10 describes the Battalion's adventures there.

3RD BATTALION 1959 – 1961

After three years in Cyprus the 3rd Battalion completed their third post-war tour of active service. In August 1959 they moved to Wellington Barracks, commanded now by Lt Col AG Way MC with Capt JP Smiley as Adjutant and A Stevens the RSM. They were quickly committed to Public Duties.

In April 1960 General de Gaulle paid a State Visit to London. At Buckingham Palace the Guard of Honour was found by the Battalion, commanded by Maj IM Erskine. Two days later, on 7 April, the General reviewed the Household Troops on Horse Guards Parade. The Battalion was on parade with the King's Troop, Royal Horse Artillery, the Mounted Regiment, Household Cavalry, and the 1st Battalions of the Coldstream and Irish Guards. This was the first and last review of Household Troops by a Head of State, other than a hereditary monarch. It was a stirring parade, and much impressed de Gaulle.

Three years before, the Army had faced traumatic cuts. Fifteen Line battalions were earmarked for reduction and twenty-seven famous names were to disappear from the infantry order of battle altogether. The Army was to be reduced from 373,000 to 165,000 – only three-quarters the size of the pre-1939 Army.

When the decisions had finally been taken on the changes and approved by the Army Council and the Cabinet, and The Queen told, Field Marshal Sir Gerald Templer, the Chief of the Imperial General Staff, summoned the Colonels of the Regiments to a series of conferences.

Most of the Colonels took it very well. The principal exception was General Lord Jeffreys, Colonel of the Grenadiers, then aged 79. He refused to accept the loss of the 3rd Battalion which, having been formed originally from the companies that had been raised in Flanders in 1656, was the repository of the oldest traditions of the Regiment. To soften the blow, it was to be placed in 'suspended animation' rather than disbanded. Lord Jeffreys insisted on seeing the Field Marshal but was tactfully told that the 3rd Battalion had to go.

At the next annual dinner of the Grenadiers' Sergeants Past and Present Club, Field Marshal Templer was the guest of honour. He spoke on the problems of the Army reorganization and received a warm and sympathetic reception. But Lord Jeffreys who was also a guest made a speech attacking the Field Marshal, and attributing to him personally the responsibility for the demise of the 3rd Battalion. The officers and sergeants were embarrassed. The Field Marshal sat through it without comment, but he was deeply hurt.

* * *

On 11 June the Queen's Colour of the Battalion, carried by 2Lt JG Cluff, was trooped for the last time. The parade was commanded by Lt Col Way: some 300 all ranks of the Battalion were on parade.

Soon afterwards, on 28 June, the Battalion bade farewell to the Lord Mayor and the City of London at a parade in the grounds of Armoury House, the home of the Honourable Artillery Company. The Battalion then exercised for the last time its ancient privilege of marching through the City with drums beating, Colours flying and bayonets fixed. Afterwards the Lord Mayor entertained some officers to luncheon in the Mansion House and provided drinks for all who took part.

The highlight of the summer was undoubtedly the official Farewell Parade before The Queen in the gardens of Buckingham Palace. This was a truly sad and moving occasion. In addition to the 3rd Battalion, detachments of the 1st and 2nd Battalions and Comrades were on parade. Her Majesty addressed the Battalion and said, 'I gladly approve your leaving behind a Company, which will stand on the left flank of the Regiment, to preserve the individual spirit of the 3rd Battalion. I think that this Company, which carries such a great responsibility on its shoulders, should bear the name of one of the most resolute and daring engagements in which your Battalion ever took part. I therefore name it "The Inkerman Company", confident that it will ever maintain the steadfastness and courage which was displayed by the Battalion in that battle.'

The last few months of the Battalion's life were as full and as varied as any before. After more Public Duties, the Battalion spent three weeks training on the west coast of Scotland. After another State Visit, and the State Opening of Parliament, the Battalion provided a detachment at the funeral on 22 December for General Lord Jeffreys. He was unquestionably the most notable Guardsman of his day with a life full of military activity and achievements. Known as 'Ma', he had mainly dedicated his life to the Regiment. He was in a way the direct successor of the Duke of Wellington, as he was the next Colonel to succeed him outside the Royal Family.

Lord Jeffreys' successor was Maj Gen Sir Allan Adair Bt CB,CVO, DSO, MC who commanded the Guards Armoured Division from 1942 to 1945. Field Marshal Montgomery said of Sir Allan at the famous Farewell to Armour parade in June 1945: 'I do not know whether the officers and men of the Guards Armoured Division realize how much they owe to General Allan Adair . . . He trained the Division for battle in England; he took it to Normandy and commanded it in the great battles there; he then led it through France, Belgium, Holland and Germany . . . throughout all this time he never failed you. . . . You owe to him more than you can ever repay. . . . The Brigade of Guards was lucky to have had such an officer to handle this armoured warfare for them . . . I wish to congratulate him on having brought the matter to such a successful conclusion.'

Those who have read Sir Allan's autobiography, *A Guards' General*, will understand the genuine affection his men had for him.

*   *   *

The 3rd Battalion mounted its final Guard on New Year's Day 1961. On dismounting, the Colours were marched to the Guards Chapel for a Thanksgiving Service. So ended the active life of a historic Battalion. The Regiment was fortunate that it had not lost its identity as Grenadiers, as had so many other famous regiments on amalgamation or disbandment. Nevertheless those officers and men who were transferred to the 1st and 2nd Battalions will still say, 'Fings ain't wot they used to be.'

In January The Inkerman Company went to the 2nd Battalion in Germany as the living visible sign of an invisible spirit belonging to the 3rd Battalion.

CHAPTER 10

THE BRITISH CAMEROONS 1961

STEPPING STONES TO INDEPENDENCE

In early 1961 the Commanding Officer of the 1st Battalion, Lt Col DW Fraser, was briefed in utmost secrecy: 'Your Battalion is going to West Africa – the British Cameroons.' Within an hour a Pakistani contractor arrived at Tidworth to see the Adjutant with the same information, asking to come too.

On 12 May 1961 the Battalion embarked at Southampton on the troopship *Devonshire*. The Regimental Band played on the quay. As the coastline slipped slowly away, there were many large lumps in the men's throats. Within a matter of hours some were being sick but the voyage was fun. Training continued on board with lectures and runs around the decks. As the troopship sailed south, porpoises and flying fish could be seen in the water.

On arrival at Las Palmas a number of Guardsmen were late reporting back to the ship and finally arrived in taxis to the annoyance of RSM A Dobson. At Lagos most were not allowed ashore. Six Queen Alexandra Royal Army Nursing Corps nurses accompanied the Battalion. Their matron was reputed to check on their location each evening. Everyone was intent on enjoying themselves except the RAF ship's Adjutant who hid the Battalion's Colours in the ladies' lavatory. They were not found until all the Guardsmens' kit had been searched.

On 27 May the peaks of the 13,000 foot Mount Cameroun became visible. That night the *Devonshire* anchored in a heavy swell off the little town of Victoria.

The Camerouns, originally a German colony, had been occupied by the allies in the First World War and then became a League of Nations

125

British Cameroons, 1961

mandate, divided between France and Britain. The British Cameroons consisted of a long, narrow strip some 400 miles long and never more than 100 miles wide. It lay between Nigeria on the west and the French Cameroun Republic on the east. It was divided into two provinces known as the Northern and Southern Cameroons respectively. It was to the Southern Cameroons that the Battalion was sent, and to which this chapter refers when reference is made to the British Cameroons. Britain administered her mandate as a province of Nigeria while France ruled her much larger colony as a separate country.

Three months before the Battalion arrived the British Cameroons had voted in a plebiscite as to whether they wished to be united with Nigeria or the Cameroun Republic. There was an overwhelming victory for the supporters of the latter – to take place on 1 October 1961. The plebiscite had been largely and emotionally anti-Nigerian; it had been misleadingly publicized with the general promise, assiduously propagated by the Premier and his party, that British methods of administration would remain unchanged, and that no disagreeable consequence of unification would ensue. Had the Southern Cameroons voted to join Nigeria, as did the Northern Cameroons, they would have seen Nigeria torn apart in a civil war over the Biafra secession in the 1960s, and considerable high level corruption which is endemic in much of Africa. And so, with the benefit of hindsight, they probably made the right decision.

The Africans in the British Cameroons were delightful – English-speaking, peaceful, easy-going and friendly. However, they had a grave potential danger on their doorstep: a full-scale rebellion by the communist-penetrated terrorists of the Armée de Libération Kamerounaise (ALNK) was underway in the Cameroun Republic, and particularly in the area adjoining the British Cameroons where the Bamaleke Tribe were especially disaffected. There was every likelihood of the rebellion spreading to the British Cameroons after unification.

The Cameroun Republic had become independent of France in 1960. Capt IA McKay, Scots Guards, was serving there as a Liaison Officer.

The terrorists conducted raids in the Republic on plantations, assaults on Gendarmerie posts, ambushes, attacks on missions and villages. They committed murder, robbery, arson, kidnapping and rape. As the Grenadiers were to discover at first hand, the terrorists' camps were normally in inaccessible positions, difficult to approach and almost impossible to surprise. Intelligence estimated that the terrorists numbered about 2,000. Some neighbouring villages covertly supported them out of sympathy, self-interest or terror. A few terrorists had received training in North Africa or China. They were directed by a central committee of exiled members in Accra.

There were, therefore, two main threats to the security of British Cameroons. First, if it was decided that British troops withdrew and the Cameroun Republic Security Forces replaced them, the people would feel themselves betrayed over the plebiscite; civil unrest, arising from fear and resentment would follow. Secondly, terrorism would spill over into British Cameroons.

With these threats in mind, the Commanding Officer deployed his Battalion group in three locations.

Battalion Headquarters remained in the south at Buea, next to the seat of Government at Victoria and near the only port. Also there were Headquarter Company (Maj PT Thwaites) and No 4 Company (Maj JRS Besly). No 3 Company (Maj N Hales Pakenham Mahon) was at Kumba, over 50 miles to the north, close to terrorist camps in the Cameroun Republic at Mongo near the important centres of communication and

within easy striking distance of numerous plantations. Over 1,000 terrorists were believed to be there.

In the north-east, 280 miles from Battalion Headquarters, lay Bamenda
where The Queen's Company (Maj PH Haslett) and No 2 Company (Maj
CCW Hammick) were deployed. Bamenda had its own NAAFI, hospital,
bakery and workshop. The Companies occupied a half-tented camp close
to a small derelict racecourse on the top of the escarpment. Terrorists operated across the border, over 16 miles away in inaccessible camps in the
Bambutos Mountains.

The Battalion Group, apart from the Grenadiers, consisted of almost a
field squadron of Royal Engineers and a movement control detachment
RE, a troop of Royal Signals, and a detachment of Royal Army Service
Corps, Royal Army Ordnance Corps, Royal Electrical and Mechanical
Engineers, Royal Army Medical Corps and the Royal Air Force. The Royal
Pioneer Corps, Royal Army Pay Corps, Intelligence Corps and Army
Catering Corps were also represented.

Geographically, the area for which the Battalion was responsible can be
divided into two halves. The southern part consists of forests, plantations
and jungle. The northern rises to a plateau about 5,000 feet above sea level
where the forests give way to rolling open grasslands. The Battalion's tour
coincided with the rainy season. The country has one of the highest rain-
falls in the world. The state of the few roads was bad. It took well over 18
hours to reach Bamenda from Buea. A Royal Air Force squadron of three
twin-engined Pioneer aircraft was based at Mamfe. It had the only airstrip
capable of taking Beverley aircraft. The RAF was responsible for three
weekly runs, collecting mail and essential stores from Bali and Tiko, but
bad weather frequently led to the flights being cancelled.

Having taken over from the very friendly King's Own Royal Border
Regiment, the Battalion set about the primary task of Keeping the Peace,
ostensibly acting in support of the police who were seldom seen. This role
necessitated becoming familiar with towns and villages, liaising with the
police and civil administration and, above all, patrolling to gather intelligence, knowledge of the terrain and how to live in it. In addition rifle
companies had to be prepared to undertake counter-terrorist operations,
such as cordoning and searching villages, deep patrolling in both jungle
and savannah; and of carrying out 'snatch' parties and raids based on intelligence. Companies were not specifically trained for such tasks since the
priority at Tidworth had been preparing for conventional war. At platoon
commander level, this lack of training was not taken unduly seriously; it
was felt that knowledge gained at Sandhurst or Mons, and the platoon
commanders' course at Warminster would suffice.

Everything seemed so peaceful that within two weeks of arrival priority was given to the Queen's Birthday Parade, the last to be held by a British Battalion in West Africa. The towering mountains overlooking the tin-roofed, palm-tree-studded town of Bamenda were still covered by clouds as the Africans flocked towards the football pitch on 10 June. Long before the parade was due to begin the ground was surrounded by a chattering mass of Africans, similar to the excitement and atmosphere of a West Indian test match. On the left of the saluting base – an upturned water-tank – sat the native kings. In 1883 the two principal chiefs were Eyo Honesty VII and King Duke Ephrain IX; the latter once called on a British official wearing nothing except a top hat. Progress had not been very great. The Minister of Posts and Telegraphs in 1961 had been known to walk around half-naked. However, on this occasion the chiefs, called Fons, were resplendently dressed in brightly coloured robes. On the right sat the small European community; missionaries, police drafted from Nigeria, Government officials and coffee planters' families. While The Queen was riding on to Horse Guards Parade in London, The Queen's and No 2 Companies marched on, led by the Corps of Drums.

Tonal communications by drums was said to be more advanced in the district than anywhere else in Africa. Two careless German officers killed in an ambush in 1917 still 'talked' to each other, for their skins adorned the drums of the Fon of Banso. The parade formed up facing the Roman Catholic Mission. The arms drill was greeted with squeals of delight and little cries of 'wunnerful'. 'Right Dress' was regarded with serious contemplation. Never before had the Africans seen white men shout at each other so loudly – never would they again.

After the arrival of the High Commissioner, the parade marched past in slow time. The scarlet sashes of the sergeants were a flash of colour against the jungle green of their tropical uniforms. They were led by 1 Platoon of The Queen's Company which then had an average height of 6 feet 4 inches; the minimum height of the Company was only two inches less. After negotiating both goal posts, the parade broke into quick time. The Royal Standard was then broken, the crowd rose, the rifles crashed into the final 'present' and the National Anthem was played. As the Companies marched off, the crowd applauded. Tiny school children in little blue frocks from the missions stared with enormous eyes. An hour later the crowd had still not dispersed.

Some patrols were equally memorable. One five days earlier consisted of eighteen Grenadiers of 4 Platoon of The Queen's Company, accompanied by a medical orderly, a drummer from the Intelligence Section, an African policeman and six porters. They set off towards a village called Bamumbo on a mountain. It was to have been a short patrol to obtain intelligence and 'show the flag'. The village had never been visited by white troops before. It was soon apparent why this was so. The approach, over dry, open, hilly country, presented no problem until the Grenadiers reached the edge of an unexpected ravine. A village on the far side, glistening in the sun, could just be seen above a sea of thick mist. What lay between was unknown. No map existed. And so the patrol descended through thick undergrowth in the mist's dense gloom. Since the patrol was still descending they knew not where, to meet they knew not whom, the patrol commander suggested to the policeman that they should put their loaded magazines on their self-loading rifles. But the policemen appeared full of self-confidence and laughingly replied that this was quite unnecessary. He was right; at that moment bare-breasted women appeared, doubling up a path carrying heavy loads on their heads, much fitter than any soldier. Their voluptuous bodies rather encouraged the Guardsmen to keep going. After several hours it was impossible to return the way they had come due to the precipitous slope or to climb the opposite side to Bamumbo. After a drop of 1,600 feet a swampy river was found at the bottom of the ravine. How to return to Bamenda now became the principal concern. One Guardsman sat down and burst into tears: it was all too much.

The villagers hid when they first saw the giants with white faces who had appeared from the bottom of the cliff-face, an approach which nobody in their right mind would have used. One overcame his fear and led the patrol to the local chief. On meeting him, the policeman, overcome by the occasion, his self-confidence evaporated, collapsed on the ground, blowing furiously on his knuckles – a sign of great respect. The patrol later slept, quite exhausted, in the small mud huts, oblivious to the lack of any facilities such as illumination or water. The village turned out to be Bamumbo after all: it was at the bottom of the valley, rather than on top of a mountain. The following morning the chief's oldest son, and the schoolmaster who was the only educated person in the neighbourhood, provided useful information. Forty terrorists had been seen at a market the previous day. Their policy, apparently, was not to attack targets in the British Cameroons but instead to withdraw if threatened.

The patrol slowly climbed 1,500 feet up a rocky slope in a circular sweep towards a road. They stopped every 20 minutes to drink and then refill their water bottles from a cool stream, always inserting the sterilizing tablets. Half way up the slope a bloody message arrived saying the Nzden village was being attacked by people from Asang. Many were said to be dying. The patrol increased its pace. Three huts were indeed found to be on fire at Nzden. Arrests were made, but quite who was responsible was difficult to discover. Nobody was hurt and the crimson blotches on the message proved to be red ink. Several exceptionally fit NCOs had been despatched ahead and brought transport from Bamenda to meet them. Such patrols, after prolonged Public Duties in London and Tidworth, made the tour in Africa so very exciting.

No 3 Company in tin huts at Kumba was on the edge of the town, surrounded by bush and opposite the prison. The prisoners wore white shorts and chains and cut the grass alongside the road. One of the more popular places for the Guardsmen was the company canteen, run by L/Sgt F Corrigan and Gdsm D Westbury, helped by Watson, an African cleaner who wore a red shirt and claimed to be a Presbyterian rather than a communist. He was paid one shilling a day for bringing them hot water for shaving. (At the end of the Battalion's tour L/Sgt Corrigan arranged for Watson to buy the armchairs and table from the canteen which were transported to his hut, thus enabling him to become the only African in Kumba with a three-piece suite.)

The canteen sold everything. 'A report on sales and returns had to be sent to Buea every morning,' recalls Gdsm Westbury, 'so this meant stocktaking every night until 2.0 am, often by candlelight as the generator had shut down, and putting in our own money to balance the books until we had the hang of it. We may have had a cushy number but it had its dangerous moments when we had to shut up the bar under the eagle eye of CSM "Flash" Huggins: the irate Guardsmen threw empty beer cans at Frank Corrigan and me as we tried to drag down the bar's grill.

'I was occasionally sent out on patrol. Once, with 9 Platoon, we were taken by truck until the road finished. We were then despatched on foot deep into the bush. We were soon hopelessly lost. Eventually we came across a remote village where the Africans still spoke pidgin German from pre-World War 1 days. We went on and suddenly saw a white man in front of us. "Thank God for that," said the officer, whose name I have forgotten and who is now probably a General. Racing on ahead, the officer said, "Excuse me, Sir," but when the man turned round he was found to be an albino, a black man with a whitish skin. Fortunately he was the local school teacher and showed us the way back.

'The food in camp was terrible. We had tinned ham for breakfast, lunch and dinner, in every conceivable permutation, sometimes in batter, sometimes fried; in the end we decided to boycott the meals and congregated in the cinema where CSM Huggins appeared with the Queen's Regulations and read the passage that says mutiny is punishable by death. "But we don't shoot people now," he said. "You just serve a life sentence." We dispersed pretty quickly, especially as I only had three months to serve. Fortunately the cook was moved and the food improved.'

The Officers' Mess at Kumba made do initially, as at Bamenda, on inadequate tinned army rations, and avocado pears which grew everywhere and were eaten at most meals. But 2Lt the Earl of Lichfield produced excellent prawns. To the Company Commander's delight on 12 August a grouse arrived from Fortnum and Mason. However, he was away that day on some crisis and returned to find that the Senior Major, the Hon Paul Freyberg MC, had eaten it. Kumba was close to Lake Barombi where a Swiss timber company enabled them to water ski for one penny a circuit. The Swiss in turn attended the Grenadier film shows at a penny a seat, and so both parties could satisfy their authorities that facilities were paid for. 'Mind the crocodiles,' the shout went up, when the alarmed water skiers were dragged into deep water, but the lake contained nothing more dangerous than the prawns.

Across the border the threat was more sinister. Maj Hales Pakenham Mahon, on returning to Kumba after a visit, was met by a distraught planter who reported that twelve of his men had been murdered at his plantation twenty miles away at Tombel on the border with the Republic. The Company Commander sent Capt MF Hobbs that night to try to get the remaining terrified and intimidated plantation workers back to work. He found the murdered men bundled together in groups of three. An ingenious and successful method was made to entice the absent workers to return; they were shown a 'Wild West' film: such a novelty proved irresistible and all turned up. Who had murdered the twelve men was obscure. Some thought that the Cameroun Republic Security Forces may have been responsible.

Many patrols found the going as difficult as that described at Bamenda. On one, north-east of Kumba on the slopes of the 9,000 foot Mount Kupe, it was so steep and muddy that it proved necessary to hang on to creepers to avoid slipping into ravines.

To reassure the Africans near the murders at Tombel, a platoon was billeted in the village's huts. The Guardsmen found it quite an experience: the African workers were woken at dawn by the beating of drums and then went out to pick bananas. The platoon went deeper into the bush. 'We were put into a palatial but abandoned plantation owner's bungalow,' recalls Gdsm Westbury. 'We could see smoke in a deep ravine which was believed to come from a terrorists' camp. When it came to our turn to do a two-man patrol around the bungalow, the night was black and the rain was coming down so hard you would not have heard elephants charging. We were supposed to patrol separately and meet in the middle, but after nearly shooting each other twice, when coming on each other out of the dark, we decided to go round together back-to-back. On return to camp after such patrols, we were scarcely recognizable, being covered in large red blotches on our swollen faces, after being bitten by banana flies and everything else in creation.'

There was little time for sport apart from infrequent ad hoc football against the local villages, and one notable Kumba Rugger Club match against Douala in the Cameroun Republic. The team, which included eight Grenadiers, set off in pouring rain in a banana wagon advertised as the Blue Train. A launch then took them to Douala where they were met by 'shattering, strong-looking Frenchmen with beautiful wives,' Capt Hobbs wrote later. 'Placards announced the game as a "Grand International". We all felt flattered but not a little apprehensive.' The match was extremely rough but the Douala side was unfit and lost 6-0. The smell of garlic in the scrum was overpowering. At one of the numerous parties which followed a Grenadier Sergeant found himself chatting up a girl in pidgin English. She later told him that she came from Croydon and knew the Guards Depot well.

The Grenadiers were struck by how advanced the Republic was compared to the British Cameroons. The French approach to their Empire had resulted in investment in better education, good roads and cultured, well trained senior African officials whose wives bought their dresses in Paris. Some officials' wives in the British territory were still in grass skirts.

A few Grenadiers took the opportunity to shoot big game. On one occasion Capt DV Fanshawe was charged by two rogue elephants. Guardsmen Cottingham and Baker saved his life by each firing 20 rounds from their self-loading rifles. When Capt Fanshawe eventually married, he ensured they both were invited to the wedding, regarding them as the most important people after the bride and parents!

On another occasion the locals at Kumba asked Maj Hales Pakenham Mahon for a battalion of troops to kill an elephant devastating their banana plantations. 2Lt R Luddington shot the elephant and brought back the tusks, and a piece of trunk to eat (which was very tough and rather high.) He had taken a bottle of champagne with him and drank it while the locals hacked up the elephant.

A Royal Pioneer Corps officer attached to the Battalion organized a successful climb of Mount Cameroun. The Grenadiers erected a plaque on the top in concrete to record the event.

Meanwhile, important political and constitutional developments were under way. Tripartite talks between the British Delegation from London, the Republic and British Cameroons were held in Buea in June. The talks ended in a complete absence of agreement on any point between anybody. But it was a turning point because the Africans at last appreciated that the British Government would definitely withdraw all their troops and administration on 1 October. Defence and security thereafter would have to become the responsibility of the new Federal Cameroun Government. Lt Col Fraser believed that if this transition could be harmonious the Battalion might play a real and constructive role in reconciling the population of British Cameroons to their future, and in neutralizing their fears. He therefore visited General Briand, head of the French Military Mission to the Cameroun Republic, and gained his enthusiastic and whole-hearted support for the concept of an orderly and harmonious relief. All the senior French officers were impressed with the necessity of expunging the hostile impression of their Cameroun Republic Security Forces. There was no major joint operation with these Cameroun Forces, although the terrorists enjoyed sanctuary in their inaccessible camps on the frontier.

The Commanding Officer was fortunate to have direct access to the Military Operations Branch in the War Office. He occasionally proposed that more active operations should be initiated in the British Cameroons, but the British Government disagreed.

One exception was to be made. Perched high on the Bambutos Mountains, 22 miles from Bamenda and precisely on the frontier with the Cameroun Republic was a terrorist camp. During their tour the King's Own Royal Border Regiment had been close to it, but the occupants had slipped away into the bamboo jungle that covered the mountain's lower slopes and re-entrants. The Cameroun Republic forces had fought an inconclusive battle there, with casualties on both sides.

It was thought that an operation by the Grenadiers from the British side might achieve some surprise. No helicopter was available, and indeed if it had been the altitude and weather conditions would have made it of little use. The best chance of success lay in a long approach march by night, with stops put out, while it was still dark, on both sides of the frontier. The camp should then be attacked at first light.

For such an operation against a considerable number of armed terrorists a declaration of a State of Emergency was necessary to give the troops power, not only to shoot in self-defence, but also to shoot anyone who ran away after being challenged.

The British Government gave their assent (to the interest of the Chief Whip, Martin Redmayne, whose son Lt NJ Redmayne was the Battalion's Signals Officer). The State of Emergency was prepared in secret and promulgated the night preceding the operation on Sunday 9 July, to the dismay of Bamenda's District Officer who said that nobody would learn of the Battalion's new powers in time. He had a good point.

A week earlier, two recce patrols, led by Majors Haslett and Hammick respectively, had sought information on the enemy and terrain. Maj Haslett was accompanied by Lt OJM Lindsay and two NCOs. They followed a track which climbed up the side of the mountain two miles from the enemy camp. While moving across the ridge, four well-equipped terrorists were observed in the distance watching them. The patrol found a recently occupied sentry post some 500 yards from the main camp. While there, the Grenadiers saw eight terrorists moving tactically in single file approaching them from the camp. When the two parties were only 30 yards apart the Grenadiers stood up to show their presence and that they were ready to shoot in self-defence. The terrorists immediately went to ground and both forces hurriedly withdrew. The terrorists were wearing a dark brown battledress-type uniform, carrying equipment and all may have been armed.

Maj Hammick's patrol, consisting of three platoon commanders and two sergeants, was also observed by the terrorists' sentries and lookouts, all of whom appeared to be armed with rifles or sub-machine guns. The patrols, on going over each crest, were immediately greeted from their flanks or front by a call which sounded like a wildfowl decoy whistle. These calls were always answered from the enemy's camp area. It was apparent that there were observation posts on all prominent hills overlooking the approaches to the camp.

The aim of the forthcoming operation was to capture the terrorists in the camp. Stop parties were to be placed at first light on both sides of the mountain range while an assault party of forty-six approached along the ridge from the south, assaulting the camp when it was light enough to shoot. The whole force was to march in from the roadhead during darkness on the night 9/10 July. The stops on the Cameroun Republic side were found by No 2 Company while The Queen's Company provided the remaining stops and assault force. Security was difficult because troops had to be withdrawn beforehand from several detached platoon bases beyond Bamenda. A deception plan was made through Special Branch agents who leaked the information that an operation would take place elsewhere.

The Queen's Company despatched a small party ahead to act as guides on the 12-hour night approach. It was led by Lt Lindsay. An impenetrable mist came down. Each guide walked holding the shoulder of the man in front whom he could not see. One NCO dislocated his knee and another cracked his ankle as they stumbled around a marshy bog searching desperately for the tiny track. Fortunately over an hour later Lt Lindsay fell into a pond with water up to his waist. He had fallen into it before on a previous recce and so at least he knew where he was. Visibility increased to 15 yards and the track was found 100 yards away.

By 10.30 pm there was still no sign of the Companies. Had they already passed through? Were they already miles ahead? In which case should the guide party rush on to catch them up? By midnight they had still not arrived. It had been raining heavily for over 12 hours. The rain continued throughout the night. Suddenly through the mist the leading section of No 2 Company groped towards the guides. Maj Hammick told them that the drive to the roadhead had been a nightmare. One 3 ton truck carrying eighteen members of the assault party had slid over the edge of the mountain's mud track. The four worst injured had returned to Bamenda. The medical vehicle had driven into a ditch. Three hours had been lost: there was now no opportunity of finishing the approach under darkness and attacking at dawn.

The going was really terrible. Deep ravines, streams and bamboo jungles still lay between them and the vast, grey, forbidding mountain on top of which was the objective. Contact with the man in front became extremely difficult. Everyone was sliding and stumbling in the mud. Several collapsed and were left to recover. At the first halt No 2 Company brewed up hot tea despite their fires being visible. At the next stop weapons, clogged with mud, were cleaned. As dawn was breaking everyone was still on the lowest slopes of the mountain which towered above them, partly hidden by clouds and swirling mist. The terrorists' spooky calls to each other, signalling the Companies' approach, could be heard. Since time was now less important, breakfast was cooked at the third halt.

At 9.30 am The Queen's Company stopped 1,000 yards from the objective and spread out to an assault formation. The medical party, the specialist Military Intelligence Officer and the Special Branch Officer were all lost. However, 2Lt Lord Ardee and the 2 Company stops had reported on the wireless that they had been in position for at least several hours. The enemy camp was apparently still occupied.

'The final hill was very steep,' recalls Gdsm NR Harrison in the leading section of Lt JWR Larken's 1 Platoon. 'I was exhausted, but being fit and highly disciplined pushed on, expecting every bush to contain a terrorist. The summit was reached. Here the grass was dense and almost waist-high in parts. I had somehow lost my correct place in the formation and found myself in Gdsm John Lunn's position. He had married two weeks before leaving England and had expressed a personal sense of foreboding before the operation. L/Sgt Dawes noticed we were in the wrong positions and we were ordered to change places.

'The advance continued in line abreast, well dispersed with the Bren guns well forward. We could see the densely wooded copse that hid the terrorists' camp. I saw one of the enemy by a tree 50 yards ahead.' Suddenly the terrorists opened fire with automatic weapons on the leading right-hand section. Gdsm Lunn jerked upwards and crumpled. Fire was immediately returned 'Stop firing,' the leading commanders yelled. 'Bren Group prepare to fire rapid – Rifle Group – prepare to advance – rapid fire – advance.' This was the battle school drill which the Grenadiers had so often practised on Salisbury Plain, and it actually worked! The sections got up and tore forwards. A grenade exploded at the foot of one Guardsman but he emerged miraculously unscathed. After two more bounds the assault group entered the terrorist camp firing rapid into the tin shacks and likely enemy positions. But there was now no sign of anybody or anything. All firing stopped. Lunn lay where he had fallen. Blood was oozing from his thigh. A second bullet through the centre of his chest had killed him instantaneously. A drizzle descended. A thorough search of the camp was made before its destruction. Grenadiers manning the stops on the Republic side had seen in the distance several small parties of men, women and children.

On the British Cameroons side a stop party of two snipers, L/Sgt Binns and L/Cpl Pollard, at 4.25 pm challenged two terrorists who were emerging from the bushes below the camp 200 yards away. Both Africans turned and ran back. They were shot dead. The following morning one body had been removed by the terrorists along with packages the men were carrying. The other was photographed and buried. He had been carrying a home-made hand grenade.

Maj Haslett had planned to leave Lt Larken's platoon with a 62 wireless set in the enemy's camp when The Queen's Company withdrew. But this plan was abandoned due to the unsuitability of the ground.

At 5.pm withdrawal was ordered. It took 20 hours to carry Gdsm Lunn's body back to the roadhead. Fortunately two of Lt CT Blackwood's recce platoon's Fulani ponies were brought forward to carry the stretcher. During the exhausting return march the terrorists' sentries were again heard signalling. Virtually nothing of intelligence value was gained from the camp or the bodies of the dead terrorists, and the camp, though destroyed, could be quickly rebuilt. With hindsight, to reach the camp on its mountain peak with any degree of surprise was not feasible and the difficulties of the approach march had been underestimated. The heavy rain added to this problem.

Militarily the operation had gained little, but politically it had shown the Government of the Cameroun Republic that the British were prepared to face danger and take risks in order to assist in dealing with terrorism. This had an undoubted effect upon the general atmosphere and immediately led to more wholehearted co-operation in the political negotiations which were leading up to a peaceful transfer of power.

It was decided that Gdsm Lunn's death would not be announced at Battalion Headquarters at Buea until the following morning. But the news travelled by African drum to Buea that night.

Before leaving the Cameroons the Battalion trained a military force to supplement the unarmed local police. The British Government agreed to equip the force. Maj Freyberg and the Deputy Commissioner of Police screened the 140 volunteers. Time was too short to train those who were not former servicemen. They carried out an eight-week Guards Depot-type course of tactics and drill run by 2Lts JG Cluff and A Heroys and A/Drill Sergeant L Jeffery. Many of the Africans had served in Burma where one had won the Military Medal. The Commissioner took the salute at the excellent passing-out parade.

By 30 September the Cameroun Republic Security Forces were in position to take over from the Battalion. Relations between them and the local people were excellent. There was a general atmosphere of relief and relaxation of tension. No terrorist activity interrupted the handover. At midnight on 1 October British responsibility ceased. The following day the Battalions embarked on HMT *Devonshire* and sailed for England.

Four months earlier Lt Col Fraser's directive had read: 'I wish all ranks to understand that this may be the last appearance of British troops in this part of Africa; and by their bearing and attention to duty to create a respect which will survive the centuries. What matters most in the long term is that when the natives ask in the far future "What sort of people were the British?" the answer should be such as to give us pride.'

In 1989 a British Defence Attaché, Lt Col S Fordham OBE, Welsh Guards, was stationed in the Cameroun Republic. He reported that the British and the Regiment were, indeed, remembered with affection and nostalgia. Everything went well and peacefully for almost thirty years. However, in May 1990 there was a considerable groundswell of opinion for a move towards multipartition. This culminated in the unofficial launching of a new party in Bamenda and a massive demonstration by the Anglophobe population. Although it was initially peaceful, the police became exhausted and part of their cordon collapsed. Shots were fired and six of the crowd were killed. Despite this setback, in 1995 the Camerouns joined the Commonwealth.

'For most Grenadiers the 1961 tour was the greatest adventure of our lives,' wrote Gdsm DG Westbury. 'Most of us National Servicemen and short-term regulars had never been abroad before. The sight and smell of Africa were unbelievable and they lived with me for years afterwards.'

All who served in the British Cameroons would agree that the tour was very memorable. Such was the cameraderie which was built up that 25 years later all the officers who had served there, with those who had subsequently been commissioned, dined together. Everyone was present except for Maj PT Thwaites who was on a kidney support machine and Lt R Luddington who was sadly dead. We will remember the Africans with affection.

CHAPTER 11

BRITISH GUIANA

1963 – 1964

British Guiana, on the north-east coast of South America – once known as the 'Wild Coast' – had been a British possession since the Napoleonic Wars. Its once flourishing economy had suffered a severe blow from the abolition of the slave trade. The British Government therefore imported a quarter of a million Indians between 1844 and 1914 to work as indentured labour on the sugar cane farms, replacing the freed slaves who had earlier been brought over by European settlers.

By 1962 the Indians formed half the population and the Africans 33 per cent. The remainder consisted of Chinese, Portuguese and Amerindians, who were the descendants of the original South American Indians. The Indians were largely agriculturalists and lived in the scattered villages working on the sugar cane estates and rice farms, while the Africans were principally the factory workers and town dwellers.

British Guiana is larger than Great Britain but its population was little over half a million, all striving to maintain themselves upon the insufficient resources of a greatly underdeveloped country. Most lived in the narrow coastal belt where the only two towns – Georgetown and New Amsterdam – are situated. Most of this land lies below the sea level and is protected by a costly system of sea defences and canals. Dense forest, primeval, primary jungle covered the rest of the country apart from the savannas lying behind the north-eastern coastal belt.

In 1953 Dr Cheddi Jagan's militant leftist People's Progressive Party was elected. The British Government, fearing his marked sympathy for the communist world, suspended the constitution. Nine years later a general strike led to spasmodic rioting and the local police appealed for military assistance. The 1st Battalions of the East Anglian Regiment and Royal Hampshires were successfully deployed. They were replaced in October

British Guiana, 1963-4

1962 by 1st Battalion Coldstream Guards on a nine-month unaccompanied tour. The next general strike, during their tour, was considerably
worse. The danger of racial strife between African and Indian had been
growing alarmingly. Black mobs roamed the Georgetown streets beating
up Indians, who, in turn, attacked the Africans in the villages. Meanwhile
the 2nd Battalion of the Grenadiers, based at Caterham, stood by to replace
the Coldstream.

The advance party left England on 15 June 1963 in a chartered
Caledonian aircraft. On refuelling in Bermuda, considerable quantities of
champagne were enjoyed, supplied by Capt the Hon JDAJ Monson's
mother who had a house on the island. Well-fortified, the flight continued
through a frightening storm; the aircraft was hit by lightning and dropped

several thousand feet. The excellent pilot landed at Atkinson airfield with his passengers suffering from hangovers and the lightning.

The disturbances became so difficult to control that 2nd Battalion the Green Jackets were also despatched to British Guiana, arriving a week before the Grenadier main party. They were posted straight to various locations on the coast.

The Grenadier Battalion, commanded by Lt Col FJ Jefferson, had its Headquarters in Georgetown with No 1 Company (Maj GEV Rochfort Rae) and The Inkerman Company (Maj PM Lambert). No 2 Company (Maj ABN Ussher) and Headquarter Company (Maj the Hon DH Brassey) initially remained at Atkinson where the Americans had built an airbase on the edge of the bush during the Second World War. No 3 Company (Maj DW Martin) consisting of the Signal, Recce and Transport Platoons were spread everywhere; the rifle companies soon had some detachments in the mines, diamond and sugar estates.

Meanwhile Headquarters 2 Brigade commanded by Brig Trevor had also flown in. 'Where's my driver and vehicle?' were his first words. Lt (QM) FJ Clutton MBE MM, the Battalion Transport Officer, gallantly pointed to his own vehicle which the Brigadier kept thoughout his tour. Brig Trevor told Lt Clutton that he was to become the Garrison Transport Officer and the Battalion's Quartermaster, Capt LC Drouet MBE, was also made responsible for the Garrison, to the dismay of the Commanding Officer who lost these two key officers. But he had a strong Battalion Headquarters team in Capt JCF Magnay, the Adjutant, Capt MLK Healing, the Intelligence Officer, and RSM D Randell.

Moreover, for a change, the Battalion was trained to some extent for the role ahead, having been the spearhead Battalion beforehand, although they still had to find two guards for Trooping the Colour. They had earlier flown in Hastings aircraft around England, practising loading and unloading techniques. Cordon and searches and riot drill, as described in earlier chapters on Egypt and Cyprus respectively, had been well rehearsed at Caterham. Everyone was confident and knew their role – the note-taker to record the incidents, the bugler to draw attention to the unfurled banner reading 'Disperse, or we fire', the marksman to fire the fatal shot, the steel helmets, the bouncing of the wire towards the 'crowd', the donning of gas masks, the firing of gas, the fixing of bayonets. Such training continued on arrival in British Guiana.

There were a number of small unpleasant incidents. For example a patrol found an Indian trussed up like a chicken, with his eight-year-old son close by, cut to pieces. The black murderers ran away as the Grenadier patrol approached. The Indian's only crime was owning a shop in a black village. The boy's body was brought back in a sack.

There were a few occasions when platoons deployed against aggressive crowds. In one incident the Corps of Drums was at a large sugar factory on the outskirts of Georgetown. Most of the drummers remained with their vehicles discreetly out of sight, listening to the roar of the approaching crowd. The Platoon Commander, 2Lt JGL Pugh, with several other Grenadiers, watched the crowd alongside the police and a few Europeans who were running the factory. Long knives and machetes were waved aggressively at the security forces, but the ringleaders, aware of the Grenadier presence, were taking few risks due to a remarkable incident towards the end of the Coldstream Guards' earlier tour.

2Lt DN Thornewill had been with five other Coldstream trying to stop a gang armed with knives from attacking a crowd of cinema-goers. The gang suddenly charged the Guardsmen showing every intention of cutting them to pieces. 2Lt Thornewill, cornered in front of his men in an alleyway, shouted unheeded warnings and then fired one bullet which killed three and wounded one other. The Major General Commanding the Household Division, John Nelson CB, DSO, OBE, MC, was instructed by the Ministry of Defence to send 2Lt Thornewill back to British Guiana to face a civil trial for murder. This was probably unprecedented, although it has frequently occurred in Northern Ireland since then. Maj Gen Nelson, a distinguished Grenadier, refused to comply. 'I simply declared that the officer in question was unavailable,' he wrote afterwards. 'This refusal caused scratching of heads, and it was finally decided that I should myself go to Georgetown, inspect the Grenadiers and discuss the matter with the British Governor, Sir Ralph Grey.

'On my last evening at Government House His Excellency referred to the difficult subject that confronted us. He stated that he saw no way to avoid the predicament that faced 2Lt Thornewill. I reaffirmed my strong feelings on the subject and added that I would resign rather that order him back to Georgetown. His rather endearing comment that he did not consider that this was a case for implementing what he described as a 'last ditch' tactic gave me some confidence but the discussion continued for an hour until I began to feel that I was getting nowhere. Quite suddenly the Governor declared that our conversation had concluded the affair and I never heard another word about it.'

2Lt Thornewill's calm and sensible action undoubtedly saved the lives of his Guardsmen. Civil Servants in London offices find difficulty in appreciating such desperate situations. The one Coldstream bullet clearly had a salutory effect on trouble-makers.

To illustrate the point, Capt Monson recalls that: 'I was called to an incident in Buxton, a large sprawling village about three miles from Georgetown. It was reported that a crowd of blacks was trying to burn

20. The Bearer Party found by 2nd Battalion Grenadier Guards carry the coffin of Sir Winston Churchill down the steps of St Paul's Cathedral after his State Funeral.

21. A h
training. Member
of 1st Battalion in
Wadi Shawkah,
Sharjah 1968.

22. The Commanding
Officer, Lieutenant
Colonel D W
Hargreaves and the
Master Cook,
Colour Sergeant J
Fisher, stir the
Christmas Pudding
Sharjah 1968.

23. Captain A J C Woodrow and Drummers of the 2nd Battalion on the streets of Londonderry, 1969.

24. Dragon Boat Racing in Hong Kong with the 2nd Battalion, 1976.

25. On 14th May 1977 Captain R L Nairac, on undercover operations in Northern Ireland, was abducted by the IRA and later murdered. Robert Nairac was posthumously awarded the George Cross.

26. On 2nd May 1980 Captain H R Westmacott, while commanding his troop in Belfast, led an assault on an IRA four-man gun team. Richard Westmacott received the first Military Cross ever to be posthumously awarded. (A portrait painted after his death by John Seibels Walker.)

down the house of an unpopular Indian shopkeeper with the inhabitants trapped inside. We drove into the village to search for the house, but quickly found ourselves caught in a maze of narrow alleyways. We could hear the noise of the mob but could not find a way through for our vehicles. I dismounted to look for a means of getting through on foot and told the patrol to remain in the vehicles until I waved to them. I rounded a corner and in front of me found the house and a mob of about 50 yelling black youths, some brandishing cane cutters' machetes and other crude weapons.

'Assuming the patrol could still see me, without looking round I waved them forward. In the belief they were at my heels, I marched confidently onwards, not even bothering to draw my pistol. As I approached I shouted something like "Clear the street. Go back to your houses", and so on. The crowd began to melt away and remarkably soon the street emptied. It became apparent that no serious damage had been done to the house. The family was unhurt though very frightened. It also became clear that I was totally on my own. The patrol had not seen me wave and had simply stayed in the Land Rovers as instructed. A police superintendent told me afterwards that the crowd were probably under the impression that, like Digby Thornewill, I had a "magic gun"!'

Companies rotated frequently. A camp at Takama proved useful for training and field firing despite the heat, flies and rattlesnakes. The humid weather was a contrast with a few months previously when the Battalion ski team (Maj Lambert, Lt MR Westmacott, 2Lt DP De Laszlo and 2Lt AMH Matheson) had competed in the Army Ski championships in Scotland. Sport in British Guiana was a high priority and included swimming, cricket, tennis, hockey, weightlifting, runs, go-karting and volley ball. The rugby team, organized by Lt JED Browne and trained by D/Sgt RP Huggins, had exceptional success. The football team took pride of place, being regarded in Battalion eyes as 'the Englishman's game', but they found the local sides to be of a very high standard.

Inevitably the operational situation dominated most events. The platoon commanders in isolated detachments on the estates enjoyed considerable freedom; the roads leading to them were so inadequate that the official visitors were mercifully few. Whenever possible Grenadiers used the antiquated railway that ran between Georgetown and New Amsterdam; it was quicker by train and more comfortable.

To protect the white managers of the vulnerable estates, ambushes were laid on mud tracks but the arsonists were seldom caught. On night patrols to nearby villages a Grenadier would be discreetly dropped off to observe anything unusual while the patrol stood-to twenty minutes away.

Patrols were on foot, in open landrovers or in three tonners. Many villagers were friendly, particularly the children who often accompanied the Grenadiers on foot on the dawn patrols. Slogans on walls read 'British Go Home'. Radical Guianese believed that the whites were exploiting them. L/Sgt NR Townsend, serving with the Drums, recalls a typical patrol he commanded. 'We set off in two Land Rovers from Rosehall Company base towards the Dutch border to check on the sugar estates where arsonists had struck. After picking up a policeman and the estate foreman, who was a huge negro, we loaded our rifles. After several hours on dirt tracks we came across occasional clearings in the jungle with several dozen houses on stilts above the swamp. The acres and acres of sugar canes looked greeny brown and rustled in the wind like a bamboo forest. We saw little game as we drove slowly back to Rosehall.

'On our return we were told an arsonist had destroyed a large area in our absence and so we spent the entire night searching for him – although it was like looking for a needle in a haystack. The police were sometimes ineffective and had divided loyalties, although their Riot Squad was very good indeed. Without a guide one became quickly lost and drove into ditches. At night our vehicles' headlights reflected thousands of red dots – the eyes of countless, small grouse-like birds. There was the same effect on water – from the eyes of small alligators. No Guardsman was ever attacked by them, except after a glass of rum too many, and then there was no teeth mark for proof!'

Each platoon undertook adventure training, long before such training became fashionable. 2Lt AA Denison-Smith took his men up river. His plan was to use native canoes but they were too old: the planks shrunk in the hot sun causing them to sink. Fortunately a boat carrying milk picked up the platoon and pulled the canoes behind, thereby enabling the sodden planks to expand, making the canoes usable once more. The patrol took arms and ammunition but little food as they hoped to live off the land. Fish were caught but the piranha stripped them of flesh while they were still being hauled in. The piranha made it unwise to swim in some rivers. At night the fish could be attracted by shining a torch on the water. They were then hit on the head.

2Lt The Master of Rollo on a route march with his platoon suddenly came across a blond on a raft. He could not believe his luck until her husband, who had been looking for diamonds, surfaced alongside her.

Lt PAJ Wright's expedition was the most challenging. Some engineers on the Diamond Estate welded together 24-gallon oil drums to make an excellent raft. Ten Grenadiers sought to spend ten days on the raft travelling 110 miles down the Demerara River. They were successful despite unexpected rapids, falls, a broken oar, snared ropes and finally almost being swept into the Atlantic Ocean.

Captains Magnay and CR Acland set off for a ten-day expedition in the

Kanaku Mountains, shooting quail and duck. They crossed by landrover into Brazil. They stayed with the manager of a cattle estate who had a pet eagle, puma and baby seven-foot anaconda which become the biggest snakes in the world. The Grenadiers swam to the launch when water skiing; a fortnight later a negro was swallowed there by one such snake.

Expeditions were arranged to the Kaieteur Falls. Each party consisted of an officer and fourteen men; in the autumn such groups were setting off every ten days. A visit involved a week's round trip by sea, river, road and finally on foot. They came across amazing sights – a 'frontier' town, full of saloon bars and swing doors, typical of the cowboy films; diamond prospectors sifting streams; women in sarongs and natives in loin cloths; Amerindians baking bread in biscuit tins over paraffin fires, and finally the impressive 741-foot falls. The Kaieteur Falls are the highest in the world – four times higher than the Niagara Falls of Canada.

University students from Europe and America were carrying out scientific studies, trying unsuccessfully to locate a near-extinct 'golden' frog. L/Sgt Townsend found the frog resting on a leaf in the middle of a puddle. It resembled the top of a pineapple and had a yellow body and golden-coloured eyes. He told the local guide who was most excited and had never seen it before.

On such expeditions the Grenadiers ate typical compo-tins of spam and stew which was sometimes followed by tinned golden syrup pudding which was inappropriate for the jungle.

Rum was the same price as coca-cola, necessitating RSM Randell keeping a close eye on consumption: drinking too much was a severe temptation. Some Mess incidents come readily to mind. During the take-over from the Coldstream, demonstrators, close to the Officers' Mess in Georgetown, led to the firing of smoke shells. Lt JR MacFarlane of the Coldstream was woken from a deep after-lunch 'zizz' on the veranda and called desperately for a glass of something, whereupon a Grenadier waiter, wearing a respirator, promptly appeared with a glass of champagne on a silver tray. Maj Martin enjoyed sipping gin and tonic outside the Mess, firing his Luger pistol at large vulture-type birds which lurked in the palm trees above him. The Brigade Commander accompanied one eight-hour patrol to an isolated estate named Leonora manned by 2Lt Pugh who was celebrating his birthday. Champagne was delivered in hayboxes with ice and cold chickens. The Brigadier was most impressed and talked about it for years afterwards.

The Royal Green Jacket Battalion excelled in everything, beating the Grenadiers in forced marches and shooting competitions. The Grenadiers refused to take the competitions seriously, with the exception of one officer who took everything so seriously that he had to be ordered by the Commanding Officer to take a fortnight's leave in Barbados.

The RGJ Commanding Officer, Lt Col G Mills, used to give lectures to his officers after supper and discouraged them from visiting Georgetown in the evenings as he thought that there was nothing worthwhile there. The RGJ officers asked Lt Clutton to arrange a discreet, regular run in a closed vehicle to take them to and from Georgetown after the lectures.

There was little sickness in British Guiana and only one death – L/Cpl TM Harrison who died from peritonitis while on adventure training en route to Kaieteur in the bush. Help could not reach him quickly as there was no helicopter. He was buried with full military honours, as was the case with all Grenadiers buried abroad. The funeral was moving and very peaceful. During the padre's address a mysterious elderly negro lady appeared in floods of tears. She wore a large black hat; through her thick veil it could be seen that she had covered her face with white powder. She tried to throw herself on to the coffin. Beyond the small cemetery were abandoned, rusty Second World War Bren carriers.

The military presence in British Guiana proved so effective that the State of Emergency was lifted after the lengthy strike had collapsed. The Green Jackets returned to England.

Resupplying isolated platoons and servicing the elderly vehicles proved difficult; United States aircraft could be used for dropping spare parts by parachute. Splendid fitters of the Royal Electrical and Mechanical Engineers slept by day and serviced the vehicles by night when the transport was not required for operations. Lt JED Browne had learnt to fly at school and piloted the Commanding Officer in a small Cessna aircraft. On one occasion they smelt burning and the cabin filled with smoke. The small fire extinguisher appeared quite inadequate. The aircraft was serviced by the Transport Platoon but the instruments were in order. Fortunately the fire was discovered to be burning sugar canes well below them.

Gradually the Battalion's tour drew to a close. Most had enjoyed it. In March 1964 they handed over to The Queen's Own Buffs, The Royal Kent Regiment. Before doing so, the Chief of the Imperial General Staff, Gen Sir Richard Hull, visited the Grenadiers. He congratulated them most warmly on a job well done.

In retrospect one of the principal reasons for the performance of the Battalion during the tour was the outstanding service of Capt Drouet, Lt Clutton and RSM Randell. Nothing was too small to escape their attention and no problem was too difficult for them to resolve, a vital help for a battalion spread out over many miles of very poor communications.

Later in the year Dr Jagan lost the election, and in due course British Guiana received its independence, although the racial violence and disorders continued. But by then the 2nd Battalion was facing fresh challenges far afield. Thirty years later Guyana, as the country is now called, has not prospered. It has become one of the poorest nations in the Caribbean with high unemployment and a daily struggle against poverty, oppression and violence.

CHAPTER 12

SOLDIERING IN GERMANY

1954 – 1981

Grenadier Battalions have served in Germany on eleven tours since 1945. Chapter 5 discussed the early post-war years, while Chapter 26 covers in detail the most recent tour.

As is the case with the Northern Ireland tours, it is impossible to write about each in some detail, otherwise this History would lead to many volumes. *The Guards Magazine, The Grenadier Gazette* and the Regimental Intelligence reports in Regimental Headquarters are there for those who want to know more.

This chapter will briefly refer to each of the tours, explain why the threat of Communism necessitated the British Army's presence in Germany and summarize the evolving tactics. Finally, Berlin and the Grenadier tours there between 1954 and 1981 are covered in their own separate section at the end of this chapter.

*   *   *

In 1955 the 1st Battalion moved from Berlin to Hubbelrath near Düsseldorf where Maj PGA Prescott MC was commanding No 2 Company: 'We were still the occupying forces. Training was totally uninfluenced by nuclear weapons, perhaps because the Russians didn't then have them, I believe. We were utter pedestrians. "CSM, we've got to get our training programme to the Adjutant now. What the hell can I put in on Thursday afternoon?" "What about padre's hour, Sir?" "No! I've put that in on Tuesday." Few things stand out. First was our total frustration at being able to do nothing about the Russian occupation of Hungary. We didn't even "stand-to". But we did when the Government passed from "us" to the Germans, and we could have deployed to the Saar in case there was trouble with the "mixed" population there. Major exercises were based on the concept that we would stop the Russians astride the Rhine and so we trained in that area. Neville Wigram, our Commanding Officer, was seriously injured in a vehicle accident on an exercise and so Rex Whitworth took over.' The Battalion moved to Chelsea Barracks in 1957.

*   *   *

In 1959 it was the turn of the 2nd Battalion to serve in Hubbelrath under command of Lt Col FJC Bowes-Lyon MC followed by Lt Col AG Heywood MVO MC. 'We still operated on foot and had Second World War equipment,' recalls the then Major DW Fraser. 'The operation plan made no possible sense.' 'Training in nuclear, biological and chemical warfare was talked about, but it did not go much beyond lectures,' remembers Maj LAD Harrod. 'We were losing our 3rd Battalion and so Jim Scott, Gilbert Lamb, Sandy Gray and I had no chance of command. L/Cpl Ayleward in my Company applied to transfer to the Welch Regiment. 'The what?' I replied, rather shocked. Six weeks later I was offered a transfer to them with a view to command.' Ayleward ended up as the RSM of the Welch Regiment's TA Battalion, while Maj Gen Harrod became the Colonel of the Regiment.

Lt Col Heywood particularly remembers the 2nd Battalion's remarkable sporting success, for the Grenadiers continued the tradition by winning the BAOR athletics for three years. 'The armoured personnel carriers, called "pigs", were the most useless four-wheel vehicles with almost no cross-country performance,' he noted. 'We had the usual annual exercises including an amphibious one with the German Navy in Eckenforde Bay against the Danish Army which was fun. I was lucky to have Fred Clutton as RSM. In 1961 I brought the Battalion back to Caterham and handed over to Francis Jefferson.'

*   *   *

Between 1959 and 1962 4 Guards Brigade was commanded by Brig John Nelson DSO OBE MC: 'The whole of the NATO strategy was based on the use, or, at least, the threatened use of atomic weapons in order to compete with the superiority of the Russians in manpower and in conventional arms,' he wrote later. 'NATO had enough troops to contain a Russian attack for a very limited time but, hopefully, long enough for the threat of a full-scale nuclear war to cause the aggressors to call a halt. . . . I had under command the Inniskilling Dragoon Guards and three transported infantry battalions of Grenadier, Coldstream and Welsh Guards. As a result of exercises I thought I had a reasonably clear idea of what the correct composition of a division should be in the context of European War with tactical atomic weapons.' So much so that Brig Nelson told the Prime

Minister, Harold Macmillan, how he could save currency and army manpower by reorganizing the Corps' command structure. This discussion took place in the Dorchester Hotel at the annual Regimental First Guards Club Dinner. The Prime Minister encouraged Brig Nelson to state his views which were subsequently put to the Chief of the Imperial General Staff, Sir Richard Hull. Brig Nelson proposed to cut out one of the many layers of headquarters, thereby saving 10,000 men and increasing direct control and efficiency. The proposals came to a grinding halt because CIGS's staff did not agree.

*   *   *

It will be difficult in the twenty-first century to understand why such a high proportion of the British Army was required for Germany, necessitating vast expenditure for 50 years. Despite the benefit of hindsight, some Grenadiers who served in Germany may have been rather sceptical as to whether the threat from the Soviets and Communism really justified such an exhaustive allied effort for so long. What was the threat?

Communism has been accurately described as a revolutionary creed resembling a religion; as applied godless materialism; as a philosophy of life which seeks to dominate and control the whole of man's life. Communism is totalitarian in its demands. It seeks the whole of man's allegiance. In the post-war years it was the most dynamic political and economic force in the world, with a membership of 45 million people from 88 Communist parties, ruling the destinies of 1,200 million of the world's population. Communists then believed that they had a destiny to create a new world and regenerate mankind. To do this they had to conquer the world, shatter the Capitalist system, and, by Communist dictatorship, establish the regenerative environment of socialism.

A series of treaties of friendships legitimized the continued presence of Soviet troops in satellite countries and served as a cover for a vast network of security agents. In such countries there was strict censorship of all literature, arts and news media, and severe restrictions against the activity of the Church. Trade Unions became mere government productivity organs. Purges and murders were commonplace. Such was the brutal, dominating, monolithic structure created by Stalin. But Khruschev in 1956 proclaimed that Communists could achieve power by peaceful as well as violent means. The publication of his secret speech, revealing the facts of Stalin's terror, released the pent-up national feeling in Eastern Europe. This brought Gomulka to power in Poland and provoked the national uprising in Hungary which was put down by Soviet tanks. The political climate in East Germany was harsh for some 50 years. In 1964 more liberalized Czech policies led to armed intervention by the Soviet

Union and Warsaw Pact. The Pact was a military framework for a form of political unity imposed by the Russians. Its purpose was to secure the political subservience of Eastern Europe to Soviet arms and policies by the threat or use of force.

Communism was not studied in detail at Sandhurst, nor at Mons Officer Cadet School (for the National Service Officers). There was no attempt to indoctrinate Guardsmen about the threat posed by the Warsaw Pact. Key personnel in Battalions being sent to BAOR studied the potential enemies' equipment and tactics rather than their philosophy. It was not until platoons undertook patrols along the East German fortified border – the Iron Curtain – that they saw and understood. The East Germans and others had constructed close-meshed fences, bunkers, watch towers, floodlights, dog runs and anti-vehicle ditches, amid millions of anti-personnel mines. Houses had been pulled down and bridges blocked – all to prevent those in the East fleeing to the West to seek freedom and prosperity.

When Grenadiers served in Germany in the early 1950s there was talk of the Red Army steamroller about to be set on an unstoppable course to the Channel. By the time the 1st Battalion moved to Hubbelrath in 1963 such an invasion was seen to be increasingly improbable. On the other hand there was the chance of miscalculation. If NATO's defences were allowed to deteriorate to such an extreme that a rapid *coup de main* against part of Germany, such as Berlin or Hamburg, was possible, then who could say with confidence what the Russians might do?

*   *   *

The 1st Battalion's training before moving from Chelsea Barracks to Hubbelrath in 1963 was quite inadequate had there been a serious possibility of war. The Battalion had spent the previous two years on Public Duties. The annual training consisted of a month at Dartmoor in September for field-firing exercises and cross-country movement on foot. The final exercise against 1st Battalion Coldstream Guards was an unequal contest, for the Coldstream had very few maps and were unfamiliar with the ground. With the help of a thick mist on the second morning the Grenadiers had no difficulty in advancing through the Coldstream positions and were able to return to camp some 24 hours earlier than the London District umpires intended.

In November 1963 the Battalion, commanded by Lt Col JFD Johnston MC, with Maj NJ Clarkson Webb as the Senior Major, moved to Gort Barracks, Hubbelrath, near Düsseldorf, which the Battalion had vacated seven years before. The barracks still consisted of leaking, temporary buildings. The Battalion was equipped with old Humber armoured one-ton personnel carriers and so driver training was the top priority.

The Company Commanders (Majors BC Gordon Lennox, DV Fanshawe, JP Smiley, CJ Airy and FJ Abel Smith) were shown the deployment plans. They briefly and discreetly looked at the ground upon which they would fight should the Warsaw Pact's divisions invade. The ground was flat, completely open and often waterlogged.

The concept of operations in the early 1960s, in simplest terms, was that a surveillance force of static observation posts and patrols would report on the enemy's approach, so that nuclear strikes could be selected, and the movement of reserves planned. The Battalion's tasks were to guard the enemy's approaches, delay them and make the attacker present targets for our tactical nuclear weapons. To save manpower and money all the battalions in BAOR had only three rifle companies, although the Second World War showed that four were essential. The fourth company was to come from the Territorial Army but the extent of its training and time of arrival were unknown. The ground was far from suitable for prolonged defence, but, for political reasons, it had been decided that the main battle should be fought close to the East German border. In view of the Warsaw Pact's overwhelming strength in armour and infantry it seemed probable that the allies could not long delay them. But any delay would buy time to resolve the crisis by diplomacy. If that failed, and the enemy had broken through the defences they would be hit by nuclear weapons quickly and accurately. The whole of NATO strategy was therefore based on the use, or at least the threatened use, of atomic weapons in order to compete with the numerical superiority of the Russians in manpower and conventional arms. It was hoped that the threat of a full-scale nuclear war would cause the aggressors to call a halt.

There was no guarantee that there would be sufficient time to deploy. The Battalion therefore had designated areas close to Gort Barracks to which it could rapidly move should an enemy nuclear strike be imminent. Alternative areas were available for peacetime exercises.

During the first few months in Germany courses were run on every imaginable subject. The Signals Officer, Capt GA Alston-Roberts-West, taught the new wireless procedure on sets not seen before. The Intelligence Officer took every company through a resistance to interrogation course. A Line Battalion had run riot in one of the German garrison towns due to boredom and drink. This led to a new policy that everyone should learn some German. The Intelligence Officer, after a scratch six-week course, taught it to the Guardsmen in each rifle company. The conversation was romantically inclined, on the lines of 'Would you like to dance?' 'Yes please,' or 'No I am engaged. There is my sister.' Judging by the overloaded bus to Düsseldorf most evenings, the lessons were of some benefit.

Every company underwent a nuclear and chemical week, but, apart from gas masks, there was little or no equipment which would enable us to survive nuclear war. Training, whether with or without troops, took up most of the time. Manoeuvre areas were still extensive and provided great scope. A farmer would readily agree to the use of his barn for a bottle of whisky.

The exercises were very worthwhile. They involved considerable night movement on foot or in vehicles, rapid redeployment, and included all phases of war in a nuclear setting. On their conclusion the Commander of 4 Guards Brigade held a debrief in a pub, attended by most officers. Criticism was levelled and lessons learnt. Reputations were made or marred. Successful command of a company or battalion in BAOR was important for promotion.

There was plenty of sport and the Battalion rented a fairly inaccessible ski hut in the German Alps. Over 100 Grenadiers undertook 'winter warfare training' there for ten days. Such training was a misnomer for it was elementary skiing, taught by a few officers; we could not afford local ski instructors. Later 2Lt AMH Matheson became the principal skiing instructor.

Capt CT Blackwood represented the Regiment at the 150th anniversary celebrations at Bayonne and Biarritz, where the Regiment had a plaque in St Andrew's Church. It recorded their casualties in 1814, the last occasion on which the French fought the British on French soil.

In April the Battalion classified with their rifles and Bren guns at Haltern after which The Queen's and No 2 Companies practised river crossings.

The former Adjutant of the Battalion, DHC Gordon Lennox, was then serving in HQ 4 Guards Brigade: 'Soldiering was taken seriously but few really believed the Russians would attack us. Our plans to defeat them were studied much more seriously ten years later. One day the Brigadier responsible for Intelligence in BAOR rang up to tell me that our Top Secret Emergency Deployment Plan had been located in Moscow. It had been lost by a Gunner Regiment. The Brigadier turned up; the meeting – to discuss the implications – took place on a cricket pitch as the Russians were unlikely to have listening devices there.'

The 1st Battalion met some Russians rather unexpectedly as BC Gordon Lennox, then commanding The Queen's Company, recalls: 'We had just finished an exercise with two squadrons of armour on a Belgian training area, when I was told that the Company must give an important demonstration in three hours' time to the Soviet Military Mission in West Germany – on house fighting. The Company was far from familiar with that phase of war. But great bangs of thunderflashes and lots of coloured smoke carried great conviction. The Russians appeared to much enjoy the demonstration and the lunch which followed.'

Officers' contact with Germans was less than minimal, apart from one local civic reception. Anne Hope, the wife of the Transport Officer, was eight months pregnant and found herself seated between two German gynaecologists! The army pay was so inadequate that young married officers could not afford to eat in a smart restaurant. But the officers' married quarters at Wuppertal to where the Battalion moved in 1964 were excellent.

While there, the old one-tonners were scrapped for fifty rather unreliable Saracen armoured personnel carriers. A REME Light Aid Detachment was formed to oversee the transition. During February 1965 the Battalion, now commanded by Lt Col MS Bayley MBE, was busy preparing for a brigade exercise in North Germany. However, the training was reduced to an enjoyable Battalion exercise because, most unexpectedly, in April the Battalion was committed to six months with the United Nations in Cyprus. Chapter 13 describes this interesting tour. Two months after their return to Germany the Battalion moved to Caterham.

The Regiment continued to be represented in Germany, for the 2nd Battalion, commanded by Lt Col AN Breitmeyer, took their place in Wuppertal, before the move to Munster in 1968. 'It was difficult keeping the interest going,' he remembers. 'It was the same training areas and similar exercises. Due excitement was caused by a new infantry radar we tried out at Sennelager. As a Coldstream battalion advanced towards us, Anthony Denison-Smith and I suddenly saw the radar's screen break into little tiny dots – the equipment worked! Company training in Denmark and France were the highlights of the Battalion's tour. But we had no proper armoured personnel carriers to provide real mobility, and so there was no operational alternative to static warfare and digging in.'

Lt Col Breitmeyer's Adjutant was Capt MLK Healing who was to die of cancer four years later. The RSM was D Randell, to be replaced by RP Huggins. In 1966 Lt Col PGA Prescott took over from Lt Col Breitmeyer. These two commanding officers were the last of those who had served in the Second World War in the Guards Armoured Division.

'I found my tank training in the war a real boon,' wrote Lt Col Prescott, 'because I could sometimes tell the tankies they were talking nonsense on exercises. We were among the first Battalions to get the Fighting Vehicle 432 and for the first time had exactly similar cross-country mobility as the tank regiments whom we were to support in the battle groups. Battalion routine was markedly influenced by this change. First parade maintenance sometimes took precedence over the Adjutant's drill parade. And my Commanding Officer's weekly parade or inspection of barracks was frequently held on the 'tank park' rather than in the barrack rooms or the square. I must have been insufferable, but we were rated the best Battalion in BAOR by Tubby Butler, the Corps Commander.'

In 1969 the 2nd Battalion moved to Chelsea Barracks. From 1972 to 1974 the 1st Battalion (Lt Col GW Tufnell followed by Lt Col BC Gordon Lennox MBE) was stationed in Munster. The introduction of the tracked Armoured Fighting Vehicle 432 had led to infantry mobility at last but there was less infantry in BAOR and it was difficult competing with armoured regiments' expertise. Moreover, to save money, the number of hours spent on training, and the number of miles each vehicle could travel, were severely limited. Many German farmers were no longer prepared to have soldiers training on their land. One major exercise was brought to a halt due to German farmers protesting. But it was the important operational tours in Northern Ireland which were having the greatest detrimental effect on professionalism in BAOR. By the time a battalion had trained for Ireland, undergone the tour, followed by leave, most of the BAOR training year was over. Despite the energy of the small rear parties, many vehicles subsequently broke down due to lack of maintenance. There was little time to train newcomers to the battalion. One company commander, in BAOR for a year which was dominated by a Northern Ireland tour, despite enquiring, never discovered where the Battalion was to deploy in the event of hostilities, nor what the concept of operations might be. This worried him.

However, operational tours in Ireland helped to develop leadership at all levels and added immeasurably to a battalion's expertise in basic infantry skills. Moreover, Ireland and BAOR justified having an army of 150,000 and no cut in the number of battalions. Tactics had, like the equipment, changed considerably. Gone were the static, immobile defensive positions of ten years before. Instead, battle groups of all arms had been formed. Based on tanks and mechanized infantry, they consisted of between two and five sub-units called combat teams each commanded by a major. The balance varied, with infantry predominating in close ground.

A battle group was responsible for an area of up to eight kilometres wide and fifteen deep. They occupied a series of planned alternative positions in depth, sited in relation to natural and artificial obstacles. From these positions Grenadiers were to check, canalize and destroy the Warsaw Pact forces by maintaining close and aggressive contact, using mobility and all available fire power, including nuclear weapons. The design for battle was therefore based upon the selection of suitable killing zones in which an advancing enemy was first contained and then, by deliberate counter-attack, destroyed. There were inevitable gaps between containing

155

positions, thereby giving the combat team commander the dilemma of whether to stand firm against an enemy assault, as opposed to withdrawing to another battle position.

Allied strategy had therefore changed from an almost automatic response to nuclear weapons to a strategy based on a wider and more flexible response, appropriate to the nature of the threat, both conventional and nuclear. The tactical cries had now become 'mobility, killing zones, and riding with the punch'. Lt Gen Sir Roland Gibbs commanding the Corps in Germany optimistically announced that his men 'were fully professional and could knock sparks off anyone'. Not all infantry combat team commanders would agree with this view; as already noted, Ireland was having a detrimental effect on mechanized training and the serviceability of the vehicles. And, as we have also seen, some were unfamiliar with the ground and tactics, while others disagreed with the whole concept. Brig NT Bagnall, a future Chief of the General Staff, was commander of the Royal Armoured Corps in BAOR in 1971–72. He 'was surprised to discover there was no operational concept. The Corps was simply divided up into compartments with nothing in the way of an operational plan.' When it was his turn to command the Corps he introduced a greater switch from static defence to that of manoeuvre warfare, with much more flexibility.

The 1st Battalion moved to England in 1974. In 1979 they returned to Germany for a tour in Berlin. Also in 1979 the 2nd Battalion, commanded by Lt Col HML Smith, started a three-year tour in Munster.

Lt Col Denison-Smith, his successor as Commanding Officer, took a Battle Group to the British Army Training Unit, Suffield, near Calgary in Canada. It consisted of Battalion Headquarters, Support Company (Maj Sir Hervey Bruce), No 1 Company (Maj GF Lesinski) and two armoured squadrons from 4 Royal Tank Regiment. The training area is magnificent. There is plenty of ammunition and supporting artillery fire. The tanks and armoured personnel carriers remain at Suffield for the incoming units. Ambitious live-firing exercises take place, followed by adventure training in the Rockies. (The superb training area is one of the contributions Canada's land forces makes to helping Britain.)

Four years earlier Denison-Smith had taken his company to Suffield, under command of 5th Inniskilling Dragoon Guards. After an exercise, Grenadiers were told to clear up any shell cartridges. There was a great explosion – seven men lay on the ground, badly hurt. The section commander, L/Sgt T Mann, was one of them, but he dragged himself, blood seeping from his leg, to give emergency first aid to his men. He refused to be evacuated by helicopter until all the others had been flown out first. It was never discovered what had caused the accident: possibly a 66 mm rocket blind had been picked up.

In 1982 the 2nd Battalion returned to England.

BERLIN 1945 – 1981

The 1st Battalion has been stationed in Berlin on three occasions since the Second World War. The first tour was described in Chapter 5.

Berlin has frequently been a centre of major international crisis – in particular during the blockade of 1948–49, the Khruschev ultimatum of 1958 and the building of the Berlin Wall in 1961.

The Berlin blockade was preceded by an escalating series of restrictions on road and rail corridors between Berlin and Western Germany. These were accompanied by bitter accusations that the Western powers were prejudicing their right to remain in Berlin which the Soviets began to claim as part of their zone. West Berlin's 2.2 million people became totally dependent on aircraft for food and coal. Fourteen months later the Russians lifted the blockade. The Western Allies, who had arrived in Berlin in 1945 as occupiers, had been transformed into protectors and comrades in adversity. The bonds which had secured the Federal Republic of Germany to NATO were forged in Berlin during the airlift of 1948/49.

While the West Berliners certainly enjoyed increasing prosperity in the years following the blockade, in East Berlin and the Soviet Zone conditions had deteriorated. Communism was simply not working. This was all too evident to the 1st Battalion which was stationed in Berlin in 1954 under command of Lt Col PW Marsham MBE. A year earlier, an East Berlin demonstration, against their government ministers, quickly turned to rebellion. The Russians sent in tanks and the 1st Motor Rifle Division which opened fire on the crowds.

1954–1958 were years of stalemate. For the West the City became the tripartite responsibility of the three Western powers, – Britain, America and France – who pledged that they would maintain their forces in the City 'so long as its protection required it'.

The 1st Battalion regarded Berlin as a popular posting due to the friendliness of the Berliners. Montgomery Barracks was comparatively comfortable, although it was divided into two, with the Russians in the other half in their zone. It was possible to stroll unintentionally into the Russian zone as the Berlin Wall had not yet been erected.

When guarding Spandau prison, Grenadiers could see from their towers Hess, Hitler's former deputy, wearing a drab military greatcoat, walking in the prison compound. Hess was not then allowed a newspaper or wireless. He had to eat the same food in his bare cell as his guards. When the Russians and French were on duty the food was very inadequate. Admiral Doenitz, Hitler's ultimate successor, could be seen wearing a blue greatcoat with a peaked naval cap.

Grenadiers' married quarters were satisfactory: each family had one wooden box in their house into which their most precious possessions were placed, should the Russians attack. The boxes were partly for morale purposes because the Russians could easily encircle the allied zone in which case no family would escape. The rate of exchange was very favourable.

There was little serious training. Deploying to the practice emergency 'crash-out' location was not popular as it was a sewage farm; the smell was dreadful. All Berlin was still in ruins, although great piles of bricks indicated that rebuilding was imminent. Many shops had a front, but with only shacks behind them.

Most Grenadiers rather enjoyed serving in Berlin. 'We had very good National Servicemen,' recalls the then D/Sgt ST Felton, 'They were just as good as the regulars, being well motivated. We never had a great deal of difficulty with them; they were so well trained at Caterham and Pirbright.

'There were frequent Sergeants' Mess functions because, for the single men, it was their only home. Bingo, poker dice, cards, sing-songs, and impromptu acts went on. Wives could go into the Mess at any time with their husbands which was certainly not the case with the Officers' Mess. The Sergeants' Mess would be very crowded on a Saturday night. On entry, one would be hit by the dense tobacco smoke. An enormous amount of beer was drunk. It was a great, very friendly atmosphere. Each night the Picquet Officer closed the Mess and collected the takings to ensure that they were not stolen. After six weeks in Berlin, my wife arrived from England, with the children and packing cases, amidst the snow. I told her I had just heard that I was off to Sandhurst as RSM. She followed a few days later. This was typical of the uncertainties of Army life, particularly at Warrant Officer level throughout the Household Division: one needs a very special wife. Some splendid, very promising young married NCOs were persuaded to leave the Army by their wives who didn't like married quarters or moving about.'

In 1955, as already related, the Battalion moved from Berlin to Hubbelrath. Western Berlin remained precariously isolated 100 miles from West Germany. All its supplies had to be brought in from the West. In 1958 Khruschev proposed that West Berlin be demilitarized. Tension was high but the West stood firm and the Soviet challenge was dissipated.

By 1961 the East German State was on the verge of collapse. Refugees were crossing into West Berlin at the rate of 1,000 a day. At 1.30 am on 13 August East German police and Volksarmee, supported by tanks, moved into position. Barbed wire fences, concrete and other obstacles were erected. So began the separation that was to last until the fall of the hated Honecker regime and the breaching of the Berlin Wall 28 years later. The Iron Curtain descended along East Germany's border with the West. It was mined and fortified. Those who tried to escape to West Germany were shot and sometimes left to bleed to death.

On 6 April 1966 a new and secret Soviet fighter crashed in a branch of the Havel River in the British Sector. The newly appointed British Commandant, Maj Gen Sir John Nelson, had the inside of the fighter's engine examined by British experts on a Royal Engineers' raft, amidst great secrecy, before the engine was returned. He later received a letter from the Chief of the Defence Staff stating that as a result of his action a 20-year gap in aircraft intelligence had been filled; the chance of a lifetime had been taken. The Americans were very impressed. The French, who, Sir John felt, regarded the Russians as their particular allies, were not told what had happened. The whole story appeared in Sir John's exceptional book *Always a Grenadier*.

Towards the end of his tour Sir John Nelson was offered the post of General Officer Commanding Hong Kong but he decided to leave the Army to everyone's regret: he had lost confidence in his political bosses. He spent his last night staying with the 2nd Battalion at Hubbelrath, 'a thoroughly happy ending to my experiences as a serving soldier,' he wrote.

West Berlin continued to be a flourishing oasis in the arid desert of East Germany, a desert deprived of freedom and occupied by twenty-two Soviet divisions of Communist controlled troops.

The next Commandant, Maj Gen James Bowes-Lyon, was also a Grenadier. He saw an intensification of the war of nerves against Berlin. Tension increased markedly with the Soviet invasion of Czechoslovakia. Fortunately the ratification of the Quadripartite Agreement in 1972 provided a turning point which produced stability and a more normal situation in West Berlin. The four governments undertook 'mutually to respect their individual and joint rights and responsibilities'. Detente was severely tested on a number of occasions, such as the Soviet invasion of Afghanistan in December 1979 and the declaration of martial law in Poland but Berlin was no longer the focus for international tension. This, therefore, was the background to the 1st Battalion's tour there in 1979–1981, under command of Lt Col MF Hobbs MBE and then J Baskervyle-Glegg MBE.

The opening months were both hectic and exhilarating with the guarding of Hess still in Spandau prison, Flag Tours in East Berlin and border patrols. The crash-out procedures, test exercises and company training ensured that the Battalion became thoroughly professional. There was certainly no opportunity to be bored. Training in May 1980 in Schleswig

- Holstein was marked by beautiful weather. The Queen's Birthday Parade followed, on the Maifeld where Hitler had harangued his followers. It was the most colourful spectacle with all three Berlin Battalions and their Colours, Ferret Scout Cars, Land Rovers and a helicopter fly-past. The Battalion participated in CRUSADER 80 – a series of exercises which culminated in the biggest manoeuvre that the British Army had experienced since the war. (On another exercise an American infantry company was seen to be high on drugs.)

The Battalion always had to be ready for battle. No more than 20% could ever be away. An example of a brief test-exercise occurred one evening during a dinner party. Lt Col Baskervyle-Glegg was suddenly ordered at 10.30 pm to carry out a night attack on a deserted village occupied by another British battalion. He wrote his orders in a Land Rover as the Grenadiers deployed. By dawn the operation was completed. Another success was the Brigade inter-platoon competition which was won by 2Lt AJ Waterworth's men.

There was an excellent relationship with the Germans; the Battalion was well up to strength and the training opportunities were superb. In short, it was a wonderful posting. In July 1981 fighting in a built-up area was studied in detail. At the Brigade Study Day several officers presented a revised concept of operations. Majors JM Craster, PR Holcroft and 2Lt HGW Swire outlined their plans which were then tested in a Brigade five-day exercise.

In 1981 Battalion training again took place in Schleswig-Holstein. It culminated in an escape and evasion exercise. Captains AD Hutchison, DNW Sewell and AD/Sgt T Rolfe played the part of female VIPs to be rescued by the companies.

One highlight of the Berlin Tour was a visit by the Colonel, Prince Philip. He was met at the airport by the Commanding Officer for he had come to see only Grenadiers. No General or Ambassador was present. Prince Philip chatted to almost every Grenadier and their families. It was an intimate affair throughout. Everyone was very proud to have him to themselves once more.

The Battalion returned to England in 1981. Berlin had proved to be an exceptional tour under two fine commanding officers both of whom became Generals.

We return to BAOR and the Berlin story at Chapter 26.

CHAPTER 13

WITH THE UNITED NATIONS IN CYPRUS

1965

In early 1965 the Commander of 4 Guards Brigade in Wuppertal, Brigadier VF Erskine Crum CIE MC, summoned his three Battalion Commanders. He asked them if their Battalion would welcome a tour with the United Nations in Cyprus. The Coldstream declined on the grounds that they had recently returned from Aden. The Royal Highland Fusiliers also turned down such an opportunity. Lt Col MS Bayley, commanding 1st Battalion Grenadiers, asked last, thought that such a tour would be ideal. And so it proved.

Five years earlier Cyprus's constitution had led to an uneasy peace between Greek and Turk Cypriots. The complex system of checks and balances might have worked had there not been a lack of determination to succeed and unwillingness to compromise on both sides. The machinery of Government quickly broke down. The following year President Makarios agreed to have United Nations peacekeeping units. The Security Council quickly voted to create the United Nations Force in Cyprus (UNFICYP) which became operational in June 1964.

The 1st Battalion's advance party left Germany on 26 March 1965, having exchanged their berets for the light blue and grenades for badges showing the world. Parkas and Wuppertal's grey skies were also replaced by shirtsleeve order and Cyprus sun. The advance party received a very warm welcome from the Cheshire Regiment who many years later were to make such a name for themselves in Bosnia. The main body deployed straight to their operational areas. No time was to be lost. The inter-communal fighting had led to armed confrontation between Greek and Turk Cypriots. A disengagement policy was therefore the top priority in the Grenadiers' Limassol area where the Greek Cypriot National Guard faced the Turk Cypriots who had not agreed to any defortification of their strongpoints in Limassol or Episkopi. There was no relaxation of the official Turk Cypriot policy of keeping the two communities apart. Suspicion on both sides was paramount.

The two races had their separate cultures, religion, language and traditions. They had, and still have, little in common. Up to a third of the Turk Cypriots had fled their villages. Not much progress had been made in re-settling them: some still lived in great poverty. Many fled to an area north of Nicosia, the boundary of which was known as the 'Green Line'. It was marked by a series of United Nations observer posts and roadblocks and lay among decaying, empty houses which had been hurriedly abandoned two years before. What was left of meals were still on the tables, the chairs askew as the inhabitants had fled in panic.

President Makarios, the leader of the Greek Orthodox church of Cyprus, did not advocate *Enosis* – union with Greece – to avoid inflaming Turk Cypriot opinion; there was very little public support for such a policy. However, many Turk Cypriots sought partition.

When independence was granted to Cyprus, two treaties were made restricting that independence. The Treaty of Guarantee, between Cyprus on the one hand and Britain, Greece and Turkey on the other, gave the three powers the right of joint or even unilateral intervention, for the purpose of re-establishing the state of affairs created by the London and Zurich agreements on independence, should there be trouble. The Treaty of Alliance between Cyprus, Greece and Turkey entitled Greece and Turkey to station contingents of their own forces in Cyprus.

The aim of the United Nations forces in Cyprus was to pacify, observe, report, restore confidence and act as a quick reaction force to prevent incidents developing into war. UNFICYP was not meant to fight. Fire could only be opened if attacked or if vital property might be destroyed, and only then as a last resort.

The force consisted of five principal nationalities. The Canadians, Danes and Finns numbered 1200; the Irish had 1000 and the British 800. Each had their own area of operations; the British UN force being in Limassol District in the south-west – an area of about 600 square miles – stretching from the sea to the Troodos mountains, from Zyyi in the east towards Pissouri in the west. To their south was one of the two large British bases, Akrotiri, with its RAF airfield, large headquarters and British Battalion. Stores, depots, two hospitals and administrative units did much to support UNFICYP.

The Battalion was based in Polemedhia camp, a hutted encampment on the northern outskirts of Limassol. It was shared with a large number of dogs. Beyond the wire was a Greek Cypriot unit in an adjacent camp. The more senior Grenadier officers lived together in a modest bungalow once occupied by Kitchener – who incidentally had designed and built the original camp.

The Battalion initially deployed with The Queen's Company (Maj BC Gordon Lennox) finding a night guard at the High Commissioner's house in Nicosia commanded by 2Lt HF Hall Hall. The other two platoons, commanded by 2Lt JWH Buxton and Lt PH Cordle, manned outlying villages.

No 2 Company (Maj DV Fanshawe) was responsible for an observation post in a flat with a prominent balcony in the centre of the Turkish quarter, a mobile vehicle on continuous patrol and more outlying villages. No 3 Company (Maj FJ Abel Smith) mounted liaison patrols in vehicles and on foot in villages in the District. They also provided a stand-by platoon and manned various observation posts.

The outlying villages included Mallia in the north-west – a mixed village from which most of the Turk Cypriots had fled, although a few had since returned to tend the vineyards. Episkopi, to the west of Limassol, was another mixed village where tension was high due to Turkish refugees and a difficult Greek Cypriot police sergeant. At Troodos, in the mountains, a mobile platoon patrolled the north of the District.

In principle, two rifle companies found these duties for two weeks, followed by one week for training. Platoons changed tasks within companies every five days to avoid boredom.

Ajax Squadron, 2 Royal Tank Regiment, in armoured cars, was under command at Zyyi, responsible for the District's eastern end, liaison with other battalions and to act as a general reserve. On 5 May A Squadron 14/20 Hussars took over from them. However, all these plans changed later when the Battalion was given an extra District to run, as will be related shortly.

On 17 May the situation suddenly deteriorated. A strong platoon of the Greek Cypriot National Guard, probably on orders from Nicosia, broke into and searched a row of fortified Turk Cypriot houses within the Turkish quarter of Limassol. They fired a few shots, took away some weapons, and left a section covering the area. This unprovoked, pre-planned assault was a clear breach of the Greek Cypriot agreement with the Indian United Nations Force Commander, General KS Thimayya DSO.

Lt Col Bayley was quickly on the scene. Lt CJE Seymour's platoon deployed between the two forces. For a period there was acute tension in Limassol. All Turk Cypriot labour was withdrawn and the shops closed. The Turks announced that they would wear uniforms and carry weapons on their National Day two days later. The Commanding Officer did all he could to dissuade them, but they insisted on the uniforms regardless of Greek Cypriot protests. Fortunately the event passed off peacefully without National Guard or police interference.

Despite the seriousness of the Greek Cypriot attack, the Turks agreed to the defortification of Episkopi. Shallow fresh trenches, overlooking the main road, were filled in. This paved the way for discussions on further defortification. Lt Col Bayley attended the weekly conferences between the Greek and Turk Cypriot leaders. The former was Mr Benjamin, the District Officer, who was not markedly pro-British, but he did have in his office a picture of The Queen with the Commonwealth Heads of Government who included, of course, President Makarios, standing close to her. Limassol's Turk Cypriot leader was Mr Ramadan Jermil OBE, a diabetic, who was a personal friend of Mr Benjamin and a restraining influence on his deputy, Kemal Pars.

Large quantities of arms and ammunition continued to be imported into Cyprus via Limassol for the Greek Cypriot National Guard. This lowered Turkish morale. Nevertheless, operationally, the situation became reasonably quiet. The UNFICYP Chief of Staff, Brig AJ Wilson, a future member of the Army Board, while recognizing that the Grenadiers were a first class Battalion, felt that they were not fully stretched. There was the opportunity for a wide variety of recreation, and to make use of the perfect conditions for swimming, athletics, mountain climbing and polo.

Although the 2nd Battalion often had an outstanding athletics team in the 1960s, '70s and '80s, the 1st Battalion's team in Cyprus achieved commendable results. On 19 May in the Army (Cyprus) Individual Championships Grenadiers won seven events and came second in five, although it was the first occasion that they had been on a running track that season. At the Inter-Unit Athletic Championships on 10 June the Grenadiers faced the three non-UN units – the 1st Battalion The York and Lancasters, the 2nd Battalion the Royal Anglian Regiment and 9 Royal Signals Regiment, all of whom had been training hard. However, the Regiment triumphed, winning nine of the fourteen events. 2Lt EH Houstoun, running in the 120 metre hurdles for the first time, set up a new Army (Cyprus) record. CSM GR Whitehead won the shot and discus.

The main opposition in the polo came from the Royal Air Force and A Squadron 14/20 Hussars, both of whom took it immensely seriously, and none more so than the Commander British Forces and Senior British Officer in the Middle East, Air Chief Marshal Sir Thomas Prickett. Some of the Grenadier ponies, imported from Jordan, were extremely aggressive. Capt the Hon GCD Jeffreys' pony in particular needed to be muzzled for matches. On one occasion the muzzle had been forgotten. Sure enough the inevitable happened. The pony devoured a large lump of leg belonging to the opposition. As luck would have it the victim was the captain of the RAF team – the much respected Air Chief Marshal. Subsequently, by way of a peace offering to the injured very senior officer, who had been immensely cheerful and tolerant of his misfortune, the Grenadier team despatched to his bedside in Flagstaff House a case of champagne by hand of the duty Drummer. On entering his bedroom, the Drummer was said to have found the Commander British Forces groaning in bed. The Drummer came to a dramatic halt, stamping his feet to the ground, Guards Depot style; this dislodged the fancy chandelier in the drawing room beneath. It crashed to the ground. Sally Rose Gordon Lennox, visiting her husband, was persuaded to give the Air Chief Marshal a box of chocolates which he accepted rather wryly.

In the final of the Inter-Regimental Competition, the Regiment won an extremely close match against the Hussars team which contained their regimental players. The Grenadiers consisted of Majors Gordon Lennox and Fanshawe, and Captains GA A-R-West and Jeffreys.

On another occasion Major Gordon Lennox, while playing polo, was told that a murder had been committed nearby. He galloped off to investigate, still wearing his United Nations blue cravat and carrying his polo stick. The startled villagers had never seen anything like it.

Many Guardsmen were introduced to the thrills and skills of rock and mountain climbing. Lt A Heroys, a renowned Army climbing instructor, plus a team of other leaders, took various parties from all the Rifle Companies the hard way to the summits of St Hilarion and other parts in the Kyrenia range.

The leave programme began early to ensure that everyone received some. Famagusta on the east coast was the most popular. The Guardsmen stayed in luxurious hotels overlooking beaches. 2Lt JWH Buxton took a group to Jerusalem for four days. Many also visited Beirut which was cheap, attractive and peaceful. Some returned to Germany and others to England, while a few wives came out to Cyprus to enjoy the hospitality, amongst others, of the Findlays at the Harbour club in Kyrenia. United Nations uniform was worn when off duty and United Nations soldiers could not 'walk out' alone.

It is fortunate that much sport and leave was completed early in the Battalion's tour for in July an Irish battalion was withdrawn from Cyprus without replacement. Grenadiers redeployed to look after both Limassol and Lefka District in the north-west beyond the Troodos mountains. No. 3 Company Headquarters was set up at Limnitis on the coast. The Irish had left their camps and observation posts in a most unmilitary state of cleanliness. A lot of work was required to make them hygienic and comfortable. One Guardsman of The Queen's Company was found to be selling water to the locals. His explanation on being sent to Battalion HQ in close arrest was a pitiful: 'The Irish told me they always did it, so I thought I had to sell water too!' Maj Gordon Lennox asked his departing opposite number what information had been gathered on the local Turk and Greek forces. The only point of interest that the Irish could provide was that the local Turk leader had a weak heart and so preferred gifts of Drambuie to whisky.

The main task was to man observation posts between the two sides. They were usually on the top of steep mountains. Nevertheless, when Maj Fanshawe was commanding there, all visitors were taken on a grand tour – on foot. The number of visitors quickly dwindled. Only the fittest came twice.

Lefka District was more troublesome than Limassol. It consisted of Turk villages containing their armed, uniformed regulars encircled by Greek Cypriot National Guard units. The Company, with A Squadron 14/20 Hussars, were kept very busy in their efforts both to promote under-standing between the two hostile factions and also to maintain freedom of movement on the roads. Administering the District provided many prob-lems. Fortunately the excellent Battalion drivers were thoroughly professional and there was no serious accident on the perilous bends of Troodos's nerve-wracking mountainous tracks. Cyprus was a hectic tour for Lt (QM) PWE Parry, the Quartermaster, who had at varying times on his charge a total of seven camps, three Districts and innumerable OPs. Fortunately a weekly resupply helicopter run was arranged.

The Greek Cypriots appeared to be conditioning the Turks to the rule of the Cyprus Government. This policy extended to running armed convoys of National Guard units through the more humble Turk Cypriot villages, risking the Turks retaliating. These provocative moves increased tension. Grenadiers, visiting the villages shortly afterwards, tried to reas-sure the frightened, despondent Turks that the provocation would be reported to the United Nations Headquarters in New York, but it was diffi-cult to sound convincing for the HQ was too remote and the events too trivial for them.

The wider political scene was in turmoil due to the outbreak of political unrest in Greece and the apparently coincidental repudiation of the Zurich Agreement by the (Greek Cypriot) Cyprus Government. Everyone was eagerly awaiting the results of the Security Council debate at the United Nations. Grivas was still a threat to stability.

In July the influential, rich, fat, urbane Ramadan Jemil, the Turk Cypriot leader, left Cyprus, handing over to Kemal Pars. This was unfortunate: Kemal Pars was apt to roll his eyes to the sky, muttering optimistically about a gift from heaven – a reference to the Turkish national air force 40 miles away on the mainland. Little did we know that this, precisely, would take place a few years later.

The Battalion Headquarters operations room was manned by fairly senior officers, each working in three eight-hour shifts. Each evening the Intelligence Officer, Capt OJM Lindsay, compiled an information summary for the UNFICYP headquarters in Nicosia. It reflected the previous 24 hours' activities and tried to forecast future trends. No intelligence was ever sought or obtained from the British Forces Headquarters at Akrotiri.

However, they twice sent Capt Lindsay on irrelevant enquiries. 'The Greeks are putting crosses in paint on rocks overlooking the main Nicosia-Limassol road, denoting future defensive positions,' he was told. The crosses were indeed there, but all an identical distance apart. This was not surprising; it quickly transpired that they were simply indicating where telephone poles were in the process of being erected. The second alarm was potentially more dramatic: 'Greek tankers are approaching Limassol with long missiles clearly identifiable on the decks.' This report was equally wide of the mark; the 'missiles' turned out to be shadows cast by the ships' hatches.

'Intelligence' was a dirty word; the United Nations were there to report, impartially and helpfully, not covertly or deviously. Slidex or other codes were not permitted; all wireless reports were in clear. However, a series of numbered reference points was issued so that the National Guard, if they were monitoring the Battalion's radio net, would be unsure of who was reporting and from where. On one occasion the reference points so exasperated Maj Gordon Lennox that he heatedly transmitted: 'This is UNO – going UNO-where.' Nobody was the wiser.

Some Grenadiers, notably Lt Seymour, became exceptionally sunburnt. When he reported that he was surrounded by Greeks and fearful of being shot, he received the unhelpful guidance on the radio that he should satisfy them that he was not the Turk whom he so closely resembled.

The recce platoon, commanded by Lt J Baskervyle-Glegg (of similar colour!), was useful in checking on the unpredictable and provocative National Guard armed convoys. Little information was available to identify Greek and Turk national units; they maintained excellent security, and the United Nations could not spy. The Intelligence Officer resorted to calling on their units, inviting them to volley ball matches or requesting a list of their officers who could be invited to a cocktail party, but such invitations were always refused and few useful facts were gleaned.

Several discussions were held with Colonel Episkopou, a retired Greek national army officer who was responsible for the Greek Cypriot forces in Limassol. He had an argument with Lt Col Bayley on UN night patrols.

The Commanding Officer was at his sternest. Progress seemed impossible until Lt Col Bayley's charming smile unexpectedly broke the ice. The Greek Colonel immediately apologized and all was well.

Being responsible for Lefka District as well as Limassol no longer left time for sport, leave or socializing with the UN forces or the hospitable British in the Sovereign Base areas.

There were some compassionate cases in Germany necessitating some men having to return to BAOR. The rear party, 100 strong in Wuppertal, was commanded initially by Maj CJ Airy who then changed places with Maj JP Smiley in Cyprus. Only one officer, Capt CJ Hope, the Transport Officer, never reached Cyprus.

When the Senior Major, DW Hargreaves, joined the Battalion his green Mercedes was impounded at Nicosia docks. Nobody else had private cars.

On 30 August the Turk Cypriots in Limassol held a uniformed parade against the advice of the United Nations and in opposition to Greek Cypriot wishes. The Battalion deployed patrols with the excellent New Zealand police. Although the National Guard sent armoured cars through the Turk quarter, the day passed without incident.

The Commanding Officer, looking back on the tour, regretted that he never made a contingency plan for internal security in Limassol in case the Battalion had been called upon to control the entire town. He felt, in retrospect, that it had been unwise that this risk was taken.

Grenadiers serving with UNFICYP believed that they were trusted by all sides; the blue berets were welcomed in the villages although the language barrier was a problem. Gestures over Turkish coffee and a backgammon board in the village square led to smiles and friendliness. There was no apparent animosity to British soldiers, certainly not to the Grenadiers, although the Greek Cypriots had scarcely been treated with every courtesy during the 1956–1959 Emergency. The Battalion developed a genuine affection for Cyprus and its people, and many have since enjoyed holidays there.

In late September, only two weeks before leaving the Island, Lefka District was relinquished to the Canadians. In its place the Battalion was given the whole of Paphos District to the west, thereby being responsible for almost half of Cyprus. Another Irish battalion, this time the 42nd, had withdrawn at short notice without replacement. The Queen's Company took over Polis and the notorious Kokkina bridgehead which had been bombed by the Turkish Air Force the previous autumn after bitter fighting. The continuous lines of confrontation ran very close to each other. The Grenadiers manned seven OPs between them. The area looked and felt under siege with many Turk Cypriots crammed into tents and makeshift shelters. Without the UN presence a massacre was a possibility. And so the tour ended on an exciting operational climax. It was from such positions that the Battalion handed over to the Royal Highland Fusiliers in early October 1965. They were promptly greeted by the first, and violent, rain storm of the autumn, reminding the Grenadiers how fortunate they were to have enjoyed six months of glorious summer weather.

On no occasion had it proved necessary to open fire. Compared to Germany, the young officers and NCOs in particular, had ample opportunity to develop their leadership and military skills. Tasks were varied and interesting, although there were, inevitably, some dull moments. The Guardsmen found themselves patrolling, escorting, observing, persuading and explaining several times a day. Duties were ultimately heavy; only one complete night in bed a week was not unusual.

In retrospect the tour was a most interesting, enjoyable and worthwhile experience. The Greek and Turk Cypriots were usually friendly and cooperative, if devious at times. The Battalion developed a close affinity to UNFICYP. They wore with pride the blue beret and, for the remainder of their service, the white and blue medal ribbon of the United Nations.

CHAPTER 14

NORTHERN IRELAND: THE OPENING PHASE

1969 – 1970

Some seventy-two hours after dismounting the Queen's Guard at Buckingham Palace, Grenadiers deployed for the first time on the streets of Ulster. A few weeks earlier, in early August 1969, the 2nd Battalion was unexpectedly placed on Spearhead. The Commanding Officer (Lt Col PH Haslett MBE), Adjutant (Capt J Baskervyle-Glegg) and Quartermaster (Capt A Dobson) flew to Wilton by helicopter to be warned that the Battalion could be sent anywhere except Northern Ireland. Khaki drill was drawn up and everyone prepared for a hot winter abroad. But it was Ireland after all.

In the Report of the Commission appointed by the Governor of Northern Ireland, Lord Cameron concluded in 1969 that in previous years there had been a rising sense of injustice and grievance among Catholics in Northern Ireland. This was due to undoubted discrimination in housing and jobs, he said. Local Government electoral boundaries were deliberately manipulated to deny Catholics influence and to preserve Unionist control. In 1967 the newly-formed Northern Ireland Civil Rights Association organized marches and demonstrations. The Association was infiltrated by subversive left-wing and revolutionary elements to exploit grievances and foment trouble. On the other side the interventions by followers of the Rev Dr Paisley increased the risk of violent disorder. The police handling of a demonstration in October 1968 was ill-co-ordinated and inept, Lord Cameron reported. The Government's announcements on the reform of local government franchise and the fairer allocation of Council houses were too late.

By 13 August 1969 women in Londonderry's Bogside were making

Northern Ireland

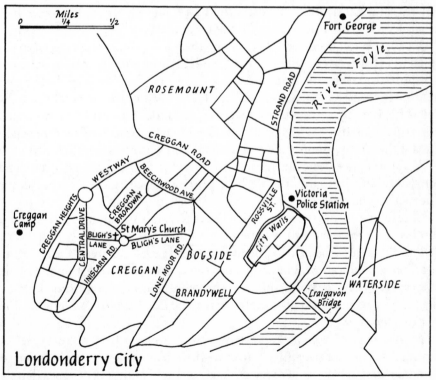

Londonderry City

171

petrol bombs; children were ripping up paving stones and piling them in wheelbarrows. The Protestant majority was astonished and angry that the momentum of riots had kept the police out of the Bogside. The following day Prime Minister Harold Wilson and his Home Secretary, James Callaghan, ordered the troops to intervene.

*   *   *

Over the following twenty-five years Grenadier Battalions served fifteen tours in Northern Ireland. This Regimental History would be much too long if adequate space was given to many of the tours which are anyway recorded in the *Grenadier Gazette* and the *Guards Magazine*. Instead, therefore, this book will concentrate on the first tour of 1969, that of 1974 when the shooting was at its height in Londonderry (Chapter 17) and on the recent tour in South Armagh of 1993/94 (Chapter 30); a brief summary of the intervening tours between 1976 and 1986 is at Chapter 23. To avoid repetition, similar incidents are not described twice.

*   *   *

Early on 22 August 1969 final confirmation was received that the 2nd Battalion was to deploy to Northern Ireland by air or sea. The Major General said that he did not expect the Battalion to be there for more than six weeks – but the tour was to last five months!

An officer purchased some maps from the Automobile Association to see which Irish towns were where. At three hours' notice the Battalion was told to report to Liverpool docks. Staff officers in their bowler hats from London District watched the coaches leave Wellington Barracks. Lt PR Holcroft's convoy took a wrong turning in Hyde Park and nearly circled the barracks again. Several coaches broke down but LSL *Sir Tristram* sailed at 2 am the following morning. It seemed an exciting adventure although the Battalion was quite untrained for Internal Security. It had returned from Germany four months earlier, since when virtually all the senior officers had changed.

On reaching Londonderry docks The Inkerman Company, 128 strong, marched into the City and was deployed immediately under command of 1st Battalion the Queen's Regiment in 24 Brigade. The remainder of the Battalion settled into Magilligan Camp thirty miles to the east on the Foyle estuary. It was a weekend, summer camp and so there was some dismay when it was eventually learnt that it was to be the Battalion's home for the next five months.

The Inkerman Company's task was to watch, wait and reassure. A Catholic Bogside barrier fifty yards to their front marked their forward boundary and the River Foyle, the size of the Thames, lay immediately

172

behind. At least map-reading was no problem! At each corner of their small complex barbed wire and knife-rests were held ready to block off militant extremists from either side. The platoon commanders (Lt the Hon Philip Sidney, and 2Lts DAS Tredinnick and JFM Rodwell) were each responsible for a few streets. Extremists were expected to burst through; but nothing happened. This was just as well for there were too few batons and shields. Some Internal Security banners had been received from Ordnance which read: 'If you cross the white line you will be shot' – a totally inappropriate message to meet the restrictions of peace-keeping in Northern Ireland. There was no intelligence, no police presence, and no accommodation apart from two filthy derelict houses. The Inkerman Company Commander (Maj OJM Lindsay) was not allowed to approach the Catholic barriers to see what lay beyond. One officer who had joined the Company the previous week felt as well prepared for Ulster as he had been when he once arrived at Rhodesia Airport to be ADC to the Governor General. Clutching a Herbert Johnson hatbox, one of Mr Wilkinson's much-advertised swords and some high-heeled evening shoes for the prime minister's wife, he was informed that the rest of his baggage had flown on in error to Ethiopia.

On 5 September Maj Gen Sir Allan Adair, the Colonel of the Regiment, visited Magilligan Camp. Speculation on the Battalion's future role was suddenly terminated. The Colonel watched No 1 Company (Capt The Master of Rollo) depart at an hour's notice to Belfast to help 1st Battalion the Royal Hampshire Regiment who had been involved in two consecutive nights of street fighting trying to separate the two communities now firmly entrenched behind barricades of paving stones, lorries and builder's rubble. Two days later the remainder of the Battalion followed them. The nights in Belfast were spent negotiating between the two factions and dispelling the wild rumours.

As in Londonderry, the soldiers were usually made welcome, particularly by the Catholics who feared Protestant extremists. In some streets old women served hot drinks and sandwiches during the night.

It was in Belfast that Guardsmen saw the seriousness of the situation. Some factories, houses and shops had recently been burnt down. The soldiers became greyer with fatigue as insufficient time was available for sleep. Minor riots ensured that reserve platoons were frequently deployed.

On the night of 13/14 September, for example, some platoons stood between two surging crowds two streets apart. Broken bottles, paving stones and other projectiles littered the streets which had been plunged into darkness. On one side the Protestants were watched by the Royal Ulster Constabulary who were *persona non grata* in all Catholic areas. On the other side the local Member of Parliament was trying to shout down

Catholic hot-heads. Innumerable lawyers, clergymen, local leaders, vigilantes and drunks in turn clutched at officers' elbows to explain how the incident had started. Explanations always differed and simply added to the confusion. Soon reports came in that schools and churches in the mixed areas were being threatened. To the delight of a TV crew who wanted action, the sections were hurriedly redeployed.

One officer, interviewed unexpectedly by BBC TV, foolishly said that there was no sign of the IRA. Fortunately the sound-recorder's equipment was faulty and the broadcast was never made. The hostile crowds quietened as the sky gradually lightened. The stars gently faded. Another day dawned. As the Companies thinned out everyone was too tired to sense any relief or elation that bloodshed had been avoided. In any event it hadn't been. That night within the Brigade Group one REME soldier had inexplicably committed suicide and another had accidentally shot dead his friend in a telephone box. The number of accidental discharges, usually when unloading weapons, became a matter of considerable concern.

Peace-keeping became increasingly frustrating and annoying, especially to see grown men behave like school children. The position became so bad that a Peace Line was built by the Army to avoid the constant raids by extremists on both sides. But the Army had still not found it necessary to open fire or use gas. An American told a Company Commander that one day snipers would pick off British soldiers on the streets. He was not believed.

Belfast became quieter. This enabled No 2 Company (Capt PML Smith) to be despatched back to Magilligan to mount a cordon operation around Dungiven. The remainder of the Battalion returned to Magilligan ready to move into Londonderry where tension was rising, although Protestant leadership remained dormant. The Derry Citizens Defence Association agreed to remove the barricades on 24 September. As the last of them came down there was a violent riot. Splinter groups of supporters stormed into the City Centre. A confrontation led to vicious sectarian close combat. A semi-curfew, Grenadier reinforcements and the issue of powers of arrest to the military all helped to impose a temporary calm.

Senior officers in the Ministry of Defence had warned that the honeymoon between the Army and the local population would not last. Soldiers had no training to be policemen. Once they started arresting troublemakers and searching houses for arms their popularity would drop and in time they would be seen as the allies of the Government of Northern Ireland at Stormont.

In October Maj BC Gordon Lennox took over as Senior Major from ABN Ussher. He had been recently involved in quelling serious riots in Hong Kong where he had witnessed excellent police/military liaison and

27. Members of No 2 Company, 2nd Battalion during an incident in Londonderry, 1980.

28. Lieutenant Colonel J Baskervyle-Glegg leads the 1st Battalion on the Armed Forces Day
Parade, Berlin 1981 with the statue of Winged Victory in the background. Those behind
are from left to right Colour Sergeant F Franklin, Captain A H Drummond, Major J M
Craster, Lieutenant Lord Michael Cecil, Major The Hon Andrew Wigram and Captain
G V A Baker.

29/30. Regimental Headquarters in 1959 and 1995. The vast reduction in the size of
Regimental Headquarters' staff illustrates its diminished role within the Army since
the rundown and its consequent reduced capacity to run Regimental events.

Back Row   Sgt N Taswell, Mr E Cummins (Archive Clerk), LCpl R Tidswell, Miss L Hughes
(Typist), LSgt L Martin, Mr A Kear (Company Clerk), LSgt P Jupp, Miss L Leppard
(Storeperson), Sgt C Burgoyne.

Front Row   Maj P A Lewis, WO2 C Stocker, Capt The Hon J G N Geddes, Brig E J Webb-Carter,
WO1 J Lenaghan, Lt Col C J E Seymour, Lt Col Sir John Smiley Bt, Capt B D Double

joint action against revolutionaries. 'I saw the dividend of a very strong Special Branch,' he recalls, 'and really firm action taken by a Gurkha Brigade on the Hong Kong border. It was crucial to deploy forces five times the strength of the enemy. None of these things had been achieved in Ireland where we lacked resources and troops. I asked a platoon commander of another Regiment during a riot where his men were. "My Battalion Commander is commanding one section and the Brigade Commander the other," I was told. Police and Army cooperation was minimal. The IRA could build up their strength undisturbed because we couldn't remove the barricades at the beginning.'

However, in Londonderry a 'softly-softly' approach seemed to be working. Before dawn on Sunday 12 October Royal Military Police foot patrols, totally unarmed, entered the Catholic Bogside. They were the first forces of law and order to be seen there since August. The Recce Platoon (Capt JGL Pugh), formed the previous day, and mounted in vehicles with loaded weapons at the ready, kept out of sight close by. 'At this time Battalion Tactical Headquarters moved into St Cecilia's Girls School and an empty adjacent warehouse,' recalls the Commanding Officer. 'We first got agreement from the local priest and then the headmistress to occupy an empty wing of the school. There were some very surprised schoolgirls when they started arriving on Monday morning, but all went well. Later the headmistress told me that the children in the other local schools were asking "Why can't we have the Army, too?" It was indeed a honeymoon period.'

After two weeks the way was paved for the re-entry of the unarmed Royal Ulster Constabulary. Thereafter a pattern of policing and patrolling was established. In the ensuing weeks lost children were found, stolen cars recovered, traffic accidents untangled, break-in and robberies investigated and drunks carried off to the nearest cell to recover. Slowly the burden of routine police work reverted back to them. The easing of tension enabled the Battalion to return to Magilligan.

Meanwhile rioting was steadily increasing in Belfast. The publication of the Hunt report which was critical of the RUC and led to the disbandment of their B-Specials, inflamed the Protestant extremists. They saw the Callaghan reforms as a sell-out to the Catholics. During the climax of the riots the Protestant Loyalist paramilitaries shot dead a Protestant policeman. Two civilians were also killed. More battalions were hurriedly flown to Belfast.

Tension throughout Ireland thereafter subsided. The Battalion became responsible for Londonderry County. As weeks flowed by, life gradually returned almost to normal. More barricades were pulled down to be replaced by troops who, in turn, were gradually thinned out and moved elsewhere. No platoon ever stayed in the same place for long.

New tactics were introduced to disperse menacing crowds. The leading platoon, shoulder to shoulder in extended line, bounced coils of dannert wire towards the crowd. Behind them were several armoured vehicles and the Company Commander with his small tactical headquarters. The vehicles were a deterrent and tactical shield behind which soldiers could withdraw and regroup in the face of a hail of missiles, or snipers who had still not been seen in Londonderry. Next on the flanks came two sections of 'toughies' who acted as snatch squads. They ran round the wire into the crowd to make arrests. Finally there were the two rear platoons who moved into cover if sniper fire was encountered. Selected marksmen could be sent to the rooftops if necessary. 'These tactics worked well,' recalls Maj JVEF O'Connell who was now commanding No 1 Company. 'We arrested youths throwing pennies at us and saw them tried and convicted.'

Grenadiers had helmets and gas respirators, sentries mounted with a charged magazine, in pairs, and with the weapon tied to the person to prevent it being snatched. Only the 'toughies' carried batons, and flak vests were worn by sentries. Gas was available and was used in Belfast, but the need for it had not yet arisen in Londonderry.

In November The Queen's Company of the 1st Battalion, commanded by Maj CJ Airy, replaced No 2 Company who returned to London to prepare for their tour in Sharjah, under command of 1st Battalion Scots Guards. Their adventures are recorded in Chapter 15.

On 7 December the Battalion moved into Londonderry to police the City while the Queen's Battalion handed over to 1st Battalion the Gloucestershire Regiment. Battalion Tactical Headquarters was in Victoria Barracks RUC station. The remainder of the Headquarters was based on St Cecilia's Girls School in the Creggan, beyond the Bogside. The Queens and No 1 Company were in the City, while The Inkerman Company was in a derelict, unpleasant gaol.

Christmas was a subdued festival with few of the traditional activities. All the Battalion attended a carol service in St Columb's Cathedral, Londonderry. An all ranks' dance was held on New Year's Eve which was visited by the Colonel and Lady Adair who lived near Strabane. Some officers' wives visited their husbands and a few took local leave in Ireland, there being no threat to their safety. Girls from the local shirt factories delivered Christmas presents to all members of No 1 Company.

There was now little to do and so the Battalion undertook some training, sport and community relations. Instructors were provided for the local Army Cadet Force units, and for youth clubs and sports teams. The Corps of Drums in tunics and the Battalion choir, under Capt PA Lewis, visited childrens' homes, hospitals and old peoples' homes. Assistance was given at scout meetings and The Inkerman Company Headquarters provided 'Meals on Wheels'. The Regimental Band, immensely popular as ever, played in five towns.

The 'honeymoon' period with the local population in Londonderry still existed. The Civil Rights Movement saw the injustices being put right. The undertrained, understrength, demoralized RUC were planning to build up their capabilities. There was no evidence to the soldiers on the ground of any IRA or Protestant extremist organizations. Was this an opportunity, therefore, for the Army to be largely withdrawn from Ulster? But significant force reductions was a gamble which the Government would not accept. In any event, the situation in Belfast was still precarious. Those who then expected the troubles to drag on with increasing bloodshed for a further twenty years or more were in the minority. However, one realist may have been the Brigade Commander, Brig PJH Leng MC. He was a former Scots Guardsman and a future member of the Army Board. When visiting a company in peaceful Coleraine he appeared in uniform in his staff car. He was furious that he was admitted without his identity card being checked. He had a point, for extremists successfully 'bluffed their way' into several armouries a few years later.

A Grenadier asked the senior RUC officer in largely Protestant Coleraine about his concept of operations against the IRA. The policeman replied that after an incident road blocks would be placed on mountains to the south-west to capture IRA flying columns returning to Londonderry or Southern Ireland. It sounded rather like accounts of the 1920's. The IRA eventually bombed Coleraine many years later. Such violence was scarcely imaginable in 1970, although the political parties were planning bigger and better demonstrations in the more provocative places.

The abiding memories of Grenadiers' first tour in Northern Ireland were the enormous relief which greeted our arrival: the few embittered Irish faces spitting hatred and intolerance; the uncertainty which greeted the removal of the barricades and the hospitality of those who could afford it least. The odd memory of confused steel-helmeted soldiers scattering barbed wire in the faces of the approaching crowds had long since been overtaken by the monotony of watching empty streets. Irish eyes were still not smiling, but the Army and some RUC patrols could go where they wished in Ulster without fear of shooting or bombing. The barricades were down.

On January 1970 the 2nd Battalion returned to Chelsea Barracks, being replaced at Magilligan Camp by the 1st Battalion commanded by Lt Col N Hales Pakenham Mahon. The 1st Battalion had a fairly quiet period to start with. They thought that everyone had seen sense and that all would be peaceful. But at Easter the Battalion had to cope with three days of violent rioting. Over 3,000 Catholics marched to the City Centre. Some suddenly tried to break into the RUC Barracks. Bricks, rocks and bottles were thrown at No 3 Company (Maj ATW Duncan MVO). 2Lts TH Holbech, the Hon RV Cecil and I Reid led their platoons into the fray. Almost everyone was hit by some sort of missile as other platoons were hurriedly deployed. The Battalion advanced deep into the Bogside, facing vicious crowds of screaming rioters. Lt JS Scott-Clarke's platoon of The Queen's Company did a sweep through the Bogside with almost every man injured. Well over half the Battalion suffered from cuts and bruises. Twenty-six arrests were made; most received up to six months' imprisonment. Thereafter everything was fairly quiet and uneventful in Londonderry. Officers could attend church in service dress. In April 1970, after five months in Ulster, the Battalion returned to England.

Even the worst pessimist did not anticipate that the conflict would so grow in intensity that over 3000 men, women and children would be murdered in Ulster over the following twenty-five years. What a tragedy.

Ten weeks later soldiers searched a house in Belfast. As they left, rioting began. The Army Commander, watching from a helicopter above, suddenly saw grenades thrown from the crowd into a platoon – the first time such an attack had been made by a Catholic crowd on the Army. A curfew and house-to-house search produced a large cache of arms and a torrent of complaints. The opening phase was over.

CHAPTER 15

SHARJAH

1968 – 1970

The British connection with the Gulf goes back to the days of the East India Company. The protection of trade and, later, of the Indian Empire and its communications with the West gradually led Britain into establishing a position of exclusive influence by means of a series of agreements with the Sheikhs of the Gulf from the Trucial Coast to Kuwait. By these agreements we undertook to protect the Sheikhdoms from aggression and assume responsibility for their external relations. These agreements were made on British initiative and primarily to serve British interests. In the process we taught the Rulers to rely on British support and protection, and they undoubtedly benefited from the relationship, sheltering under British protection.

By the late 1960s the politically fragmented Gulf area was in a power vacuum. The external pressures were numerous and conflicting. Persia, Iraq and Saudi Arabia all had claims and ambitions, but perhaps the most dangerous threat came from revolutionary movements supported by Cairo. The areas bordering the Gulf contained two-thirds of the non-Communist world's proved oil reserves and accounted for a third of its total production. Western Europe, including the United Kingdom, obtained more than half its oil supplies from these countries. Oil had thus given the Gulf a new, intrinsic importance for Britain. The Russians were made aware that our vital interests were at stake there: we and the West would go to war to defend those interests, as was evident when Iraq invaded Kuwait in 1990.

In 1968 the 1st Battalion, after two and a half years on Public Duties at Caterham, welcomed the approach of their nine-month unaccompanied tour in Sharjah. Their last overseas tour had been with the United Nations in Cyprus described in Chapter 13, prior to which they had served over the previous decade in Germany, UK and the Southern Cameroons.

179

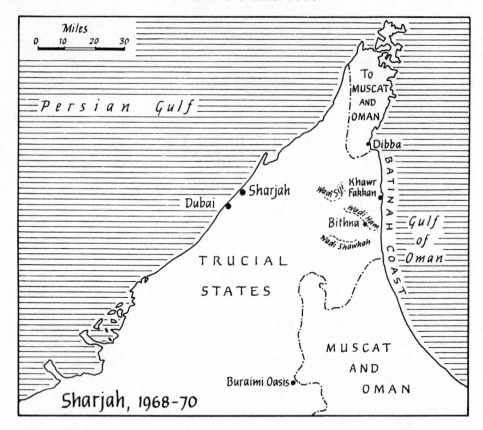

Sharjah, 1968-70

The advance party, commanded by Maj JPB Agate, left Brize Norton for Sharjah on 24 July 1968 to begin taking over from the 3rd Battalion Royal Regiment of Fusiliers. Sharjah is on the west coast of the Arabian peninsula at the mouth of the Persian Gulf. It was one of the seven sheikhdoms forming the Trucial Oman States, later to be known as the Union of Arab Emirates. They were to the north and west of Muscat and Oman with which they had no political connection.

On 16 August the Battalion, commanded by Lt Col DW Hargreaves, was complete in Sharjah and assumed its operational role from the Royal Fusiliers who surreptitiously turned off the air-conditioning in the barracks as a parting gesture. However, the Grenadiers had the last laugh as the Fusiliers left their Colours behind.

The first three days were spent swimming, touring the bustling towns of Sharjah and Dubai and listening to lectures on local politics and customs given by the Trucial Oman Scouts whose officers were largely British on secondment. Their role was to police and maintain the security of all the Trucial States – some 30,000 square miles.

The Battalion's daily routine consisted of physical training at 6 am, working through to lunch, followed by a siesta. August was the hottest time of the year. Initially, a two-mile march in light equipment in the loose sand and under a burning sun seemed quite a feat. But everyone gradually became acclimatized. During the first months, the Battalion concentrated on section and platoon training, with the emphasis on desert survival and first aid. Intensive driver and signal courses were run.

The Battalion had to be capable of carrying out internal security duties, in support of local forces, or acting independently. This meant being ready to protect British service installations in Sharjah as well. Secondly, the Grenadiers had to be prepared for counter-insurgency operations, or to resist external aggression. One company was to be kept at six hours' notice to support the Trucial Oman Scouts or protect British lives and property anywhere in the Trucial States. No 2 Company (Maj PH Cordle) was the first to leave camp for training in the desert, away from civilization, relying upon resupply by air and road from Sharjah. They spent a fortnight in September near the shanty port of Khawr Fakkan on the east coast. The temperatures and humidity were formidable. 'It is to everyone's credit that they pulled through so well,' it was noted in the Grenadier Association handbook. 'The Company was training in temperatures of 120° and above. The need for guts and self reliance quickly became apparent. It was more than encouraging to see young men mature very quickly.'

The platoon commanders, 2Lieutenants RHBI Cheape, CH Walters and CRJ Wiggin, had to keep a close eye out for heat exhaustion which was enough to kill. 'Men were capable of just wandering off on their own almost delirious on a march, impervious to orders, because of their condition,' recalls one Guardsman. 'Hence a very tight buddy-buddy system was required to drag them into the shade and, using up all the section's precious water, to revive them.' Seeing someone suffering from heat stroke was an alarming experience. NCOs had to react quickly to prevent the soldier dying.

In October The Queen's Company (Maj GW Tufnell) set up a camp for the Liverpool-Scottish Territorial Army Company at Wadi Shawkah. No 1 Platoon (2Lt JS Scott-Clarke) acted as the TA's advance party. No 3 Company (Maj HS Hanning), meanwhile, had spent a week on Bahrain Island, before training at Bithna, a small picturesque town with a mud fort. They had under command throughout the tour a platoon of the Welsh Guards commanded by 2Lt REH David. They were all quite excellent. So often small detachments of another Regiment excel in the environment of competition and pride; the Welsh Guards platoon was no exception.

No 4 Company, commanded initially by Capt NA Thorne and then Maj ATW Duncan MVO, consisted of the Signals Platoon (Capt

AC Chambers), the Mortars (Capt ET Hudson), the Anti-Tank Platoon (Lt CXS Fenwick), and the Transport Platoon. The Signals had won the London District Concentration for the previous three years. They provided a communications network throughout the Trucial States and ground-to-air communications with the Royal Air Force.

The Mortars were fired at Buraimi Oasis adjacent to Oman, while the Anti-Tank Platoon, finding there was no tank threat, quickly converted to a machine-gun platoon, to avoid, it was said, having to revert to foot-slogging as an ordinary rifle platoon. Equipped with the General Purpose Machine Gun and its sustained fire tripod, it was unique to the British Army. The Motor Transport Platoon under Lt (QM) JR Dann had inherited the salvaged remains of the Aden garrison, but, with superb REME support, managed to provide sufficient transport under extremely difficult conditions. All the vehicles were incredibly heavy because of the steel welded to the underneath and sides to counter mine damage in Aden. The Land Rovers were not good in sand and frequently got stuck, but the Bedford 4 tonners were excellent and always got through, mineplating and all.

Headquarter Company (Capt CJE Seymour) found that the only practical way of resupplying isolated detachments in many cases was by air-drop. So an air freight team was established under the Master Tailor, C/Sgt Harrhy. They soon became experts in shifting almost everything from forty gallon drums of helicopter fuel to 'Polar Packs' of enormous cubes of ice. The heat made conditions in the Cookhouse terrible, but the Master Cook, CSM Fisher, and his staff coped admirably.

On 9 December the whole Battalion embarked on a major air-supported exercise. As the first convoy left Sharjah it started to rain hard. It was the first of the season and well before time. The route was over salt flats which flooded; some drivers narrowly avoided driving into the sea. Memories of Sennybridge and other UK exercises which had been rained off came readily to mind. The RAF declared the exercise airstrip unsafe. However, the sun reappeared and the Companies were flown in to skirmish with the enemy and establish their positions.

Battalion exercises in Sharjah were exciting affairs due to the provision of supporting arms which were never seen in London District. The Battalion Group consisted of two Saladin troops, and a gunner battery of light howitzers. For air support they had one Beaver aircraft and four Sioux, two Scouts and four Wessex helicopters, with three Hunter attack aircraft on call. Resupply was by Andover and Argosy aircraft. The exercise, which was set and directed by Lt Col Hargreaves, involved clearing 40 miles of the Wadi Siji and the Wadi Ham of an insurgent enemy found by a company of the Cheshires flown down from Bahrain and three Ferret Troops.

The country was very mountainous and difficult to clear. But cleared it was thanks to the overwhelming air support. Lt TTR Lort-Phillips joined the Battalion as one of the Sioux pilots. The recce platoon, commanded by Capt EH Houstoun, found a route through the jebel where the Machine Gun Platoon and 9 Platoon (2Lt AGH Ogden) prevented the enemy from escaping. The Queen's Company and No 2 Company pressed forward with No 3 picqueting the heights. This set the pattern for the remainder of the exercise: the last day ended with three Company helicopter assaults.

The Battalion was in high spirits when it went into close leaguer on the red sands of the Battinah Coast from where they swam. The following day the transport was embarked on the landing ship logistic *Sir Bedivere* to return to Sharjah while the rest of the Battalion flew back in Andovers.

It was not unusual for a Company to train with a troop of artillery and Saladin armoured cars: this was an exciting experience for a young officer.

Christmas 1968 was celebrated with the usual parties and games which are traditional in the Regiment. One novel feature was an ingenious inter-company 'It's-a-knock-out' competition organized by CSM R Thompson on Christmas morning. It was hilarious for the spectators and strenuous for the competitors. The Sergeants' Mess arranged to play donkey polo against the officers but the donkeys never appeared.

After Christmas the Companies left Sharjah for training once more. Maj GW Tufnell decided that The Queen's Company should march from Sharjah on the west coast to Dibba, 70 miles away on the north-east coast of the Trucial States. They set off on 10 January during an exceptionally heavy rain storm. The following morning the sun returned and they could see, marching across the desert, greenery and flowers appearing. It was an amazing sight. At the first wadi the water was flowing so strongly that they could not ford it for two hours and only then by roping themselves across.

Lt EJ Webb-Carter joined the Battalion and was flown by helicopter to await the Company at an oasis, but it was the wrong one. He had rather an alarming wait. Meanwhile the Company struggled on with CQMS J Smith arriving by helicopter each day to replenish them.

'Frankly, it was bloody hell,' recalls Gdsm M Greenberry. 'Some wag had calculated that 25 miles per day was perfectly feasible. In that heat? Through many tracts of soft sand? Yes, the Company made it, but, oh dear! that journey is burned on everyone's soul.'

The Company stopped at Manama on the Trucial Oman Scouts' ranges and gave them a firepower demonstration. Maj Gen John Nelson visited the Grenadiers at this stage. 'Unbeknown to us,' wrote Maj Tufnell later, 'our presence prevented a fight between two villages. The local Resident told us on our return that a party of villagers was on its way to attack a rival village when, in a wadi, they saw The Queen's Company marching in full battle order. They were amazed at our quick reaction and returned quietly to their village.'

On arriving at Dibba the much-loved Colonel of the Regiment, Maj Gen Sir Allan Adair, was there to welcome them. His energy, cheerfulness, encouragement and personality, as always, were an inspiration to many.

As the Companies returned from their respective exercises, adventure training became the priority. The Regiment has always encouraged young officers to take their men far afield. Sharjah proved to be no exception.

The precedent had already been set by Capt EH Houstoun who had earlier driven 5000 miles in 16 days from Caterham to Sharjah at the beginning of the tour. He took three Land Rovers and seven Grenadiers. On arrival at the Dover ferry Drummer Blagden had driven straight into a caravan: there were bigger excitements in store. Their route was Nuremberg, the Austrian Alps, and the University City of Zagreb where the camp site had a dance floor. 'The entire female population of the university was available to us,' he reported. 'The night was spent giving "English lessons" and so on.' Then on they went to Belgrade where they found a kilted Scotsman thumbing lifts. He originated from Birmingham but had already travelled twice round the world on the strength of the kilt. They continued down the Ægean Sea by Salonica, then east to Istanbul, Ankara, Erzurum and Iran. At Bandar Abbas they crossed by dhow to Sharjah, feeling a great sense of achievement at such a thrilling journey.

2Lt CH Walters and RSM PA Lewis took parties to India. Others went to Iran and Singapore, while 2Lt JS Scott-Clarke took Guardsmen on LSL *Sir Bedivere* to Masirah Island and Salala. Skiing in Cyprus was also very popular. 2Lt J Luddington took fourteen Guardsmen to Muscat, covering 200 miles visiting the local villages, and 2Lt SMA Strutt brought a party out from the 2nd Battalion at Chelsea returning via Afghanistan.

By now the Grenadiers were thoroughly familiar with local customs. In towns they had to be properly dressed with shirt and trousers for the Arabs regarded some nakedness as a sign of poverty and disliked it. The Commanding Officer approved. 'The Empire was won by people in long trousers and lost by those in shorts,' he was heard to exclaim, having decided that all uniforms must have long trousers.

Soldiers were not permitted to camp alongside a well as the women would not draw water while they were there. Only one Guardsman let the Battalion down – by throwing his shoe into a well, fouling the water, in a fit of pique over an Arab girl. The women were very shy and inclined to run away if a European approached. 'You should appear oblivious to their existence,' the Guardsmen were told. 'Any attempt at familiarity could lead to a very ugly situation.'

And what did the Guardsmen make of Sharjah? Gdsm M Greenberry, quoted earlier, had set his ambition on being in The Queen's Company since his father had served in The King's Company. 'Having made a great fuss, I joined the Company in Sharjah, while the rest of my Depot Squad went to the 2nd Battalion,' he wrote later. 'CSM "Nobby" Clark had heard of my impudence in advance. 'Demanding to join my Company are you?' he said to me before sending me to my section commander. My father warned me that they were tall, and I'm six foot four myself, but when I knocked on the door it was opened by L/Cpl Ray Lowe who was six foot eleven! I thought they must all be that height. After the CSM's verbal barrage and then meeting the land of the giants, all I wanted was my Mum!

'So what do I best remember? The pre-breakfast run, a murderous routine; the boxes of paludrine tablets and salt tablets in the cookhouse: a regular "wind up" was to get the new whities from Blighty to take two salt tablets with their egg and bacon. Of course after a short delay they were sick as dogs, and only then were they told that the tablets were for the spare water bottle. I have never forgotten the fantastic hailstorms with lumps so big; we wore steel helmets to the loo during the storm; the frustration of receiving notice of my first born with no leave available; all I had to cuddle was a telegram. I sent a signal saying "please send sex" which was misunderstood by everyone in Sharjah, but the reply came back "a boy". Then there were the "dhobi wallahs" who used to starch the laundry so much you had to force your way into it; as soon as you went on the square, it all went limp at the first hint of sweat.'

As usual, the Grenadiers tried to make everything as pleasant as possible. A golf course was built around the camp's perimeter and grass was planted around the Officers' Mess although nobody was allowed to walk on it. A football pitch was also constructed for the local school. The Sheikh of Dubai came to the Mess for supper occasionally, as did the local Europeans for the Sunday night film show.

Relations with the RAF, who shared Sharjah Camp with the Battalion Group, were never quite as good as they should have been. The Regimental Sergeant Major made matters somewhat worse when speaking on the local British Forces Broadcast Station's celebrity spot. He was asked what he most disliked. He replied: 'Long hair, short shorts and desert boots.' The point was not lost on the RAF whose uniform had just been described.

The Sheikh of Dubai asked the Commanding Officer if one of his officers could teach his son English. He had evidently heard of the reputation of the Royal Army Education Corps. Lt CXS Fenwick was selected much to the disappointment of his brother officers. Fenwick disappeared for days on end only to return to regale them with tales of harems. These stories were probably untrue but he was given a white Mercedes Benz for his efforts.

Clay pigeon shooting was brought out from England; new engines were obtained for the motor boats, making water skiing possible. Archery, football, baseball, tennis, squash, swimming and cricket were all popular.

No Grenadier succumbed to malaria in Sharjah as they all took paludrine, but many RAF fresh from England failed to do so and fell ill. However, one Grenadier officer contracted malaria in Berlin nine months later. He had not taken the requisite three weeks of paludrine on leaving Sharjah.

Confiscated, contraband cigarettes were issued free to the Guardsmen. The dangers of smoking to soldiers' health were not appreciated. (It was not unusual to have 'smoke breaks' in the Army and most soldiers smoked. Cigarette firms made their occasional appearance to promote their brands – for example at an Army concert in Londonderry in 1969. By the 1980s the risk of cancer was better understood.)

In February 1969 intense exercises resumed. Companies left barracks again for the sand and jebel to train from self-contained Company camps. On 31 March the Battalion flew down to Buraimi in Andovers and Argosies for the final test exercise. Supported by B Squadron, 3 Royal Tank Regiment, 13 (Martinique 1809) Light Battery Royal Artillery, 78 Squadron (Wessex) RAF and 208 Squadron (Hunters) RAF, they located the enemy, ferreting them out of the wadis and chasing them into the hills. The exercise covered 800 square miles, with companies and platoons well dispersed. On the last day a successful battalion attack, high in the jebels, destroyed the enemy's headquarters in the scorching heat. Headquarters Land Forces was more than satisfied. It was a fitting climax to the Battalion's tour. In May they handed over to the Queen's Own Highlanders to return to England. Some Grenadiers pulled the legs of the incoming Battalion by appearing at the airport wrapped in blankets, complaining about the cold.

The Battalion had found its nine-month tour thoroughly worthwhile and rewarding. High standards of individual military skills were achieved, while the professionalism of the Quartermaster, Capt FJ Clutton MM, was outstanding. The opportunity for training with other arms and the RAF was unique. Much was accomplished. All that was missing was a serious military threat and a real enemy, but the presence of the British Army and the Trucial Oman Scouts would probably have been sufficient had such a threat developed. The Battalion's tour was a splendid change from Public Duties.

*   *   *

The Regiment felt that there would be no opportunity to serve in Sharjah again. However, in 1970 the 1st Battalion Scots Guards, destined for a tour there, was well below strength. No 2 Company of 2nd Battalion Grenadiers, commanded by Maj DH B-H-Blundell, (a suitable choice since his father had been in the Scots Guards), therefore joined them for a nine-month tour, becoming the Battalion's third rifle Company.

'Your flight to Sharjah will be delayed until it stops snowing,' the mini-kilted Caledonian air hostess told the Grenadier and Scots Guardsmen who sat huddled in greatcoats at Gatwick. After taking off eight hours late, everything went according to plan. Indeed, that could be said of the rest of the Company's tour, for an excellent relationship was maintained between the two Regiments. The Scots Guards were commanded by Lt Col MP de Klee, while the Grenadier hierarchy included Capt CWJP Langdon-Wilkins, and 2Lts SMA Strutt, RM Scott-Hopkins, the Hon AFC Wigram, CSM Hunter and CQMS J Gowers. The Battalion routine had not changed from Grenadier days, apart from the physical training starting fifteen minutes later at 6.15 am.

The Battalion exercises were as worthwhile as those during the Grenadier Battalion's tour almost two years earlier: training in the desert in arduous conditions was excellent.

Regimental tradition amongst the Grenadiers and Scots Guardsmen was, if anything, heightened by the close association. But the rivalry remained friendly and the competition entirely healthy, even when a Grenadier very nearly won the tossing the caber event at the 'Sharjah Games'.

The final test exercise in October 1970 involved a landing from the sea at Muscat, followed by the searching out and destruction of the 'enemy' which consisted of B Company, 2nd Battalion Royal Irish. So many were captured that the exercise finished a day early, to the dismay of Brig P Ward, a former Welsh Guardsman.

Ten days later the Battalion gave its 'Farewell to Sharjah' in the form of a Searchlight Tattoo which was directed by Maj IA Ferguson who later ran the Royal Tournament in London so successfully. The tattoo had everything from the fireworks accompanying the *1812 Overture* to the Ceremony of the Keys; from a musical ride in Landrovers to an assault course; from massed bands to a drill squad found by No 2 Company Grenadiers, and Right Flank Scots Guards. They gave a faultless display, finishing with advance in review order to (after some discussion) the 'British Grenadiers'. The firepower demonstration culminated with explosions of such force that the police station in Dubai, ten miles away, was besieged by terrified telephone enquiries. Just after midnight the bands played the Evening Hymn, followed by 'Retreat'. The guests left for home in their ships, aeroplanes, motor cars, on donkeys or on their feet.

It was all over in more senses than one, for these were the last days of the British military presence in the Gulf, the states of which were now benefiting from the newly found wealth. In mid-November the Grenadiers returned to Caterham and the Scots Guardsmen to Pirbright. They had all enjoyed each others' company for there has always been great bonds between the Regiments of the Household Division.

The success of the tour, at all levels, was marked by the presentation by the Scots Guards to the Grenadier No 2 Company of a Pipe Banner and a specially composed pipe tune.

This was the first occasion since the Second World War that a Grenadier Company served a full tour with another Regiment. In the years to follow the Household Division proved its flexibility by occasionally detaching platoons or companies to other Battalions.

CHAPTER 16

BRITISH HONDURAS

1971 – 1972

British Honduras, as Belize was then known, had been a British colony since the Anglo–Spanish treaty of 1786. Its eastern coastline is on the Caribbean, while the country is bordered by Guatemala to the west and Mexico to the north. To the south, a short length of coastline belongs to Guatemala before Central America curves round into the north-facing coast of Spanish Honduras. Mexico City is about 1000 miles away, and Miami about three hours flying time to the north.

Both Guatemala and Mexico had maintained claims to British Honduras since the early 19th Century. Although Mexico dropped her claim, in 1945 Guatemala published a new constitution which referred to British Honduras as an integral part of the State of Guatemala. Three years later the Guatemalans threatened to invade British Honduras in pursuit of their claim. In response the British Government established a permanent garrison based on a rifle company group.

As is frequently the case, it is easy to put troops into a country only to find thereafter how difficult it is, militarily and politically, to withdraw them.

In 1970, as a result of further Guatemalan aggressive statements, the force level in British Honduras was increased to battalion strength for a short time until the situation became calmer. In 1971 the 2nd Battalion Grenadiers became the designated 'Caribbean' battalion. This commitment required one company group of about 180 men to be deployed in British Honduras, with the remainder of the Battalion available to reinforce if required. The Caribbean commitment was for 18 months. The first six-month period was undertaken by S Company Scots Guards (Maj MB Scott) which had been formed from the disbanded 2nd Battalion Scots Guards and came under command of 2nd Battalion Grenadier Guards.

189

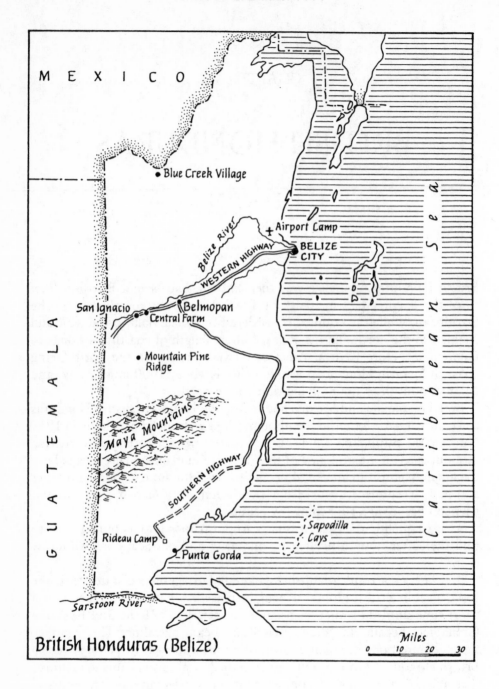

British Honduras (Belize)

On 23 August 1971 The Inkerman Company Group, 180 men strong, started taking over from S Company as the resident company in British Honduras for a six month tour. It was commanded by Maj RS Corkran and stationed at Airport Camp, some six miles from Belize City. When the locals heard that Grenadiers were arriving they assumed that the troops were from Grenada and must be all black. The group consisted of three rifle platoons, commanded by 2Lts JPC Hawkesworth, RG Cartwright and Sir Hervey Bruce Bt, and a support platoon, consisting of mortar and anti-tank sections, under Capt HJ Lockhart. The Company 2IC was Capt JM Hirst, while Lt RG Woodfield acted as both Garrison and Company Administrative Officer. CSM Sherlock with CQMSs Holmes and Sharman completed the hierarchy.

British Honduras was, and still is, terribly poor. Apart from its mahogany, which once proved an essential asset to the Royal Navy, there were few natural resources. Belize City, the original capital, later replaced by Belmopan which was well inland and less vulnerable to hurricane damage, was located on the Caribbean coast and consisted largely of wooden buildings on stilts to avoid the flooding which accompanied the hurricanes. The city suffered from the obnoxious odour of its many open sewers. It was built in 1920 after a hurricane had destroyed the previous capital. 50,000, a third of the country's population lived there.

The country is beautiful with magnificent mountains and jungle. Lovely beaches are found on the offshore islands, known in the Caribbean as cays. The coastline is mangrove, the interior forest and the climate sub-tropical. The inhabitants were friendly and easy-going. They gave the most extraordinary names to some of their villages. One was called 'Never Delay' and another 'Trousers Down'. The country's beauty queen orig-inated from a village with the unromantic name of 'Double Headed Cabbage'!

On 7 September Hurricane Edith, rather than Guatemalan tanks, became the primary threat. A hastily-contrived hurricane alert practice quickly became overtaken by events and became a real alert, minds concen-trated by the knowledge that it was a fact of history that the worst hurricanes tended to occur at ten-year intervals, and in 1961 Hurricane Hattie had killed 400 people and made many homeless in Belize City and the surrounding area. By 9 September it was evident that Edith was approaching fast. Wind strengths of 160 miles an hour were registered. Red One Warning phase was announced. 10 September was a national holiday and the locals were more interested in celebrations.

By now the Grenadiers had pre-positioned ration packs, filled emergency water tanks, tested assault boats, anchored roof-retaining hurricane wire on buildings and lashed down vehicles. Maj Corkran put on his heaviest boots and the Officers' Mess croquet hoops were removed to a place of safety. 2Lt Hawkesworth's platoon was positioned in the Palloti convent which was well fortified behind stout walls and much prayer. The platoon was tasked to perform anti-looting duties in the city centre if required. An operational headquarters also in Belize City served a very depressed cabinet with ice cream soda. The Governor, Sir John Paul, who became a great friend of the Company, received the occasional whisky and soda.

At the convent the Grenadiers could not believe their luck when the dancing girls from the Bamboo Bay Night Club clamoured to get in, but the nuns did not consider their souls worth saving and they were turned away. L/Cpl Carnell, however, was persuaded by the Mother Superior to stay up until 4 am, sampling her excellent brandy. Two hours later the shelter doors were opened on a bright and almost cloud-free day; Edith had unexpectedly turned north, striking Texas where it killed 100 and made 60,000 homeless. Although rather an anti-climax, the Grenadiers felt that they had completed the transition from Public Duties.

Maj Corkran got his platoons away from Airport Camp as much as possible – not only was training better and more relevant 'up-country' but the humidity and heat were oppressive at sea level. The Platoons trained initially at Mountain Pine Ridge at an altitude of 2000 feet. Jungle courses, training in navigation and field-firing were a welcome change from fatigues and guards at Airport Camp.

A platoon went to Blue Creek village on the Mexican border to liaise with the Mexican Army. Expeditions went into remote areas on 'hearts and minds' patrols, usually receiving a warm welcome from the Indians. On one occasion L Sgt K Schofield and his patrol had to apprehend a local who had gone berserk with a shotgun. 2Lt Hawkesworth, who spoke Spanish, visited Guatemala unofficially with L/Cpl Carnell and Gdsm Caldow to see what the Guatemalan Army was doing, but learnt little of interest.

✳ ✳ ✳

The remainder of the 2nd Battalion was stationed at Caterham carrying out Public Duties. They still had under command S Company of the Scots Guards, with whom an excellent relationship was established. There was much rejoicing when the Government announced that overcommitments and lack of infantry necessitated reforming the 2nd Battalion Scots Guards, amongst others. In January 1972 they left the Grenadiers to join their re-formed Battalion.

It had always been planned that No 1 Company (Maj JVEF O'Connell) would replace The Inkerman Company following a major Battalion exercise in British Honduras planned for February 1972. No 1 Company was to remain in British Honduras, replacing The Inkerman Company which would return to England at the end of the Battalion training period. The Commanding Officer, Lt Col JRS Besly, and the Battalion Advance Party had already left for the exercise when the Brigade Major, Lt Col BC Gordon Lennox, telephoned the Battalion at Caterham to ask how quickly they could move. The Foreign and Commonwealth Office had reported that Guatemala seemed about to invade. Maj HML Smith, the Senior Major, was asked how many mines the Battalion held!

Capt EJ Webb-Carter, the Adjutant, was having lunch with the Major General (James Bowes-Lyon) 'when a message came through telling me to report to the Brigade Major,' wrote Capt Webb-Carter later. 'Reeling from the port I went to Horse Guards, then to the Ministry of Defence where I was asked, "How long will it take you to jack up the Battalion with anti-tank weapons and mortars and emplane? You have got three days to get moving." It was a Friday and everyone had been sent on weekend leave. Moreover, it had been planned that the support weapons would remain at Caterham. Lt Col Besly had previously been warned that it could be a long tour in British Honduras but he could not tell anyone this. He decided to tell his wife and the Quartermaster, Capt PA Lewis. Meanwhile at Caterham it was decided to 'play it for real'. The police called on leave addresses to recall the Guardsmen. Within 36 hours everyone was back; gas masks were fitted and wills made. There was scarcely time to zero weapons. Ammunition and mines were drawn up. Plans changed so frequently that one unfortunate NCO said a tearful goodbye to his wife no less than three times.

In British Honduras, meanwhile, the minute garrison was preparing for war.

Maj Corkran recalls that 'Sir John Paul had left and until his successor arrived the Second Secretary was the senior diplomat in the country. It was after midnight and we had been drinking too much whisky together in my sitting room when a FLASH signal from the Foreign and Commonwealth Office was delivered to him. He showed it to me. The crucial words were: "Guatemalan attack could even be imminent" – I said to him, "What do your lot mean by imminent? Next month, next week, or tomorrow?" After some thought the Second Secretary replied, "They could mean tomorrow". I said that in that case we had better get moving, and after the Second Secretary had showed the signal to the Garrison Commander, Col JN Shipster, I summoned my orders group.'

Maj Corkran remembers that 'Col Shipster invited me to give out orders and promised that he and his Garrison HQ would do everything they could to help. It was all up to us as the police force was small with little equipment, as were the British Honduras Volunteer Guard which I had only seen on a Remembrance Day parade. The Guard had the annoying habit of defecating in running water upstream of our Mountain Pine Ridge training camp.

'I deployed the Recce Section of six men 40 miles to the west on Western Highway which runs from the Guatemalan border to Belize City. A rifle platoon, together with two anti-tank guns, were a little closer, supported by the mortars, while the rest of us deployed around the airport and camp. My plan was to fight a withdrawal battle, causing the maximum delay to the Guatemalans by opening fire at long range, forcing the enemy to deploy off the Highway onto soggy, unmotorable ground. We would then pull back and do the same again. We had practised this some weeks earlier and so went to war using exactly the same exercise instruction, apart from changing the words "blank ammunition" and "Exercise Secret" to read "live ammunition" and "Secret". I had a private fall-back plan, in the event of our being pushed back before reinforcements could arrive, of holding a dropping zone in the north in the bush, hopefully with the Mexicans, perhaps conducting guerrilla operations against the flanks of the Guatemalan division. I had little confidence that the airport could be held against parachute forces and air attack taking into account our equipment limitations. I remember a Guardsman behind a Browning machine gun at the airport saying: "If I get a Guat pilot in these sights, he's going to die – laughing".' (The Brownings seldom worked.)

Sgt Eddy, the Pioneer Sergeant, assured Col Shipster that his box of explosives would blow up the Hawkesworth Bridge on the main route to the border. It later took two Royal Engineer troops several days to prepare the demolitions. Sgt Eddy may have been a trifle over-optimistic.

Maj Corkran sent the only FLASH signal of his military career asking for extra anti-tank ammunition because he had only twenty-six rounds while the Guatemalans were understood to have thirteen tanks. That night another FLASH arrived from England. Expecting confirmation on the ammunition, Maj Corkran was dismayed to learn that it concerned a change of date for the Second Secretary's wife's forthcoming dental appointment. The Foreign and Commonwealth Office seemed to have different rules concerning signal precedence. The extra ammunition never arrived.

*　　　*　　　*

The Battalion Advance Party, meanwhile, was flying to British Honduras. Some Grenadiers took it in turns to sleep on camouflage nets in vehicles in the cold RAF aircraft. On Lt Col Besly's arrival, Maj Corkran clambered aboard the aircraft and reported, 'The Inkerman Company is deployed to meet the threat, Sir'. Col Shipster added that the situation was very serious: the Battalion must dig in around the airfield immediately.

The rest of the Battalion arrived earlier than planned. They were followed closely by reinforcements, including an RAF Regiment Air Defence Squadron equipped with the untried Tiger Cat missiles, later replaced by Bofors – much more practical for that part of the world – as well as two troops of combat engineers who were most welcome since they were able to build splendid camps up-country. The Battalion also received a large Royal Signals detachment. The Chief of the Naval Staff made a point, just as one aircraft carrier was on its way to the breaker's yard, by sending the old *Ark Royal* which deployed Phantoms. They made a salutary pass over the Western Highway. The frigates HMS *Phoebe* and *Dido* appeared offshore, providing the forces on land with most welcome helicopter support. The Officers' Mess at Airport Camp, built for seven officers, soon housed seventy. A nervous Guatemalan Government sent a protest to the British Government describing the deployment as an unfriendly act. Initially the United States refused over-flying rights so the troop-carrying aircraft had to take the longer route via the Azores. This had one advantage in that during stopover in the Azores the Guardsmen were given a meal which included a bottle of red wine per man. Guardsmen, preferring beer, gave all their wine to the officers!

By 30 January the Battalion was complete at the airport and camp, largely in marquees on concrete hard standings. The blazing sunshine and stifling heat were a welcome change from a coal strike in freezing England.

No 2 Company (Maj the Hon Samuel Coleridge) moved to Central Farm, about 13 miles from the border on the Western Highway. Maj Corkran handed over command of The Inkerman Company to Maj HS Hanning and became Senior Senior Major for a short spell (Maj Smith became the Junior Senior Major) and soon afterwards departed rather gratefully for Singapore. The Inkerman Company remained at Airport Camp, together with the Defence Platoon commanded by Lt JFM Rodwell. No 1 Company trained at Mountain Pine Ridge which was 17 miles south of Central Farm. Battalion Headquarters deployed nearby: the area was similar to Pirbright ranges, especially when it rained. Each day Battalion HQ enjoyed a swim in the nearby rapids. The Battalion Intelligence Officer, Capt CXS Fenwick, like many others, was highly sceptical as to whether the Guatemalan forces really had reached the border in strength. Despite the best efforts of Lt MGA Drage, the Battalion Signals Officer, radio communications were hazardous.

Lt Col Besly was conscious that he had a very untrained Battalion and his NCOs were short of experience. New drafts from the Guards Depot had joined in such a hurry that they had not all completed their basic training. However, Lt Col Besly did not believe that the Guatemalan Army was about to invade, although according to the local newspapers their tanks and field guns had 'taken up battle positions facing the east'. The papers considered that it was the most serious confrontation in Guatemala's long claim to British Honduras.

Lt Col Besly's priority was to get into the training areas, rather than maintain static guards around the airport. He felt that he could react quickly if necessary with the companies and his Headquarters well forward, close to the only approach to Belize City. A standing patrol was out every night on the likely enemy approach. There was some patrolling along the border. Although it would mean pulling the wool over Col Shipster's eyes to some extent, the Commanding Officer's aim was to train the Battalion for their next Northern Ireland tour. Having served there the previous year, he knew that the pre-Northern Ireland training for a Foot Guards Battalion in London would be inadequate. The Deputy Commander-in-Chief United Kingdom Land Forces, Maj Gen FD King, visited the Battalion during this period. Although he was somewhat sceptical about the Battalion Group's rudimentary plan to halt an invasion, he agreed entirely with the priority of training. Later, when he was GOC in Northern Ireland, he saw the benefit of this policy.

The Battalion soon learnt that their six-week exercise in British Honduras was to become a seven-month tour. They had brought no civilian clothes or sports kit and the inevitable compassionate cases arose amidst some frustration. The 300% increase in the force level caused many problems.

Fortunately everyone was kept busy. For nine weeks the companies rotated around the camps, each taking its turn at training, forward surveillance and the defence of the airport and camp.

Some of the Battalion's vehicles still had the London District sign of a sword and castle on them. This upset Col Shipster who wanted the Belize swordfish on them instead. Battalion Headquarters disagreed. A compromise was reached whereby both signs were displayed, but the London District sword was placed at such an angle that it was piercing the swordfish. This caused some umbrage.

An intensive patrolling programme was designed to establish a military presence countrywide, gathering intelligence and discovering the negotiability of the numerous unmapped roads and tracks. Two companies covered the centre while the Recce Platoon (Lt Hawkesworth) took the north. The Mortar Platoon (Sgt Marshall) and Anti-Tank Platoon (Capt the Hon Philip Sidney) took the south. Even the Adjutant and RSM WR Clarke took to the jungle for six days. Everyone took part in joint exercises with the Royal Air Force Regiment Air Defence Squadron, which by now was commanded by a splendid character known as Squadron Leader February because he habitually awarded 28 days detention to wrongdoers.

Sometimes he made it 27 days – to avoid the four days remission for good conduct which those serving 28 days could earn. Needless to say he ran a very good squadron! The Battalion helped to train the Police Special Force and the British Honduras Volunteer Guard.

In March The Inkerman Company, known locally as the Inca Men, returned to Caterham for Public Duties. This left 318 Grenadiers in British Honduras and 208 in Caterham. Nos 1 and 2 Companies alternated between the airport at Belize and the surveillance commitments close to the border. Nevertheless this provided time to train which is so seldom the case for a Public Duties battalion in England. Courses were run covering tactics and drill, support weapons, signals, driving and intelligence, and as many individuals as possible were sent on courses in the United Kingdom. As the tour continued, all platoons attended a jungle course. A detachment of the Royal Hussars Recce Troop joined the Battalion in April. The expertise in reconnaissance work which they passed on to the Battalion was greatly appreciated. The Air Troop of the 14/20 Hussars, who were equally popular, also supported the Battalion.

Representatives of the Organization of American States later visited British Honduras to see if either side was preparing for war. The team was lead by a Columbian, General Valencia, and a high-ranking Chilean lawyer. They were all lavishly entertained in the Grenadier Officers' Mess and played tennis and croquet with the Commanding Officer, Maj Smith and Col Shipster. The visit was such a success that the situation was defused and the team wrote a very flattering report on the Battalion.

By now even fewer were taking the threat from Guatemala very seriously. Col Shipster had hired an aircraft, piloted by an ex-Naval officer who flew hundreds of miles over Guatemalan roads. Neither the pilot nor a Military Intelligence Officer who accompanied him saw anything suspicious. But the Joint Intelligence Committee in the Ministry of Defence was still uncertain and Canberra aircraft were sent out to fly photographic missions. Col Shipster later visited Guatemala, accompanied by his wife as cover, pretending to be tourists. They met, as planned, Americans at a secluded cafe, who gave Col Shipster the entire Guatemalan forces' order of battle. It revealed that the Guatemalans had no radios in their tanks. Their troops seemed more interested in sleeping under banana trees.

L/Cpl B Inglis was one of the few Grenadiers to see Guatemalan soldiers on the border: they were dragging their rifles in the dust behind them. He remembers sitting with 2Lt Hawkesworth on the bridge of that name: it had been built by one of the latter's ancestors. They would have had mixed feelings about blowing it up as planned should the invasion occur.

2Lt Cartwright had earlier walked over the border. This led to a complaint reaching the Governor. He later attended an American jungle warfare course in Panama but was unimpressed by it. The American forces were in a demoralized state as the Vietnamese war was reaching its unhappy conclusion.

On 21 April No 2 Company took part in a parade to celebrate the Queen's Birthday. The Governor took the salute. Several commented on the dignity of the Mayor although his top hat was green with mould. The parade was commanded by Maj Smith who suddenly felt ants climbing fast up his legs: he was standing on an ants' nest. When at last he could give the order to march off, he almost ran, certainly moving faster than Light Infantry pace.

Belize City still had little to offer apart from rum, which was cheaper than coca cola. Then there were two particularly notorious ladies called Gangbang and Kipperfeet who lived in a shack near Central Farm. Maj Smith later visited No 2 Company in its country role. He found Maj Coleridge undertaking a cordon and search of a local village. The Company Commander was arresting the women because a number of the Guardsmen were getting venereal disease. The Battalion doctor gave all the women compulsory injections of penicillin before their release. The operation was carried out with impeccable good manners.

Sports equipment had now arrived and soccer, rugger, basketball and cricket were played against local sides.

A Mexico leave scheme was very popular. Parties of twenty-five spent a week sampling the luxury of Hispanic civilization at Merida. Although not many were interested in architecture, the Mayan ruins proved fascinating, some being built between 1390 and 1440.

A very successful Battalion exercise was held in July on Mountain Pine Ridge. No 1 Company decided to march back to Airport Camp although the distance was 86 miles and it would take four days. The roads were very bad and everyone received dreadful blisters despite being very fit. To encourage the Company, a landrover went up and down the platoons playing regimental music on a gramophone. The platoon commanders were Lt CNR Brown, 2Lt HARO Tweedie and Lt RM Festing and the platoon sergeants Sgts Mundy, Winch and Ball. All the population living on the Western Highway turned out to see those sent from England to defend them.

One of No 1 Company's platoons, commanded by Lt Festing, made a name for itself when a fire broke out at San Ignacio, four miles from the border. The platoon rushed to the scene in time to rescue some inhabitants. Almost thirty wooden houses were destroyed. Pulling people from them was not easy as the large refrigerators were run by calor gas containers which exploded, as did the bottles of rum. The platoon lost only one bush hat and a burnt shovel, but the Quartermaster received claims for a platoon's worth of new equipment. Belize's Prime Minister visited the village and sent a personal message expressing his deepest gratitude and congratulating the Grenadiers on their prompt and effective action.

On 11 June the Battalion athletics team took on the Belize Athletics Association. It was a great success largely due to the efforts of Maj O'Connell and CQMS A Hughes. The combined garrison team won in the closing minutes by a narrow margin. This showed the strength of the Battalion's athletic team for the Inkerman Company, back at Caterham, won the minor units athletics at the London District meeting and also the Southern Command championship. They came third in the Army minor units, despite being heavily committed on Public Duties, having to find a guard on the Queen's Birthday Parade. By now Lt RM Scott-Hopkins, 2Lt NJ Coates, 2Lt EPG Hay and 2Lt CAH Wills had joined the Battalion. Capt JWH Buxton's marriage in England was duly celebrated by his absent friends in British Honduras.

The social life in the Sergeants' Mess, based on Airport Camp, was fairly hectic. The numerous dances were accompanied by steel bands and 'Reggae' music groups. The mess seniors were RSM WR Clarke, RQMS E Mitchell, AQMS M Griffin, D Sgt BT Eastwood and AD Sgt D Ashworth.

No 1 Company gave a dance on Bird Island when they spent a phenomenal amount of money and achieved a ratio of four girls to every Guardsman.

Few were sorry to see the white faces of The Devon and Dorset Regiment who took over from the Battalion in August. The Grenadiers returned to Caterham for four weeks' leave and Public Duties once more.

The Commanding Officer had succeeded in his aim of turning a young, inadequately trained London District Battalion into a fit, more experienced and professional body. Within seven months, as he had anticipated, the Grenadiers were to find themselves on a very different tour in Northern Ireland. The opportunities to train in British Honduras had not been neglected. The six-week exercise which stretched into a seven-month tour had been put to good use.

CHAPTER 17

NORTHERN IRELAND: THE SHOOTING WARS

1971 – 1974

In August 1971 the 1st Battalion, commanded by Lt Col GW Tufnell, deployed at a week's notice to Belfast, five days after dismounting Queen's Guard. (The background to the troubles in Northern Ireland was covered in Chapter 14.)

The Queen's Company (Maj J Baskervyle-Glegg) and No 3 (Maj AA Denison-Smith) were in the North Ardoyne area. No 2 Company (Maj PH Cordle) was detached, under command of 1st Battalion Scots Guards, to Ballymurphy to the south; they later deployed to Londonderry to act as Brigade reserve. While there they took part in some of the most bitter fighting ever seen in the City, being frequently deployed in the Bogside and Creggan. The rest of the Battalion, meanwhile, saw plenty of activity in Belfast.

L/Sgt Vernon, following a suspect into a house at night, had three shots fired at him. The Grenadiers returned fire. This was the first occasion in which a soldier in the Battalion had fired a shot in action since the Cameroons in 1961 (Chapter 11). There was some sniping and lots of tension. No Grenadier was killed, due, in part, to luck. Maj Denison-Smith's vehicle, for example, was hit by Thomson machine-gun fire, but the rounds passed between those in the vehicle.

There was mutual distrust between the Army and the trouble-makers. Suspected terrorists had been interned without trial since August. This policy was questionable and attracted a greater hatred for the security forces.

In October 1971 the Battalion returned to Chelsea Barracks. A week later they had resumed Public Duties. The Northern Ireland tour had curtailed their conversion training for Germany where the Battalion moved in February 1972. Ireland was not forgotten. Such was the lack of infantry that it was necessary to call upon Gunner, Engineer and Cavalry Regiments to fight the IRA. The Battalion provided instructors to train 19 Field Regiment RA and gave a presentation to the Life Guards. Fortunately a Northern Ireland Training Team was set up to train all units going to Northern Ireland. It was commanded with outstanding success by Maj J Baskervyle-Glegg who was more than familiar with operations in Ulster.

In November 1972 the 1st Battalion left Munster for Londonderry once more. The Battalion was responsible for the Creggan Estate and Rosemount areas of the City and the southern half of the enclave between the City and the Border. Within hours of arriving, Lt Col Tufnell, still commanding the Battalion, made a spectacular catch. On turning a corner he found a Russian-made RPG 7 rocket and launcher which the terrorists had hurriedly abandoned.

Battalion Headquarters, The Queen's Company (Maj PH Cordle) and No 2 Company (Maj JM Craster) were in Creggan Camp which was on a high slope called Piggery Ridge, overlooking the estate and most of Londonderry. No 3 Company (Maj AA Denison-Smith) was in a factory at Bligh's Lane and Support Company (Maj DHC Gordon Lennox) was based on Brooke Park Library. The IRA sometimes target newly arrived battalions; No 2 Company had three engagements with gunmen within the first few days. L/Cpl Anderson was shot in the foot but quickly recovered.

Maj Denison-Smith compiled a list of statistics to cover their four-month tour in the Creggan: 'We had one RPG rocket attack at night against our camp. There were two mortar attacks on us; L/Sgt Dehnel's quick follow-up located one home-made mortar and bombs. There were eleven sniping attacks against our patrols – only one was successful: Gdsm M Clegg received a flesh wound. Gdsm K Beckett's rifle was hit by a sniper but he was uninjured. Two of our patrols were blown up by claymore mines triggered by trip wires, but there was no casualty. We had twenty-three serious riots to deal with. This necessitated firing 165 baton rounds, gas and smoke grenades. There were six bomb hoaxes and numerous bomb scares. We had to remove twelve barricades, some with support from the Royal Engineers who were quite excellent – one never knew if the barricades were booby-trapped. Removing them necessitated deploying bulldozers, cranes and the Company to deter snipers. Numerous houses were searched, and all the gardens at least twice. We arrested twenty-three men, six of whom were passed to the Special Branch. It was an exciting time for a twenty-eight-year-old Company Commander.'

Other Companies had an equally difficult time. The Queen's Company fired a startling quantity of rubber bullets at stone throwers. Gdsm D

Clarke was shot in the leg while Gdsm Leary was seriously injured by a cleverly booby-trapped metal detector which had been planted at his vehicle check point.

The Recce Platoon, based close to the border, was particularly active due to the character of its commander – Capt Lord Richard Cecil. He used himself as a bait, driving in an open Land Rover along the border hoping to draw the IRA fire while the remainder of the platoon was concealed nearby. He took the risk because he thought that the IRA were bad shots. He later had GPMGs mounted on vehicles like the SAS in the Western Desert in 1943. This was frowned upon as it was too provocative and warlike. The Recce Platoon twice came under rocket and small arms fire from across the Border. One terrorist was killed, and another leading one was captured in a separate incident. The Brigade Commander rewarded the Platoon with a crate of champagne.

On 8 March a plebiscite was held in Northern Ireland. The day passed off quietly, but barricades were erected each of the following three nights. They were dismantled by Grenadiers in time for the visit of William Whitelaw CH MC PC who was Secretary of State for Northern Ireland. He had won the Military Cross serving with the Scots Guards in 1944. The Commanding Officer remembers: 'I drove Willie Whitelaw down Central Drive in my landrover and we got heavily stoned. I was able to show him what the Guardsmen experienced every day. My aim was also to dominate the area which we managed to do in the end.'

Three distinguished Grenadiers also visited the Battalion – Lord Carrington PC KCMG, Secretary of State for Defence; Sir Ian Gilmour PC, Minister of State for Defence; and Lord Windlesham PC, Minister of State for Northern Ireland. Many other former Grenadiers have also been involved in politics and became Ministers. Among them are Lord Balniel PC, (subsequently the Earl of Crawford), who was a former Minister of State, Defence, 1970–1972, and the Earl of Arran, (formerly ADC Gore).

The Battalion returned to Munster on 28 March 1973, unaware, of course, that they would be back in Londonderry for an equally difficult tour within twelve months. They had achieved an immense amount – finding fifty-eight weapons, 9,128 rounds, 692 lbs of explosives and twenty-five terrorists.

* * *

The 2nd Battalion, meanwhile, had been deployed to Belfast for two weeks in March 1973 to cover the plebiscite. It returned to Victoria Barracks for more Public Duties including Guard Mounting from Horse Guards and the provision of two Guards for the Birthday Parade, with the rest of the Battalion, including the Commanding Officer and Adjutant, lining the

Mall. The Battalion only had time for two weeks' pre-Ireland training which, according to the Chief Training Officer, United Kingdom Land Forces, was the shortest time any Battalion had ever received. However, they did manage to win the Lawson Cup and Prince of Wales relay, and reach the final of the Army basketball cup – which was duly won in July. After a week's leave the Battalion moved to Belfast on 5 July, and lined the Shankill shopfronts for the 12 July Orange Day Parade. Fortunately, as related in Chapter 16, the Battalion had trained for Northern Ireland during their time in British Honduras (later Belize), thanks to the foresight of the Commanding Officer, Lt Col JRS Besly.

He provides an interesting insight on how a Commanding Officer saw a tour in Northern Ireland in those days: 'We took over the Shankill and Ardoyne areas of Belfast. The latter had been a "no go" area; the previous but one battalion had lost half a dozen killed or seriously wounded and the last Commanding Officer had had a nervous breakdown. We took over from 3rd Battalion the Parachute Regiment who were extremely tough but quite excellent, although they made few friends. Every patrol in the Ardoyne during our first three weeks returned to base covered in spittle.

'I coped because I had three outstanding Company Commanders in Martin Smith, Sam Coleridge and Henry Hanning. It was important to get them in the right place. The Bone area of Belfast was very tough, and so Martin went there as he stood no nonsense whatsoever. Sam went to the potentially more difficult Ardoyne because he was extremely experienced, having served with Special Forces. He had also shown himself to be a very good trainer in Belize and taught his men to operate effectively in small packets. His schoolboy looks surprised the locals. The Protestant Shankill area contained lots of criminals so I put Henry Hanning there; his very astute but charming diplomacy defused most situations. He also kept a firm grip on what had been a very difficult area for 3 Para. He was assisted by D/Sgt T Farr, who was responsible for an ad hoc quick reaction force in the Protestant Bone area.

'Bob Woodfield ran a special patrol group, spending most evenings in the Ardoyne, chatting up locals, getting the best out of them, helping on "hearts and minds", such as ridding the place of a large rat infestation. Because of his experience in Borneo with the Guards Parachute Company, Bob also made an ideal patrol master in the Ardoyne, helping many an inexperienced patrol commander.

'Hubert de Lisle ran an excellent intelligence cell; all wore uniform and had hair cuts – rather than being James Bond types. Hubert was considered the best Intelligence Officer in Belfast at the time and thereafter became a regular lecturer at the Intelligence Corps Centre. Good intelligence work enabled the Battalion to arrest a bomber in the Ardoyne within 24 hours of his arrival from the South.

'Robert Nairac did a lot of patrolling and was tasked to get to know the youth. The local Catholic Fianna knew him as Robert and respected him. He was in Tom Lort-Phillips' backup No 1 Company; we also had a company of 1st Battalion Light Infantry with the delightful John Morgan under command. Peter Lewis, the Quartermaster, combined the old-fashioned high standards of the Regiment with modern thinking, ensuring the Guardsmen got what they needed. For example he developed microwave cooking, as in St James's Palace, for the observation posts, thereby getting rid of fry-ups and junk food.

'My Adjutant, Captain Evelyn Webb-Carter, became a first class Operations Officer, handing his Adjutantal duties to his assistant, Nigel Brown. Finally I should mention the Senior Major, Andrew Duncan, who had taken over on our return from Belize and who played a significant role throughout the tour in over-seeing the administration – organizing some very successful hearts and minds exercises and helping me on the operational side.' (It is possible that many Battalion Commanders, given half the chance, would have been equally appreciative of the officers and men whom they commanded on such tours. It can be seen from this account to what extent a Commanding Officer relied upon his team.)

'We had no padre,' continues Lt Col Besly, 'and the doctor was only with us for the tour; had we had the doctor earlier we might have weeded out several Guardsmen who were medically unstable and nearly caused some major catastrophes. Over the years I had seen how better-educated Guardsmen had become. This is why I felt it was important to keep them informed of my intentions and expectations. In England this was achieved mainly by arranging a series of talks from people who had served in Northern Ireland in recent years. In Belfast I spent a considerable amount of time discussing the day-to-day situation with my Company Commanders and others.

'The Army too had become much more professional; there were excellent individual courses and a first class pre-Northern Ireland training package. My wife Dinah, and the Families Officer, Lt Colin Jenkins, were greatly helped by the Lieutenant Colonel, David Hargreaves, and Regimental Headquarters. Colin and other wives kept morale ticking over in the rear party in England. Several people came up to me afterwards and said "your wife saved our marriage"!

'The Battalion in Belfast may have appeared to be more relaxed than others, but the statistics belie this. During the tour we found fifty-eight weapons, 9,056 rounds of ammunition, 693 lb of explosives, and we put 104 terrorists/criminals in prison. This compared very favourably with our predecessors and successors. Incidentally a good proportion of our arrests and finds were from the Protestant areas which did not endear us to the police and local authorities. Luckily no Grenadier got seriously hurt. I am also glad that we didn't shoot any gun-toting IRA youngsters, for our priority was to arrest them and put them, together with the Godfathers behind them, in prison.

'Before returning to England in November I was sent for by the head of the local Catholic community who said, "Our people have found that your Guardsmen have behaved in a gentlemanly and courteous fashion, and we thank you for it." Although the Company Commanders and others did their bit, it was the individual NCOs and Guardsmen who won this accolade: one relied on them making the right decisions at the right time. When an incident occurred they went in hard. I was fortunate. If I'd had a Guardsman killed during one of the many shooting incidents, I might have felt differently and we could have had problems.'

Having quoted the Commanding Officer in some detail on how he saw his Battalion's tour, let us turn to one of his Company Commanders, Maj HML Smith: 'When we took over from the Parachute Battalion it was too dangerous to drive down Old Park Road in Belfast; we could only patrol on foot. I never allowed more than four Irishmen to group together. At night I switched off all street lights with the master key so the IRA couldn't see us. But they had a key too, so switched them on to set up ambushes for us. I twice shot out the lights, just as we shot at petrol bombers. There were a lot of butterflies in tummies as there was on average a shooting incident in the Battalion's area each night.

'My policy was that if we were shot at from a house, the Company took the house apart within a few hours. Floor boards were pulled up if necessary and walls that sounded hollow were knocked down. Royal Engineer mechanical diggers dug up the garden. I was known as The Reverend Bastard. After the search, fair and full compensation was paid for the damage we caused. Some Irish are illiterate and so we helped them fill up the compensation forms. If we had broken windows with our rubber bullets or knocked down doors, we would repair them immediately using our Pioneer Platoon. It was a very tough policy but fair. The locals soon got the message. We followed the Paras' example of checking everything; were the same people in the streets? – the same number of milk bottles? – the usual clothes on the line? It was a successful tour. When we took over there were lots of shootings and killings; on leaving all was quieter because the terrorists had moved out into an easier military area in which to conduct their atrocities.'

Luck, as always, played its part. Gdsm Smith was in the Ardoyne area: 'The IRA were watching one of our patrols and listening in to our wireless,

but they "lost" us, perhaps because we had stopped to have an illegal cigarette in a derelict house. As luck would have it, the IRA opened fire on an
observation post, unaware of our presence nearby. We heard the shooting,
quickly surrounded the house and found an M1 carbine in the coal yard;
a sixteen-year-old had fired it and was arrested.'

On another occasion intuition by a Company Commander enabled the
Grenadiers to find a loaded Armalite rifle in the house of a prominent
member of the civil rights movement. She was sent to Armagh prison for
a long time.

All NCOs and most Guardsmen had received first aid training in Belize
or the UK. A number were trained paramedics. Thus when a man collapsed
at the feet of a patrol in the Ardoyne the NCO realized he was having a
heart attack and gave resuscitation including heart massage. Despite a very
hostile crowd, he called the armoured ambulance and got him to hospital.
When the man returned home he spread the word that his life had been
saved by the military and he even thanked the NCO concerned in person.
The attitude of the locals changed considerably. Later in the tour, when a
Guardsman was injured, the locals helped to direct the ambulance to the
scene and a well known Republican controlled the crowd.

Finally let us hear from Lt RG Woodfield who had served in almost every
rank in the Regiment before being commissioned in 1971: 'Robert Nairac
was fantastic with the Irish youth. He had been to Dublin University. I used
to visit the Sporting Club in the Ardoyne with him. He introduced me to
all the Club's committee and sang Irish rebel songs with them, knowing
more about Irish history than all of them. He also knew most of the
fourteen–fifteen-year-olds in our area, and how crime was in their blood.
He used to give talks to their mums and dads on where their children were
going wrong.

'After the Battalion had returned to England, five of us were left behind
to see in the 1st Bn the Argyll and Sutherland Highlanders. On their first
night a young Argyll officer looked into a car in the Crumlin Road and saw
a fuse burning. He just had time to get his patrol and us into cover before
the massive explosion broke all the windows in the neighbourhood and
took off four roofs. None of us was hurt: Robert was on the scene until 3
am. On the second night we were still both patrolling with the Argylls,
wearing their headdress, with blackened faces. A local dignitary told us
that he had just seen one of his lads being abducted: unless he was reported
safe by 10.30 pm all hell would break loose. So we went into the crowded
pubs to get information. We were cursed and abused but I knew many of
them by now. We carried our weapons cocked so could have fired within
seconds. Robert covered me, and six Argyll platoons were deployed as

31. Officers of Regimental Headquarters, 1st and 2nd Battalions in London, 17th May 1978.

Back: 2Lt N J C Wilson, Capt B T Eastwood, Lt B C Symondson, Capt W R Clarke, Capt G R Whitehead, 2Lt P J Lees Millais, 2Lt G V A Baker, 2Lt J G Heywood, 2Lt C H Manners, Capt W Williams, Capt D R Kimberley, Capt B E Thompson, Capt N Collins, Capt D Mason.

Middle: Lt M C Y Madsen, Lt A H Drummond, Capt J S Lloyd, 2Lt A J Fane, Lt T E M Done, Lt M I N Brennan, Capt D M Davis, Lt A D Hutchinson, Capt A C Ford, Lt E T Bolitho, Capt R E H Aubrey-Fletcher, 2Lt H Hobhouse, 2Lt S H P Hay, Lt J R Lloyd-Jones, Capt P J S Allen, 2Lt C T King.

Centre: 2Lt T W G Dennis, Maj J M Hirst, Capt T Attwood, Maj P R Holcroft, Capt C E Elwell, Maj C S S Lindsay, Capt Sir Hervey Bruce Bt, Capt G W M Chance, Capt J F M Rodwell, Capt D C L De Burgh Milne, Capt R C Wynn-Pope, Capt Lord Valentine Cecil, Capt J P Hargreaves, Capt R J C Hawes, Maj J S Scott-Clarke, Capt M G A Drage, Capt D L Budge.

Seated: Maj The Hon Philip Sidney, Maj D M Braddell, Maj J M Craster, Capt G F Lesinski, Capt C R J Wiggin, Maj N Boggis Rolfe, Lt Col D H Blundell-Hollinshead-Blundell, The Colonel, The Queen, Col G W Tufnell, Lt Col M F Hobbs, Maj O J M Lindsay, Maj P A J Wright, Capt J F Q Fenwick, Maj The Hon Samuel Coleridge, Maj T T R Lort-Phillips, Maj H A Baillie.

32. A group of Officers, 2nd Battalion, British Honduras, 1972. *In rear*: Lieutenant C N R Brown, Major The Hon Samuel Coleridge, Captain E J Webb-Carter, Captain The Hon Philip Sidney, Second Lieutenant H A R O Tweedie, Second Lieutenant N J Coats, Lieutenant M G A Drage, Lieutenant A J W Powell (RH), Second Lieutenant R M Festing, *In front*: Captain T T R Lort-Phillips, Captain G A M Lonsdale-Hands (RAPC), Lieutenant R M Scott-Hopkins, Lieutenant-Colonel J R S Besly, Major J V E F O'Connell, Captain (QM) H C Jenkins, Second Lieutenant N S F Browne.

33. After the United Nations Medal Parade. Members of 2nd Battalion, Cyprus, 1983. Sgt D Smith, Sgt C Cox, Sgt K Knight, CSM S Marshall, Sgt A Goddard, Sgt H Booth, CSM J Marshall, Sgt J Wigg, Sgt C Angel, Sgt R Bedford.
Seated: Lieutenant Colonel A A Denison-Smith, Captain E T Bolitho.

back-up. We discovered that the lad had only been playing snooker and had him back by the 10.30 deadline, so all was well.'

*   *   *

By now those posted to Northern Ireland were familiar with the aspirations of both the IRA and the Protestant Loyalists. Ideologically the IRA is a descendant of Wolfe Tone's United Irishmen and of the ideas and ideals of the French Revolution. Over the last two centuries few strategies and tactics have not been tried by the Irish Republicans. The IRA were waging war for the ideal Republic – a united Gaelic Ireland. Despite the murders, theft, extortion, blunders, betrayals and schism, new volunteers have always been forthcoming to further their dreams. Their faithful seek to destroy the English connection. Most volunteers are young, working class, increasingly from the north, very Irish, deeply nationalist and from Catholic backgrounds.

In 1970 the IRA Army Council had decided to move into their armed struggle by three stages. First to organize and arm a secret army to represent themselves as the defender of the nationalists. Secondly to provoke the British Security Forces so that they would alienate the nationalists and appear to be occupying forces. Finally to engage the British in an urban-rural guerrilla war. All these objectives were achieved within two years.

By July 1972 hundreds of bombs had reduced the centres of many towns to rubble; shooting incidents were daily events; the police could no longer enter many nationalist areas; the Northern Ireland Government at Stormont had been dissolved and their Orange system dismantled; the IRA representatives had met with British Ministers in London. When the truce collapsed, the war resumed. Operations were directed against targets throughout the United Kingdom and Europe; arms and funds were obtained from the United States and Middle East.

In 1972 103 British soldiers, twenty-four RUC and three of their reservists were murdered, whereas during the 1956–1962 IRA campaign only six RUC were killed and no soldier lost his life. Their campaign had failed then partly because of Ulster's increasing prosperity.

As far as the Protestant extremists are concerned, all have been prepared to fight Britain to stay British. By 1971 the Ulster Defence Association was the most influential paramilitary group. Since they murdered as indiscriminately as the IRA, particularly civilians, RUC policemen or prison warders, (the soft targets), referring to themselves as 'Loyalists' is outrageous. Loyal to whom? Three years later they felt strong enough to take on the British Government, objecting to the prospect of power-sharing and the Council of Ireland. They organized a strike which led to the collapse of the Government of Northern Ireland and the reimposition of 'Direct Rule' from Westminster.

In 1971 internment – imprisonment without trial – was introduced. Both the IRA and Protestant paramilitary internees were given 'special category' status. These privileges led to the belief that they were 'prisoners of war'. Despite this, internment helped to reduce street violence. However, it also became a catalyst for some to oppose the British Government, and so when the security situation improved internment was ended. The problem with internment is that unless used massively and for a sufficient period it is less effective as a security measure than as a cause of IRA recruitment.

Until 1972 the British Army interrogated suspects in depth. This involved frightening and psychologically disorientating methods rather than physical brutality. Lord Parker's committee examined the technique and reported that the information gleaned from these methods was indispensable. The Government, however, stopped such interrogation due to international opinion. The European Court of Human Rights was critical of the practice. The loss of such vital intelligence was a very serious blow to the Security Forces. Lack of evidence led to few arrests and convictions.

Questioning by most soldiers was of doubtful value since they were not trained to interrogate, nor were those arrested inclined to cooperate. A major setback occurred on Sunday 30 January 1972 when thirteen civilians were killed in Londonderry by 1st Battalion the Parachute Regiment. None was proved to have been shot while handling a firearm or bomb, but some may have done so earlier that afternoon. The Widgery Report exonerated the soldiers from the IRA charges of firing indiscriminately at a crowd running from them. The political consequences of 'Bloody Sunday' were catastrophic. Many Catholics in the North thereafter identified the Security Forces with the Protestant cause. It was not unusual to see photographs of the thirteen 'martyrs' in people's houses.

* * *

On 28 March 1974 the 1st Battalion took over responsibility for the Creggan Estate which lay on the west of Londonderry, overlooking the Bogside and Brandywell. (See map on page 171). Beyond the city centre lay the River Foyle, crossable only by the Craigavon Bridge, to the Protestant Waterside. The situation had deteriorated still further since the Battalion's previous tour. Battalion Headquarters with The Queen's Company (Maj OJM Lindsay) and No 3 Company (Maj AJC Woodrow) were based on the camp on Piggery Ridge overlooking the Estate which contained some 15,000 Catholic Republicans in small terraced houses with individual gardens (although flowers were never noticeable). The majority resented our presence and ignored us.

Creggan Camp was surrounded by a wall of concrete, wire and sand-bags with watchtowers containing narrow firing ports and periscopes. The camp was said to contain the longest sandbagged wall built since Korea. On April Fools' Day three large crumps were heard followed by fifty high velocity rounds. The camp was being mortared! A rocket hit one of the towers, while the mortar bombs exploded harmlessly 50 yards beyond the camp. In the prolonged follow-up Grenadiers searched 159 houses. The Regiment to the north found the home-made mortar, while No 2 Company located the unfired mortar bombs.

Both No 2 and Support Companies were based on Bligh's Lane, three minutes drive to the east between the Creggan and the Bogside. They were commanded by Majors MF Hobbs MBE and DM Braddell respectively.

The other Regiments in Londonderry, all in the infantry role, were the Queen's Dragoon Guards, 16 Light Air Defence Regiment, 1st Battalion the Royal Regiment of Fusiliers and 1st Battalion the Duke of Wellington's Regiment. The last two were on 18-month tours. They were in the quieter areas, but nobody envied them. We felt that four months were quite enough.

During the Ulster Workers' Council strike, the Army's overall strength was 17,000. Less than four years earlier, after the removal of barricades, the Creggan had been patrolled by L/Cpl GF Lesinski and a single Royal Military Policeman who carried a concealed pistol. They wore forage caps, buff belts and well polished boots. Even the vigilantes then seemed friendly because the Army was regarded as saviours from the Protestant hordes. How times had changed! Many of the welcoming children grew up to become terrorist sympathizers.

The Commanding Officer, Lt Col BC Gordon Lennox MBE, like most officers, had served in Londonderry before. His priorities were first to dominate the Creggan and neutralize the gunmen. Secondly to build up the intelligence picture. An important consideration was to ensure the safety of his men.

A card was delivered one night to every Creggan household. It read: 'Our business is the maintenance of law and order in your City. We are here to help. We must stop the gunmen and bombers. Help to bring about a return to peace. Telephone (anytime) Londonderry 61021'. The card was ignored by all.

Lt Col Gordon Lennox felt that he lacked the free hand he wanted. Fast, new Puma helicopters were available. When it came to the highly provocative Easter marches, he wanted to fly in soldiers and arrest the IRA leaders. This was forbidden. He also felt that common law should prevail: cars without licences should be removed. The Army Commander disagreed: 'Soldiers should concentrate on terrorism, not cars,' he was told. Lt Col

Gordon Lennox's point was that it was all part of the same thing; political will was everything; one can not control terrorists if one is not the sovereign power.

To summarize a few of the early incidents: on 3 April the senior Army Roman Catholic Chaplain's visit to the Battalion coincided with five shooting attacks. There were eight next day. Over the following weeks Drummer Saunders was wounded; the Adjutant, Capt CXS Fenwick, in another Battalion's area, lost an eye when a brick was thrown into his Land Rover; L/Sgt G Lightfoot was also seriously injured, losing a leg after entering a booby-trapped flat following up after a shooting attack on Creggan Camp. Maj Woodrow had the utmost difficulty rescuing him from violent crowds. Sgt T Mann was wounded in the upper chest by an armalite round, but continued to maintain radio contact with Battalion Headquarters. He later made a good recovery. A fourteen-year-old girl was hit by IRA shots fired at a patrol. During five shooting incidents on 22 May L/Cpl Chick was shot in the leg. The following day the Commanding Officer's escort vehicle was hit as had been three of the Company Commanders' in preceding days. All this was within the first eight weeks.

And so it might have continued had it not been for the increasing professionalism of the Battalion whose endeavours to drive the gunmen from the Creggan were meeting with success. The pressure upon the terrorists was constant. Patrolling, to dominate the area and acquire intelligence, was vital.

Let us follow one typical patrol in the opening weeks. The company commander will have planned a mixture of vehicle and foot patrols which cover the entire company area ideally 24 hours a day in an irregular and unpredictable pattern, in such a way that no patrol is left unsupported. The patrols are usually 'multiple' from the outset, covering an area of streets working in parallel in a co-ordinated manner and in radio contact. The IRA sniper will therefore never be sure of his escape routes. The pressure on the young, junior NCOs who are commanding most of the half sections is considerable. Rifle platoons usually consist of four sections of six men plus the platoon HQ.

The company group intelligence cell gives them a mission and an update on what has occurred since their last patrol. We do not want them to regard themselves as the 'duty targets'. They move out from the camp at speed. Half are in fire positions while the remainder run – never walk. The routes are varied, tracks are not followed and it is necessary to look *down* as well, to avoid trip wires or mines. The mission is to search for weapons. The IRA favour hedgerows, dustbins or sheds because the weapon can be quickly disposed of and not attributed to anyone if found. Explosives, also, are unlikely to be hidden in houses because they would lead to the arrest of the occupants, possible explosions and suspicious stains. Within minutes the 'P' (Personal Identification) checks will start, although they are usually inconclusive. An individual will be asked to identify himself or herself. The identification will be verified with the P card held by the intelligence cell which is monitoring the net. The cell will then suggest intricate questions which the suspect should be asked so that his or her story can be checked. The patrol is vulnerable while stationary and so, if doubts exist, the individual is arrested. The soldier simply says, 'As a member of Her Majesty's Forces I arrest you.' Nothing more is required. In principle an arrested person is handed over to the RUC or Royal Military Police but neither are seen in the Creggan.

To avoid a hostile crowd building up, it is preferable to arrest when vehicles can quickly collect the suspect. No vehicle ever travels singly. The individual is treated with courtesy, can take a companion along to see fair play and an explanation is given to the crowd. 'Aggro' must be avoided for it diverts us from our mission, plays into the IRA's hands – for they want an overreaction – and we are bound to meet the deliberately uncooperative, such as the woman who will not allow us to look beneath her baby in her pram, necessitating taking them both to our camp for a servicewoman to search.

The company group's operations room monitors the patrol's reports. Also listening in, probably, is the Battalion Duty Officer, and the excellent Medical Officer (Capt N Gavin) who can be with any casualty within three minutes. By now every soldier is familiar with slick radio procedure.

To prevent the IRA knowing the multiple patrol's location, very detailed maps are carried which show numbered points to which the patrol refers. They are meaningless to an outsider. When a map is lost, new numbers are issued. Security is also preserved by not wearing badges of rank nor referring to names; code words are used and it is assumed that all communications are insecure. The patrol may be watched by our sentries in the camp's watchtowers or by mobile patrols. The IRA have been known to stalk a patrol or lie in wait. Recognizing *their* signals would save casualties; none of us know them for certain although we have our justified suspicions. A cigarette tossed on the ground, a hand removed from a pocket, a football kicked in a particular direction, a youth who starts combing his hair, a group playing cards, or even someone suddenly scratching – these are, we believe, the signals occasionally used by the IRA sympathizers to indicate to their gunmen that our patrol is approaching or that they can set up a shooting for we are not in the area. The patrol run from fire position to fire position, the rifles of those on the ground are in the shoulder and aimed at likely IRA positions – a corner of an alley, an upstairs open window or the shadows beyond a shed. Are the IRA setting up an incident with their 'sentries' posted? If so, the patrol suddenly changes direction. But the initiative remains with the IRA who can decide when and where to open fire.

Suddenly the patrol hears the loud crack and thump of the bullet above their heads. It is immediately reported: 'Shot rep. Alpha:1627 hours. Bravo: source of fire at Junction of Bligh's Lane and Central Drive. Charlie: a single round. Delta: No casualties. Echo: Running to junction to set up cordon. Foxtrot: Little reaction so far.' Within several minutes the reserve platoon and perhaps the Company Commander with his small group will be on the ground; ideally a helicopter may be hovering but there are too few of them. Snap vehicle check points are set up. A 'hot pursuit' or returning fire is often impossible because the gunman is seldom seen and his location unknown. A thorough search usually fails to find the weapon. A few arrests may follow and several houses receive a brief search. Gradually the tension subsides; normal patrolling resumes; the incident is over. The knowledge that the gunman is still free to strike again is disturbing. The IRA, like us, set no pattern; they change their sentries, signalling and time of shooting. However, they don't fire from the south because their escape routes are poor and the people there may be rather less sympathetic to the IRA. Nor do they shoot at night because they had two killed within days of a previous battalion receiving night viewing devices.

On return to base the patrol carefully unload their weapons – accidental discharges are still a problem. They are then debriefed in detail. The intelligence cell patiently update their records and fire questions: 'Who was in which house? Describe them . . .' The name cards, street cards and car cards receive added detail. Meanwhile the questioning of those arrested is continuing under the Senior Major, DH B-H-Blundell, and his Battalion HQ screening team who are in close contact with the RUC Special Branch.

The following morning at 3.30 am the houses are searched by a trained team which includes a sniffer dog and a servicewoman from the Royal Military Police, Women's Royal Air Force, or Ulster Defence Regiment. A cordon outside wait patiently in the dark. The house's occupants are held in one room; they usually cooperate, perhaps because they have something to hide. On conclusion, the amount of damage, if any, is mutually agreed. The patrol is picked up quickly by two vehicles and then deposited at the next houses at 5 am. By 7 am they are back in camp to be debriefed before washing the camouflage cream off their faces, a large breakfast and further patrolling. It is exhausting work. Most lack sleep.

The typical pre-planned patrols and searches described above very seldom found a weapon or wanted person, although we held good photographs and descriptions of those we were seeking. But due to the constant patrolling and relentless Grenadier pressure, the IRA started seeking safer pastures. Between 20 April and 28 May, 11,088 cars and 257 houses were searched; 505 people were arrested of whom eleven were detained. During the same period there were sixty-eight shooting incidents; thirty-nine blast or petrol bombs were thrown at us and there were twenty-one major stoning incidents. Gradually, the number of Grenadier successes tipped the balance.

On 29 May an alert sentry in one of Creggan Camp's towers alerted 3 Platoon (2Lt DHP Luddington) to a suspicious red Ford Escort. It was seen by Gdsm Knight (6'7") who stopped the car by leaping on the bonnet. The two occupants were on the Brigade wanted list. (Knight received a large financial reward which he passed to a wounded Grenadier.) The car also contained a loaded and cocked Schmeisser sub-machine-gun, .303 rifle and twenty-four rounds. An immediate 'hot pursuit' led to the find of a Luger pistol, 172 assorted rounds, blast bombs, explosives and an incendiary device. Then the shooting started. A gun fight developed between five gunmen and all the soldiers who could be deployed. Capt CH Walters, accompanied by three Army Catering Corps cooks and five Royal Army Pay Corps clerks, chased an ambulance. The shooting went on intermittently for fifty minutes. Grenadiers fired ninety-four shots.

Three weeks later L/Cpl Dodd, also of 3 Platoon, saw a long, suspicious parcel on the back seat of another Ford Escort. The owner could not be traced – he later claimed that it had been stolen – and so the specialist bomb disposal unit was called in. An hour later the parcel was gingerly pulled by wire from the car to reveal two rifles. Despite a large, vicious crowd, the car was removed to Creggan Camp for forensic tests.

All other platoons had their successes, too. Gdsm A Thomson of 2 Company found a rifle near a post office as a result of a tip-off. 'I also saw three empty bullet cases on a window sill,' he recalls. 'I picked one up but it was booby trapped. I fainted and fell backwards down some stairs. Another Guardsman caught me as I came bouncing down. I enjoyed the tour: it was fun because it was a contest between the IRA and us.'

Important lessons were learnt. For example, after a shooting incident, L/Sgt Dowland's section of 7 Platoon saw two girls aged fifteen and sixteen wearing long overcoats. The girls agreed to open their coats but could not withdraw their hands from their pockets without dropping parts of the rifle which they were holding in the coat linings. 9 Platoon had an identical success, finding an armalite rifle together with eighteen rounds being carried by two other females. No 2 Company, meanwhile, had arrested two more girls with explosives at a vehicle check point. The IRA used women because only servicewomen were permitted to search them. This rule was often disregarded because there were too few servicewomen and they could not run round the streets with us. Another rule which was very occasionally broken was that only aimed shots were meant to be fired at identified gunmen. However, since we usually did not know their precise location, a few shots were sometimes returned, against a brick wall or whatever, so that they would ricochet harmlessly into the ground. The gunmen immediately ceased firing.

Very little definite evidence is available about IRA casualties over this period but on two occasions extensive pools of blood were found after No 3 Company had been in action. Neither side claimed many hits because some engagements were at 500 yards at very small targets. Our weapons were re-zeroed frequently but there was insufficient time to improve marksmanship. The IRA had greater problems; their weapons were stripped down after shooting incidents and so presumably had to be taken to Southern Ireland for re-zeroing. Very few telescopic sights were available.

There were occasions when companies worked together. After one shooting incident a platoon of The Queen's Company seemed about to be overwhelmed by massive crowds around St Mary's Church. A wounded IRA gunman was said to have crawled into the Church for safety. Guardsmen were struggling to hold on to their weapons as the rest of the Company deployed. Support Company fired gas. A senior clergyman accused the Commanding Officer of shooting at the Church and roundly cursed him, but, after the identification of the gunmen, Lt Col Gordon Lennox was taken aside and blessed. The Commanding Officer later saw an intelligence report which said that a pistol had been found beneath the altar.

One three-hour night operation involved three companies, with support from the Gunners and Fusiliers. Thirty-six Creggan houses were searched very thoroughly and a further thirty-eight given cursory searches. Nothing was found. It is probable that a Guardsman told his Creggan girlfriend on the telephone beforehand that a big operation was imminent, although he may not have known the area to be searched.

Grenadiers wore brown berets, which had replaced the dark blue/black ones some five years earlier, and cloth rather than the polished brass grenades. The purpose-designed helmets with visors were introduced much later. Flak jackets provided some protection against low-velocity rounds and shrapnel.

The heavy and bulky 7.62 mm self-loading rifles were designed for war with the Warsaw Pact rather than internal security, but we had confidence in the rifle. 9 mm Sterling sub-machine guns and, at road blocks, pistols were sometimes carried. A few members of each patrol usually carried batons and weapons to fire rubber bullets and gas, but they were seldom used as it was now a shooting war.

Vehicles – Alvis Saracen armoured personnel carriers – were available, together with sufficient Land Rovers covered by rather ineffective macralon light armour, and Ferret scout cars.

The Battalion's standard drill was to have two men standing in the back of a Land Rover with rifles, and the passenger in the front seat with his door fastened open and a rifle pointing out. Such tactics were questionable for the soldier was vulnerable to snipers or brick-throwers. Fire could seldom be returned, for the snipers were not seen.

As more shooting was taking place in the Creggan than in all the rest of Ulster the Battalion received more visitors than others. They included the Chief of the General Staff and, intriguingly, two groups of Members of Parliament and the House of Lords. They were led into Central Drive and a ramshackle shop where they, predictably, received obscene abuse, were spat at and there was some stoning, but no shooting because the patrols had swamped the area. It was to the politicians' credit that they came.

The Creggan's pub, the Telstar, was a difficult area to visit as belligerent drunken Irishmen, crowded together, were likely to lead to trouble. Nobody therefore entered it. However, perhaps to impress the visiting BAOR Brigade Commander, The Queen's Company was ordered at 6 pm to search it at 7 pm and again at 11pm. On the first occasion 2Lt AGR Way's platoon manned the inner cordon and 2Lt NJ Coates the outer. An arrest force stood by in Saracens in Creggan Camp while several mobile patrols toured the area. The Intelligence Cell with the Captain rushed in, but there was no real trouble as The Queen's Company Guardsmen always look tough. At 11 pm a large drunken party was in full swing upstairs. It was agreed that two Grenadiers could check their identity. Capt the Hon James Hogg, the Company Intelligence Officer, spent half an hour doing so. One man was arrested. Outside the crowds suddenly cleared. A shooting incident seemed imminent, but the cordons had twelve infantry weapon sights which provided excellent illumination in the dark; a Guardsman had turned off the street lights and the Grenadiers were in good fire positions. The IRA, if they were around, left it for another day.

The gunmen attacked the soldiers indiscriminately regardless of the danger to civilians. On an earlier tour Grenadiers received a telephone call at 9 pm. A man complained that someone was moving round his garden. A patrol visited it soon after but found nothing. The IRA had positioned a claymore mine for them in the neighbouring garden. During a church service the following morning it killed a ten-year-old. When Grenadiers entered the house a vicious crowd surrounded it; the patrol was extracted with difficulty.

Although Creggan Camp was extremely crowded – even the Commanding Officer shared a room – life was not as uncomfortable as earlier tours. An occasional film show was arranged, although the senior ranks were too busy to watch it. There was no rank distinction when it came to sharing the wash basins. The only bathroom was in the medical centre. Everyone ate the same excellent food. Guardsmen were allowed two cans of beer every 24 hours. There was an Officers' and Sergeants' Mess. The former contained silver and pictures brought over from Munster. The Battalion was very fortunate in still having Peter Denton as the Padre who held a short service every Sunday.

At the beginning of the Battalion's tour RSM BT Eastwood handed over to RSM A Holloway. As usual the Warrant Officers were a tower of strength. They were: RQMS B Hunter, D/Sgt DJ Webster, AD/Sgt R Barnes, CSMs D Sherlock, A Davenport, P Hodgkinson, B Sheen, R Holmes, and P Bell.

There was no sport or socializing with the locals. One Company Commander, for example, found only three opportunities to leave the camp, apart from operations and the few days 'rest and recuperation' in Munster. The first occasion was Beating Retreat by the hospitable Royal Fusiliers. The delightful ceremony was attended by their Old Comrades. What was particularly memorable was the machine-gun posts facing outwards on tall buildings around their barracks, the former naval base HMS *Sea Eagle*. On the second occasion the Brigade Major, JPW Friedberger, gave a lunch party in the same barracks. The three guests arrived by helicopter to avoid needing a two-vehicle escort, but he was at an orders group. Within five minutes the officers were ordered to return for an impending operation. On the third occasion a dinner party was held in a vast stately home. The butler placed the three officers' loaded pistols in the safe; the ladies wore long evening dresses and the talk was about anything other than Ireland and the troubles. The contrast with patrolling the Creggan a few hours before can be imagined.

At vehicle check points some officers very occasionally came across Irish friends whom they had known at school or met at social occasions on the relaxed initial tours. To avoid compromising them, no recognition was exchanged. Since we were in the Creggan all our efforts were directed against the IRA and their Catholic sympathizers. However, on 16 July some of us at vehicle check points saw Protestants in buses setting off for their marches, to stir up trouble. They looked middle class, elderly and smug, clutching their bowler hats and orange regalia. It takes two sides to prolong trouble.

The rear party at Munster was run by Capt (QM) T Astill. He did much to keep the families as cheerful as possible, as did Sally Rose Gordon Lennox and three Company Commanders' wives – Tessa Hobbs, Clare Lindsay and Fiona Woodrow. They regularly visited their 'patch' and helped the families with any problems. When a Grenadier was wounded, *every* family was told immediately, so there was no rumour. Families received an excellent newsletter edited by the Padre Peter Denton. (The IRA later claimed that they had all the editions.) He first censored any pin-ups in the background to the newsletter's photographs. Naked breasts had black bikini tops delicately drawn over them.

Although the situation in Londonderry was scarcely under control, the Queen's Dragoon Guards were withdrawn. The Bligh's Lane companies therefore moved to Fort George on the west bank of the River Foyle. This camp was originally named by the Coldstream after their very popular Colonel, Maj Gen Sir George Burns. Despite our own excellent recce platoon, commanded by Capt SMA Strutt, a Life Guards troop, a platoon of the Fusiliers and a Military Police detachment, there were simply not enough men to dominate the area. This was because, of the fifteen platoons, half were committed to static duties such as vehicle check points or guarding the two bases which were shot at occasionally. Confinement to static locations gave the IRA a freer hand.

Nevertheless, due to our constant patrolling, and searching, the gunmen went elsewhere. The number of shooting incidents trailed off dramatically and by now we were usually firing first. The Creggan at last seemed subdued – or at least dormant. No Grenadier was shot in the last two months.

The only known death in Londonderry, apart from two IRA who blew themselves up in a supermarket, was Capt ASH Pollen, Coldstream Guards. He and another soldier were in plain clothes serving in another unit. They were cornered by the IRA on Easter Sunday at a demonstration in the Bogside, where they were attempting to take photographs. Capt Pollen was shot dead while the other soldier escaped. None of us was aware of their activities.

It may be invidious to refer by name to the principal Grenadiers who contributed most; everyone did well. But those listed under the Honours and Awards annex received the appropriate recognition. Special mention should be made of Maj Woodrow who received one of the first Queen's Gallantry Medals to be awarded. Since the Battalion had no permanent static observation point in the Creggan he took to lying up at night, almost alone in a street, to see what took place. Fortunately his location was never discovered: the IRA would have killed him before the stand-by platoon reached the area.

On 24 July 1974 the Battalion handed over a fairly quiet parish to 1st Battalion the Staffordshire Regiment. It can be seen that the 1974 tour had nothing in common with the first one in 1969 (Chapter 14), nor with the 1993/94 tour (Chapter 30). On 28 July a Battalion Thanksgiving Service was held in Munster. There was much for which we were thankful: no Grenadier had yet been killed in Northern Ireland. This was partly because the gunmen in Londonderry were relatively amateurish. Indeed some were referred to as 'cowboys'; they had not yet mastered the use of command-detonated bombs.

The gradual attrition of the IRA was paying dividends; Grenadiers gave evidence against five men and six women caught carrying weapons or explosives; five senior 'IRA officers' were arrested and convicted. But their sentences were short and, with remission, they were free to resume their terrorism all too soon. Even so Londonderry was a safer place for a while. The number of incidents thereafter were few. Force levels were reduced accordingly. Creggan Camp could be dismantled after a further Grenadier tour there; an uneasy peace descended upon the long-suffering Estate which the Grenadiers had dominated for four thoroughly unpleasant months in 1974.

CHAPTER 18

HONG KONG

THE TOUR OF A LIFETIME

1975 – 1976

'Hong Kong will be a tour of a lifetime,' Lt Col DV Fanshawe, Commanding the 2nd Battalion, promised his men in Victoria Barracks, Windsor.

The Grenadiers arrived in the Crown Colony between 4–16 January 1975 to take over from the King's Regiment. First impressions were distinctly favourable although it was still the rainy season. The barracks at Stanley Fort were on the most southern peninsula of Hong Kong Island. They had been built well before the war on high ground around several playing fields and a barrack square. Spacious white buildings with wide verandas contained the Guardsmens' sleeping quarters, offices, stores and a few married quarters. Before the surrender of Hong Kong to the Japanese in December 1941, the final stand of British and Canadian troops occurred in Stanley Fort.

Beyond the Island's north shore lay Victoria Harbour, the New Territories and some 25 kilometres beyond Kowloon, the faintly menacing Chinese Border. Stanley was separated from the teeming city centres by a tortuous narrow road. It was secluded and remote enough to ensure peace and quiet, and not sufficiently accessible for the Guardsmen to be able to reach the nocturnal delights of the garish but enticing Wanchai District too frequently.

The tour, indeed, promised to be exciting and unique – seemingly un-limited opportunities for training, exercises, sport, travel, sight-seeing and

219

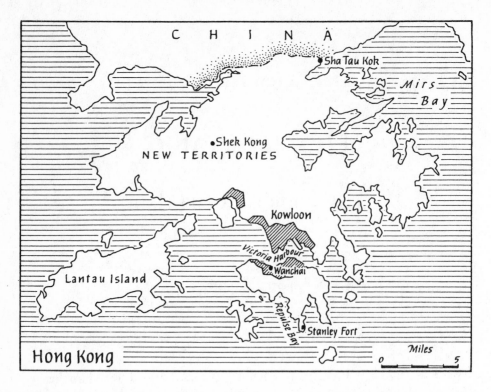

fun. However, first impressions are often deceptive. Many challenges lay ahead. Not all British battalions have prospered in Hong Kong.

On arrival the Commanding Officer reported to Lt Gen Sir Edwin Bramall who made it clear that the Battalion should obtain the very most from the opportunities to train and enjoy life. He expected the highest standards and he gave a real assurance of his support whenever it was required. The General became a much respected Commander and firm friend of the Battalion. (He was destined to become Chief of the Defence Staff and to receive a peerage.)

The tour got off to an excellent start for a variety of reasons; first because, in the many months before the Battalion's tour, Grenadiers were trained in every form of course with the emphasis on sport and recreation – including canoeing, dog-handling, combat survival, judo, sub-aqua and refereeing.

The Battalion was determined to enjoy themselves and to make their mark. Within weeks of arrival the Officers entertained a large number of strangers at a cocktail party. The Sergeants' Mess, led by RSM DJ Webster, RQMS T Farr and D Sgt P Richardson, also gave a party for representatives from all units in Hong Kong. Enduring friendships quickly built up thereafter with the civilians and military. All ranks were thoroughly briefed on their role and some of the temptations to be avoided, such as drugs. Slides were shown, during four anti-vice lectures, on victims of venereal disease and on prostitutes' austere beds.

The Battalion quickly discovered that they had to train for and be proficient in five different types of warfare, the principal one being internal security. The others were limited war, jungle warfare, counter-revolutionary war and civil assistance.

During the previous 25 years major problems had arisen in Hong Kong. In the early 1960s communist-inspired riots had threatened law and order. The 1956 disturbances were caused by Kuomintang supporters and then exploited by the secret Triad Societies. They were followed by the 1962 mass influx of illegal immigrants from China, and the 1966 riots which arose from social and economic pressures. In 1967 the Red Chinese militia overran a small hill on the border and opened fire on the Hong Kong police, killing five and wounding eleven. They were forced out by a Gurkha battalion.

In 1975 China regarded Hong Kong as Chinese territory, but they claimed that they saw no need for conflict; the historic question of who would own Hong Kong Island would be solved when the time was ripe. The British lease of part of Kowloon and all the New Territories was to expire in 1997. China, meanwhile, needed Hong Kong as an outlet for trade and as a source for almost half her foreign exchange earnings. China's policy towards Hong Kong was therefore expected to remain benign while local communists took every opportunity to widen the base of their influence within the Colony preparing for the day that they took over.

The economic pressures, particularly on the hundreds of thousands of new immigrants, were potential causes of unrest – industrial disputes, underemployment, and frustration over lack of housing. Petitions and processions might lead to illegal gatherings and sporadic outbursts of violence which could result in riots, looting and sabotage.

Internal Security was therefore of prime importance, but, over the months that followed, the Battalion was gradually trained for the other roles as well. The forces in Hong Kong were insufficient to deal with any really serious threat. A cost-cutting exercise was in force, partly because the Hong Kong Government had to contribute to the expense of having British units there. The Royal Navy was being reduced to five small patrol craft and the RAF to eight Wessex helicopters.

The Army consisted of the Headquarters Land Forces and two brigades – 48 Brigade in the New Territories and our 51 Infantry Brigade with its Headquarters in Kowloon. The principal units were 20 Light Regiment, Royal Artillery, three Gurkha battalions, the Grenadiers, Royal Hampshire

Regiment, and the Hong Kong Regiment which was the equivalent of a Territorial Army battalion, largely manned by Chinese. The Gunners and a tank squadron were withdrawn without replacement. The Gurkhas had their own engineers and signallers.

Within six weeks of the Battalion's arrival the first of many counter revolutionary warfare exercises took place. The recce platoon was despatched to Lantau Island, to the west of Hong Kong. Unfortunately, due to poor navigation, it was dropped five miles in the wrong direction in the middle of a hostile, sensitive communist village. (On the next exercise a Royal Hampshire Company was dropped by the Gurkha Engineers on the wrong island!) The recce platoon's task was to locate the 'enemy' force of 51 Infantry Brigade's (Chinese) defence platoon commanded for the occasion by Capt RG Woodfield. The Battalion was then dropped, largely in four feet of water, by landing craft or helicopters. After a number of operations and some genuine casualties due to the difficult going, the exciting exercise ended with a Battalion attack. It was an interesting introduction to the available naval and air support.

Brig D Boorman commanded the Infantry Brigade. He was a frequent visitor to the Battalion in Stanley and on exercises. He took a Grenadier orderly. 'I was chosen,' recalls the then L/Cpl DG Boucher, 'because my wife was the daughter of the Quartermaster of the Blues and Royals and so it was assumed I would know about airs and graces. One afternoon I left the Brigadier's house too early, to visit my wife. In my absence a Chinese lunatic who had escaped from Lantau Island somehow gained entry to the Brigadier's bathroom. The housekeeper assumed he was a guest. The lunatic then put on Brig Boorman's service dress and got into his bed where Mrs Boorman was said to have found him. I got the sack!'

The internal security training started with close liaison with the Royal Hong Kong Police and joint training days. The Companies each had their own area of responsibility in an emergency which coincided with police boundaries. Although somewhat preoccupied with police corruption trials, the Commissioner of Police visited Stanley with a police demonstration company. Maj the Hon Samuel Coleridge was posted to the police military liaison staff which further enhanced police cooperation with the Battalion.

It had always been anticipated that a number of families would have difficulty in settling in Hong Kong because military duties on the border, in Brunei and elsewhere would lead to excessive separation. The wife of the Senior Major, Clare Lindsay, who visited virtually all the 294 wives, having come out in the advance party, found one exceptionally young, lonely, newly married German wife who could speak no English and lived in a flat amid a noisy, filthy Chinese ghetto. Fortunately the very next on her list was another, older German wife who was immediately rushed round to cheer up her compatriot. The more senior officers' wives each regularly visited their 'patches'; activities were frequently run for wives and the older children. The local schools were very good. Two Grenadier wives, Sheila Eastwood and Pat Mason, ran a first-class primary school within Stanley Fort. The high ratio of infant mortality experienced by the previous battalion was frequently discussed with the SSAFA representative, Battalion Families Officer (Lt BT Eastwood), and the two medical officers Capt JRS Reid and Dr Seymour Jones. The Battalion was very fortunate in having Capt PB Denton as the padre. He had served with the Grenadiers in Germany and Londonderry, was a great friend of all and joined in most training. His chapel in the Fort, flying a large flag of St George, was the centre of much activity and his Sunday services were invariably well supported.

The following month, March, was spent preparing for and operating on the border with China. The Battalion was responsible for observing and reporting on the activities of the Chinese Communist Army, and capturing and handing to the police the Chinese illegal immigrants. The police were 40% understrength on the border and played a very small part.

In some ways it was a relief to take up Border Duties for they were usually interesting. Soldiers at all levels with their respective 'chiefs' could settle down and get on with their work without interference. Companies deployed with enthusiastic professionalism under their commanders. Companies were commanded by Majors J Baskervyle-Glegg MBE, A Heroys, AC McC Mather, TJ Tedder and ET Hudson. Not many illegal immigrants were expected because the communist army on the Chinese side captured most. The British Government's policy was to return them immediately to China. The ground was mountainous with dense vegetation making it difficult to catch them. Moreover, 'snake head' Chinese patrols from Hong Kong hoped to find the illegal immigrants first, to make money out of them.

Each night patrols lay in readiness to seize the unfortunate, exhausted Chinese who also were often suffering from exposure, shock, malnutrition and thirst. They were probably seeking their fortunes in Hong Kong or the West, rather than leaving China for ideological reasons.

Additional patrols found by Headquarter Company personnel were commanded by the Commanding Officer or Senior Major (OJM Lindsay). It became quickly apparent that while illegal immigrants were still crossing the border on foot many were also swimming across a 4,000 metre inlet to the east of Sha Tau Kok near Mirs Bay using every form of flotation equipment. One board was found which was kept afloat by lots of rubber gloves, the fingers of which had been cut off and blown up to provide the buoyancy. The paddles were ping-pong bats. Occasionally decomposed bodies were found which had been partly devoured by sharks. The Battalion was the first to redeploy to take account of those crossing by Mirs Bay.

Different tactics were tried on land. Ambushing paths in the rear areas was unsuccessful. Instead night viewing devices were used by observation posts to discover where the border crossings were occurring. Patrols were then silently directed to capture the illegal immigrants before they could disperse into the villages or the wilderness of the New Territories.

Ninety-five were arrested by Grenadiers, the Royal Navy and police during the three-week border tour. This compared favourably with the two captured by the Gurkha battalion in the previous fortnight, but the Grenadiers had deployed most of the Battalion on the border. The Guardsmen usually gave their captives food, coffee and cigarettes and often regretted that they had to be returned to China. Sometimes it proved necessary to disguise the illegal immigrant in a combat jacket to prevent sympathetic villagers trying to snatch him back. A few of those who had been captured a second time fleeing from China were apt to put up a fight.

Some of the illegals were highly professional, very well educated and would have contributed much to any society. The Commanding Officer tried to reason with one senior Government official in Hong Kong expressing the view that returning them all to China was an inflexible policy. He was told there would be no exception. Grenadiers understood that they were returned partly to appease China and also due to the over-crowding in Hong Kong.

With the benefit of hindsight it can be seen that the British forces in Hong Kong were kept in ignorance of events in China. They were told, should they ask, that the returned illegal immigrants would be subjected to mild counselling on their wish to leave China.

It was not until some ten years later that reliable books started to appear written by Chinese, telling at first hand of the humiliation and persecution during the Cultural Revolution. In 1967–68 the Red Guards and Revolutionaries had run wild in the towns, looting and abducting people at will, torturing them in secret courts and killing them in every way imaginable. One of the ugliest aspects of life in communist China was the demand by the communist party that people inform against and denounce each other. In the mid 1970s Mao Tse-tung was still whetting the appetites of his followers, encouraging them to ransack the homes of the 'capitalist class' and violently attack the intellectuals. Had all this been known to those responsible for returning the illegal immigrants to China, it is questionable as to whether the soldiers would have hunted them down with such determination. (True, it is not the soldier's job to 'reason why'.)

During the border tour it was interesting observing the Chinese army.

Their sentries were so close that the Adjutant, Capt PR Holcroft, could inspect their turnout through his telescope. Some of their soldiers were females. One couple held hands throughout the night. Several instances of brutality were observed. They usually involved an illegal immigrant being run down by extremely aggressive Chinese alsatians and then, on arrival of the Chinese soldiers, a dog being permitted to continue savaging the prisoner. Most dogs were ill-disciplined. On three occasions they tried to cross the border to attack the Grenadiers or to seek more fundamental attractions.

Occasionally shots were heard when it was known that captives were in the area but there was no proof to support suggestions that summary executions were taking place.

There was no confrontation between the Chinese Communist Army and the Battalion. Only once did their soldiers shout and smile at our patrols. They usually adopted a Mongolian dead-pan expression.

At the end of the border tour the Commanding Officer wrote a detailed report to Headquarters 48 Gurkha Infantry Brigade. Its many recommendations, such as the need for direct communications between the Marine Police and troops on the ground, were gradually implemented to the benefit of future operations.

Immediately after the Battalion returned from the border, over twenty officers and NCOs left for Singapore for a six-week Jungle Warfare Instructor's course. For the remainder, courses and sport were the priorities. The courses included signals, driving, canoeing, sailing, judo, aikido, assault pioneer, support weapons, swimming, waterpolo, free-fall parachuting and subaqua. All Battalion Commanders have their own hidden or well-published priorities. For one in England it might be shooting while another would decide on adventure training. Lt Col Fanshawe's two priorities seemed to be zest for life and physical fitness. Very early each morning small groups of Grenadiers were to be seen doubling off to isolated spots in Stanley Fort and beyond. The Commanding Officer's orderly room group of a dozen were led by him down precipitous steps to the sea on the southernmost tip of the peninsula, beyond the gaunt pill boxes of the Second World War. As the waves crashed against the rocks close by, each member of the group chose in turn the exercise, before everyone doubled up the cliff to breakfast. After months of such activity it is doubtful if there has ever been a fitter Grenadier Battalion.

Not surprisingly, when it came to sport, the impact was dramatic, despite strong competition. The Battalion won the Land Forces Athletics Meeting by fifty clear points, and six major titles and five junior ones at the Hong Kong National Championships. The principal heroes were L/Cpl J Taylor, CSM H Grime and L/Sgt P Vergo. Sgt T Ratcliffe later represented

Hong Kong at the Asian Games in the Philippines winning gold, silver and bronze medals. CSM A Hughes broke the Hong Kong national record for the hammer. Winning the football knockout was improbable because RAF Kai Tak had previously defeated the Army side 3-1. However, they were overconfident and the Battalion team won an exciting, tough match 5-0, due, in part, to their exceptional fitness. The inter-unit cricket was easily won the same month.

The final of the basketball competition was notable because the Gurkhas had traditionally won the trophy and been 'struck off' for months preparing for the encounter. The Grenadier team, however, had done particularly well in England the previous year and had the advantage of height, if not of agility. Some reluctant Grenadiers were detailed to support the team which set off for Shek Kong. The match against 6 Gurkha Rifles was too close for comfort but, to the wild cheers of the now enthusiastic supporters, the Battalion triumphed in the closing seconds.

However, the Gurkha battalions swept the board at the Forces Rifle Meeting in the New Territories. It was not until the Bisley Championships some ten years later that Grenadiers reversed the results. A jungle path with eighteen pull-up targets was built at Stanley alongside a thirty-metre range. This partly compensated for the other ranges being two hours away in the New Territories.

A leave camp was established on Lantau Island. In April 100 single soldiers could relax there for five-day periods. Apart from canoes, beer and beautiful walks, it simply offered an empty, clean beach which was impossible to find elsewhere in Hong Kong. Holiday chalets were available for families.

In May companies became responsible in turn for running a refugee camp for 600 South Vietnamese refugees. They were largely families who had been rescued from a sinking ship.

Saigon, the capital of South Vietnam, had fallen to the Vietcong the previous month although the war there had officially come to an end with the American withdrawal in 1973. Many American military felt that they had betrayed the South Vietnamese. After President Nixon's resignation, Congress had cut off the flow of ammunition and spare parts. We also had considerable sympathy for the South Vietnamese refugees. They were friendly, optimistic, well educated and well behaved. Although carrying large sums of money, they had few possessions. Our involvement was to prevent them breaking out and others entering illegally. Nobody anticipated the tens of thousands of refugees from Vietnam who would follow soon afterwards, nor the reluctance of other countries to accept them.

Examples of the variety of activities in Hong Kong, some exceptional and some routine, are best illustrated by some of the twenty-five entries in the Regimental Intelligence report for May:

1 May.	Activation of Colony-wide Married Families warden scheme;
5 May.	Tin Hau festival: Grenadiers attend Chinese Opera as guests of local village association;
6 May.	Prince Philip, The Colonel of the Regiment, visits the Battalion;
7 May.	Tour by helicopter of Battlefields over which the Japanese attacked Hong Kong in 1941. Five survivors describe their adventures.
8 May.	Commanding Officer's tour of barracks;
9 May.	Recces preparing for June test exercises;
12 May.	Annual stocktaking board commences;
16 May.	All Grenadiers candidates pass Lt to Capt practical promotion exam;
19 May.	Preparations begin for Capt Holcroft's Battalion Review;
20 May.	Battalion Sports meeting;
21 May.	600 Stanley Sea Cadets watch Battalion training and Corps of Drums Beat Retreat;
22 May.	Briefing for rural patrols;
22/23 May.	Night exercise with police;
23 May.	Corporals Course pass out;
26 May.	Commitment of looking after refugees ceases. Conference on June Dragon Boat races;
27 May.	Debrief exercise conference with police;
29 May.	Meeting of wives' visiting committee;
30 May.	Fourteen athletes receive South China prizes in Peninsular Hotel;
31 May.	Five officers fly to Brunei to plan forthcoming exercises.

The above summary was very typical of the extraordinary pace of life which the Battalion was to enjoy throughout its two-year tour in Hong Kong. The detail may appear mundane, but what a glorious life it was compared to London District soldiering! Every month recorded a similar variety in activities.

The first visit of the newly appointed Colonel of the Regiment, Prince Philip, to the 2nd Battalion was a great event. His relaxed, friendly manner, good humour and interest in all our activities were very marked. The Queen and Prince Philip were met by a full scale parade, commanded by Maj Baskervyle-Glegg.

In June the first of the Battalion exercises was set to test companies in their Limited War role, both defence and withdrawal. The Headquarters British Forces operational plan was that the Battalion would withdraw from the border through built-up areas towards Victoria Harbour, regardless of civilian casualties, chased by the Chinese Communist Army. Such a concept was scarcely credible but was not changed until the late 1980s.

Many Grenadiers by now had gone on expeditions in the Battalion's speed boat *Sir Allan Adair* which had been brought out from England in a container ship, funded, like so many such adventurous activities throughout our history, by Regimental Headquarters. Some officers shared a stately Chinese junk upon which they generously entertained their friends.

Each rifle company and the Wives' Club entered teams for the traditional Chinese Dragon Boat large canoe races. No 2 Company won the European section, while The Inkerman Company were runners up to the Royal Navy in the Combined Services race. (No 2 Company won it in 1976). The Grenadier ladies were less successful, being overweight, but they did not suffer the same indignity as the Royal Navy ladies whose canoe sank. The Corps of Drums, commanded by D/Maj K Green, played on the beach beforehand. They were to go everywhere – from Tokyo to the *Queen Elizabeth II* liner, from the Excelsior Hotel to Stanley Village harbour.

In July The Inkerman Company group (Maj Hudson) spent six weeks training in Brunei. They learnt to live in a jungle environment while undertaking basic offensive operations. The other rifle companies and Support Company followed them in due course. The platoons were apt to do a circuit: one on patrol techniques, while another learnt ambush training and the third mastered jungle survival. Gdsm Kendal decided to look for fruit but took no compass. He became lost in the jungle for three days, keeping on the move the whole time to avoid being eaten by animals. Fortunately he eventually walked into a Grenadier cut-off group practising ambushes.

In July the Battalion also found the United Nations platoon Honour Guard in Seoul, Korea, for two months. They had no operational role and had orders to withdraw at all speed should the North Koreans attack the South once more. They were responsible for ceremonial guards for visiting VIPs, for guarding the UN Commander-in-Chief's house and the underground UN complex. The platoon, commanded by Capt HARO Tweedie and Sgt Tillotson in 1975 (and by Lt JM Gage and Sgt J Mundy in 1976), included soldiers from Thailand and the Philippines, all forming part of an American company. All the Grenadiers were volunteers and single men. They had to adopt American drill and so became familiar with such stirring commands as: 'Stand Slack, Soldier!'

It was probably the first occasion throughout this History that a

Grenadier platoon served under American command. Despite the formidable American presence in NATO and elsewhere, the British did not usually come across the American soldiers in Germany or England. Nevertheless, in July an excellent relationship was established with eighty United States Marine Corps whose D Company visited Stanley to teach us how to play their soft ball and football, and to learn our cricket and soccer. (Grenadiers much later trained in America, and served with them in the 1991 Gulf War.)

In August the Battalion was made responsible for running a major 'Tactical Exercise without Troops' for HQ 51 Infantry Brigade. The subject was likely operations forward of the main defensive position. To catch the interest of the outsiders, it started with Capt JFQ Fenwick, representing a Chinese Communist officer, giving the Chinese concept of operations. Lt Eastwood, representing a newly arrived Royal Green Jacket officer fresh from Eton Combined Cadet Force, then closely questioned a panel of experts. By a deliberate coincidence he sat next to Lt Gen Bramall whom he resembled and from whom the Green Jacket paraphernalia had been borrowed.

Also in August Maj CWJP Langdon-Wilkins ran a section commanders' course for the two British Battalions, and Maj TJ Tedder a support weapons concentration for all Far East units.

The Battalion completed a second border tour in October which proved to be rather an anticlimax as few illegal immigrants evaded the Communist army patrols to the north. The Guardsmen manned ten static observation posts overlooking China and set ten ambushes each night. But they had a fairly dull time. A Company was placed for the first time on the Peninsula to the east where they arrested twenty-three illegal immigrants.

The principal exercise of the year took place in November. It will be described in some detail as it was the annual test exercise and unusual in many respects. The exercise was set by Brig Boorman and Maj JP Foley MC, the Brigade Major. Both were to be future Commanders British Forces, Hong Kong.

The counter-revolutionary war exercise fell into two phases. There were few exercise instructions as to which side should do what and when. The two opposing forces had no indication of the location of the other. We had obtained in a devious way some preliminary exercise instructions for the umpires, but they revealed little.

On the night of 18/19 November five recce patrols, commanded by Captains SMA Strutt, AGH Ogden, RG Cartwright, HARO Tweedie and RG Woodfield, flew by Wessex to a rendezvous in the unpopulated, hilly land mass some twenty kilometres north-east of Kowloon and south of Mirs Bay. A landing craft then took them close to the 'neutral' territory into which Chinese 'terrorists' had infiltrated. Capt Sir Hervey Bruce dropped off each patrol in assault boats. For the following three days the patrols operated covertly and succeeded in locating four of the enemy camps and several enemy ambush positions, although two of the patrols were themselves discovered. The patrols were commanded in the field by Maj Heroys. They were collected on 20 November by Maj Lindsay in assault boats and the landing craft where they were skilfully debriefed by Capt Fenwick for over an hour each, despite heavy seas. Within half an hour of our return to Stanley Fort, Capt JFQ Fenwick gave a masterly summary to Maj Gen RWL McAlister, Deputy Commander Troops Hong Kong, on all that had been learnt about the enemy.

On 24 November the Commanding Officer was ordered to destroy the enemy terrorists. The Battalion deployed by ship and vehicle the following day. The initial attack on the major enemy camp included No 1 Company (Maj Baskervyle-Glegg) putting out stops while No 2 Company (Maj CWJP Langdon-Wilkins) assaulted the camp. Some of the enemy escaped. However, over the next 24 hours virtually all the enemy were 'killed' in a succession of successful ambushes. The Inkerman Company, and a fourth rifle company which included the Corps of Drums, Corporals' Course and a 'Toughie' platoon commanded by Capt GR Whitehead, were particularly successful.

On 27 November the Commanding Officer was told to invade 'enemy' territory on a further peninsula six miles to the west of Mirs Bay. The Battalion's outer cordon was put in by Wessex helicopters and the inner cordon and assaulting Companies by assault craft. Battalion Headquarters (main), which hitherto had been confined to six landrovers, controlled the battle from a landing craft at sea. The Battalion spent the next few hours fighting through a determined enemy found by 6 Gurkha Rifles. The exercise was imaginative, realistic and fun. Brig Boorman told the Battalion that he was extremely impressed with their performance but the umpires were critical of Battalion Headquarters' lack of defensive measures.

After an enjoyable Christmas in Hong Kong the Battalion returned to the Border for a third tour of duty. One exceptional incident occurred which brought credit on them. A serious fire broke out late at night at Sha Tau Kok where The Inkerman Company Headquarters and a platoon were situated. The local fire-fighting organization and equipment proved hopelessly ineffective. Maj Hudson, CSM J Gowers and the other Grenadiers fought the fire, found water supplies, demolished a house to create a fire break and evacuated the villagers. This was accomplished amid exploding gas cylinders and the fierce fire. The villagers subsequently made a formal presentation to the Company in recognition of the skill and courage of its members.

Gdsm K Winn later saved the life of a drowning boy on Repulse Bay Beach. The boy was technically dead but was brought back to life by artificial resuscitation. Winn had to fight off two Chinese 'Lifesavers' who considered that he was molesting the boy.

All the Services in Hong Kong had to be on their guard against flirtatious ladies with the enticing names of Alice, Betty and Doris. They were typhoons, rather than Wanchai tarts, and were capable of sucking air conditioners (possessed by very few) from walls and throwing ships onto shores. Clara came very close while Ellen caused torrential rains with a record sixteen and a half inches being recorded. The resulting landslides and floods led to twenty-four deaths. At that time England was suffering from a severe drought.

Fortunately the weather was fine, apart from the wilting humidity, when the Battalion celebrated the Queen's Birthday Parade on 21 April in the Hong Kong football stadium before a crowd of 10,000. The fortnight's preparation smacked of public duties and was for many the first taste of Spring Drills. The Escort was commanded by Maj Baskervyle-Glegg and the Colour carried by 2Lt JR Lloyd-Jones. For effect a *feu de joie* and advance in review order were added.

As the tour wore on more officers had the opportunity of enjoying the 'tour of a lifetime'. Among the other newcomers to the Battalion in 1975 were Lts GWM Chance, MDH Mitchell, 2Lts JS Greenwood, HC Flood and D Sgt L Perkins. They were joined in 1976 by Capt AW Fergusson-Cuninghame, Lt CE Elwell and 2Lts OP Bartrum and BC Symondson. The more senior officers in the Battalion also changed. Lt Col DH B-H-Blundell took over command from Lt Col Fanshawe with HS Hanning as his Senior Major in 1976. The Company Commanders became Capt CSS Lindsay (No 1), Maj JM Craster (No 2) Maj HJ Lockhart (Support), Maj CJE Seymour (HQ) and Maj EJ Webb-Carter (Inkerman). The Adjutant was Capt TH Holbech. Other appointments included Capt DP Ratcliffe (Signals), Capt D Mason (Quartermaster) Capt DL Budge (Transport) and Capt BE Thompson BEM (Families). The CSMs were MJ Joyce, M Kenny, D Marshall, A Souster and D Cumming.

Border tours continued and separation for families usually ran on average at twelve days per month. Considerable effort therefore went into looking after the thousand 'dependants'. Many of the married quarters were superb and a local overseas allowance was helpful. Most had now purchased all the bric-à-brac of a tour in the Far East – the picture of junks amid glowing sunsets, the hand-made tapestries of opulent tiger hunts and the hand-carved side tables in the shape of elephants appropriately humbly kneeling. Husbands of all ranks became familiar with: 'If I hadn't bought it, it wouldn't be there when I went in again,' or 'Darling, I couldn't afford not to.' Some wives had difficulty adjusting to Hong Kong and missed England while other wives coped magnificently.

Mrs Barnett's tour in Hong Kong got off to a bad start because she flew out alone and was not expected. She spent the first three hours weeping at Kai Tak airport. Although just 18 and unqualified, she was one of the few to get a job – teaching English to Chinese at private schools. At the end of the day a parent's luxurious Rolls Royce drove her home. She was reluctant to take taxis because she lived in Cloud View Road and was too embarrassed to ask for the street: in Chinese it was called 'Wanking Doe'. She and Gdsm Barnett employed a Chinese servant as a novelty but later decided to spend their money on other things.

Six married officers lived in Stanley Fort in tall, white elegant mansions which would have well satisfied even the most socially ambitious of wives of minor colonial governors. From large balconies the twinkling of the lights of the fishing fleet far below could be watched at night. Mr Fishyman was to be found squatting beyond the kitchen the following morning selling his lemon 'Dover' sole caught a few hours before.

Chinese food was preferred when dining out. At the floating restaurants which closely resembled Mississippi steam boats, lobsters, shrimps, crabs and scallops wallowed alive in large cages beneath the restaurant, a few paces from the toilets which spilled their contents into the dark, stagnant water.

Training in Brunei and Hong Kong and numerous exercises within the Colony kept the Battalion extremely busy. All Companies took two weeks' block leave in the summer. Some returned to England but many families enjoyed the opportunity to explore the Far East. Thailand was particularly popular.

On 25 August the Governor of Hong Kong, Sir Murray MacLehose, visited the Battalion. He had endeared himself for frequently taking parties from all ranks of the Battalion on his Junk for trips around the islands.

On the first day of the September border tour Chairman Mao Tse-tung's death was announced. Although widely anticipated by the media, the stock market plunged. The Battalion adopted a low profile. Flags were flown at half-mast. His 'Cultural Revolution' was later officially declared in China to have been a national catastrophe: it had destroyed both Party discipline and civic morality; 10,000 people had died unnaturally in Shanghai alone. A month later his wife was arrested and people became more optimistic.

But the Battalion was more concerned with sport than events far afield. The six-a-side cricket final against 1st Battalion Light Infantry was convincingly won thanks to Capt DL Budge, Lt GWM Chance, Sgts Lamb and Evans and Gdsm De'ath and Hewitt. The Battalion even took on the Gurkhas at their own game in the annual Gurkha Khud hill-race. Capt

HR Westmacott and L/Cpl P Greenwood produced creditable performances. This was also the case in the various military competitions: Lt RJC Hawes and a team from No 1 Company won a military skills competition organized by the Royal Hong Kong Regiment in the New Territories while Sgt Randell's platoon distinguished itself in an inter-platoon patrolling competition. The Gurkhas, on the other hand, won the support weapons competitions.

Gradually the Battalion's tour in Hong Kong was drawing to a close. A final farewell parade in Hong Kong, Chelsea Barracks and numerous Queen's Guards were fast approaching. In December 1976 the 2nd Battalion returned to England, to be followed by several thousand packing cases of 'special offers' which were to transform many a married quarter and barrack room into a Chinese emporium.

After the Battalion had left a few Grenadiers were posted from England to the staff in Hong Kong. They found that the reputation of the Battalion was still very high indeed.

For a number of years most people remained optimistic that the Chinese take-over of Hong Kong in 1997 would go well. However, the massacre of students in Tiananmen Square in Beijing destroyed confidence in China. In 1991 the continuous reduction in the British forces in Hong Kong necessitated the police taking over the border responsibilities.

And so what will be the Grenadiers' abiding memories of Hong Kong? Everyone will have their own – for some the dawn patrols along the border villages, listening to the cocks crowing, or the pathetic family of five illegal immigrants who knelt, imploring the Grenadiers not to return them; the isolated observation posts resupplied only by helicopters or mules on the mountains overlooking China; or the darkness and humidity in Brunei's dense jungles; the beauty of the crimson sunsets, sparkling sea and the lights of the distant fishing fleet, all visible from Stanley; the cheerfulness of the Chinese, and the friendships made in a fine Grenadier Battalion. Many will agree that it was, indeed, the tour of a lifetime.

PART TWO

1945 – 1995

CHAPTER 19

THE GUARDS PARACHUTE COMPANY AND G SQUADRON SAS

1946 – 1995

Grenadiers served in the Guards Parachute Battalion, the Guards Independent Parachute Company, and in the SAS role on active service in Borneo. They are now serving in G Squadron 22 SAS. This chapter reflects some of their activities.

After the war, to help recruiting in 6 Airborne Division, each of its nine parachute battalions was affiliated to infantry groups. The 1st Battalion of the Parachute Regiment became the Guards Battalion. Lt Col EJB Nelson DSO MC was selected to be its first and, as it turned out, only Commanding Officer. He joined the Battalion outside Tel Aviv in Palestine in November 1946. He was the first Guardsman to join it.

'With the arrival of General Hugh Stockwell in May 1947 the Division reached its high potential,' wrote Lt Col Nelson in *Always a Grenadier*. By then the first Guardsmen and Troopers, all volunteers, had reached the Battalion.

The operations that winter were exciting – keeping the peace between Arab and Jew. The Guardsmen were often defending themselves against both, without a single casualty. The Palestine story is told in Chapter 3.

By March 1948 the final plan for the future of Palestine had been agreed. Almost the whole of the country was given to the Jews to form the sovereign state of Israel. 6 Airborne Division left for home six weeks before the end of the Mandate, but the GOC (Gen Sir Gordon MacMillan) especially asked for the Guards Battalion to remain behind to help during the difficult days around Jerusalem. This they did and had the honour of escorting the last British High Commissioner out of the City on 14 May 1948. After sailing from Haifa to Liverpool, the Battalion was met by Maj R Steele MBE, the Second-in-Command. The Battalion was then disbanded with the rest of the Division, but the Guards Battalion was allowed to preserve its identity in the form of 16 (later 1) Parachute Brigade Pathfinder Company. 200 men of the Guards Battalion remained to form this Company. So the Household Brigade was able to preserve its links with Airborne forces: an exciting future lay ahead of them.

It was a unique Company being the only all-regular one in the then National Service Army. A happy arrangement was made with the Parachute Brigade. The Regimental Headquarters of the Grenadiers became responsible for all 'A' matters such as promotions and discipline, while other responsiblities, including operational ones, rested with the Parachute Regiment.

1 Guards Independent Parachute Company, as it eventually became known, reached Pirbright in July 1948 and found King's Guard on 24 August to show that the Company upheld the traditions of Guardsmen. This was the first occasion that the maroon beret was seen on guard at St James's and Buckingham Palaces. The Company then moved to the Harz Mountains in Germany. It was part of the Occupation Forces and also had a role in policing the border between West and East Germany. Training was fun but rigorous because it was adapting techniques pioneered during the war to the requirements of the day.

The Company consisted of three platoons, No 1 being commanded by Capt JD Makgill Crichton Maitland, with Maj Steele the first Company Commander. He wrote: 'Our role may sound like a short cut in wartime to suicide; we precede all other airborne forces, and lay out navigational aids, to guide in the aircraft which carry the main body, in such a manner that parachutists and gliders are released at the correct moment.'

The Company was therefore responsible during its first ten years for the pathfinding role – the securing, marking by visual and radar means, and defending of parachute dropping zones.

It is not the purpose in this chapter to tell the full story of the Company, but rather to refer to some of the activities in which Grenadiers were involved. All those named in these pages were Grenadiers, unless indicated otherwise.

In June 1951 the Company as part of 16 Parachute Brigade sailed in the aircraft carrier HMS *Triumph* from Portsmouth to Famagusta to await aircraft for a possible drop at Abadan where Dr Mossadeq was threatening British oil interests. This problem was solved by diplomatic means, but directly afterwards King Farouk abrogated the Anglo–Egyptian Treaty and the Company flew with the rest of the Parachute Brigade to Egypt for

34. The 1st Battalion London District Rifle Association Shooting Team, 1984.
Standing: Maj J S Scott-Clarke, L Sgt Spencer, Lt Col A Heroys, Gdsm Hailes, Beale, Finney, Talbot, Westover, Langdon, Sgt Stone, Gdsm Westwood, Negus, CSM Dehnel, L Sgt Thompson, L Cpl Mizzi, Gdsm Wilson, Lt H V L Smith, Maj E T Hudson, RSM Halford.
Kneeling: Sgt Kinton, Gdsm Chapman, L Sgt Edwards, Sgt Watts, CSM Wilmot, L Sgt Wardell, L Cpl Hodgetts, Gdsm Seward.

35. A tactical pause. Training with AFVs 432.

36. The Queen with Grenadiers on the Queen's Birthday Parade, 1984. Major T J Tedder, Regimental Adjutant; Lieutenant Colonel GAA-R-West, Equerry in Waiting to The Queen; Lieutenant Colonel Sir John Johnston, Equerry in Waiting to The Queen; The Duke of Edinburgh, the Colonel; The Queen, the Colonel in Chief; Colonel A T W Duncan, Lieutenant Colonel Commanding the Regiment; Lieutenant Colonel A Heroys, Commanding Officer 1st Battalion; Lieutenant Colonel J V E F O'Connell, Commanding Officer 2nd Battalion; Captain C T G Bolton, Adjutant 1st Battalion; Captain G V A Baker, Adjutant 2nd Battalion.

37. 1st Battalion past Commanding Officers' and Sergeant Majors' Dinner at Cavalry Barracks Hounslow, 27th September 1985.
Front Row: Colonel D W Hargreaves 1966-69, Lieutenant Colonel R Steel 1957-60, Major General C M F Deakin 1947-50, General Sir David Fraser 1960-62, Colonel The Lord Wigram 1955, Lieutenant Colonel E J Webb-Carter 1985-, Major General Sir Julian Gascoigne 1941-42, Major General R H Whitworth 1956-57, Brigadier M S Bayley 1964-66, Colonel N Hales Pakenham Mahon 1969-71, Colonel G W Tufnell 1971-74, Brigadier M F Hobbs 1978-80.
Rear Row: Major L C Drouet 1957-59, Captain L E Burrell 1949-53, Major A Dobson 1959-63, Captain D R Rossi 1979-82, Lieutenant S R Halford 1982-84, WO1 Ling 1984- Major G C Hackett 1945-47, Major P A Lewis 1967-69, Captain A Holloway 1974-76, Captain T Pugh 1964-67, Captain W Williams 1969-71, Lieutenant Colonel G R Whitehead 1971-72, Major B T Eastwood 1972-74, Captain P F Richardson 1978-79.

internal security duties. Happily there was also time for plenty of parachute training in the desert and for exercises in Iraq and Jordan on both of which the Company dropped having emplaned in Egypt. By now Capt N Hales Pakenham Mahon had joined the Company.

While in Egypt Brig AGW Heber-Percy DSO met the Grenadiers. He had commanded the 3rd Battalion in North Africa and Italy from 1942 to 1944.

The pathfinder sticks participated in an exercise in Cyprus which was typical of many. Dropping independently at last light they laid the night 'T' for the first lift. Shortly after midnight, with the heavy drops completed, they moved across country to the next dropping zone 16 miles away. After successfully laying out the day dropping zone at 1 pm, they formed the Brigade HQ defence platoon. On returning to the Canal Zone they were joined by CSM Taylor.

Not all exercises went to plan. In October the Company was to land, mark the dropping zone (DZ) and to secure two bridges over the River Main in Germany. The leading aircraft developed engine trouble and the occupants baled out. The second aircraft could not approach the DZ due to fog. Eventually, very tamely and very cold, the Company motored up to the battle area and captured the bridges. The exercise was therefore disappointing, particularly as so much work, as always, had been put into the preparation, including the rehearsing of a *coup de main* party by swimming the Thames in full equipment.

In 1956 the Company participated in the Queen's Birthday Parade, 'keeping the ground' with 1st Battalion Welsh Guards, before moving to Cyprus for operations against EOKA terrorists (Chapter 8). Five months later, in the Suez operation, a stick of the Company dropped operationally to provide a fighting patrol for a mission with the French to the south of Port Said, as well as the vanguard for the advance down the Suez Canal causeway (Chapter 7).

The following year the Company received the Freedom of the Borough of Aldershot with the rest of the Parachute Regiment. Operations followed in Jordan.

In 1960 reconnaissance and surveillance became the primary role. With the development of new RAF techniques, pathfinding became less elaborate and more flexible, hence the introduction of a more clandestine role, equipped with machine guns and long-range wireless sets. By now Capt PM Lambert had become the Company 2IC and MTO, with Capt AGR Ellerington the Air Adjutant and commander of 1 Platoon. An anti-tank section had been added to the Company.

By 1961, when the Company received its first Colour, the overall strength was 109. Each year about 100 volunteers commenced the selection course of whom thirty were chosen for the three-year posting. Very high standards of physical fitness and moral determination made the Company an elite; great emphasis was placed on personal skills and weapon-handling which were to stand the soldiers in good stead in their next operational role.

BORNEO

Three years later the Guards Parachute Company was attached to 22 Special Air Service as an independent squadron for active service in Borneo.

For a year 'confrontation' had simmered over the sprawling 1000-mile border between Malaysian North Borneo and President Sukarno's Indonesia. Throughout this time small SAS patrols had operated in the deep jungle, partly as an early warning screen for the British and Gurkha Brigades posted behind, and partly to raise, train and direct local irregular forces. The SAS urgently needed an extra squadron. The Guards Parachute Company provided it, a testing and exciting role.

After retraining in England with Capt the Hon TRV Dixon, and test exercises in Malaya, the Company flew to Borneo in June. On arrival CSM RG Woodfield had to produce a bearer party for an SAS soldier killed in action. Bodies were sent to Singapore. It was a sober reminder of what lay ahead. CSM Woodfield was glad that he had brought out the ceremonial drill book.

'Our base was in Sibu, a predominantly Chinese river town in Sarawak,' wrote Capt JG Cluff later. 'From a forward base we were taken by helicopter, flown by 845 Naval Air Squadron pilots from HMS *Bulwark*, to be lowered into jungle clearings or sand spits on the rivers. We were re-supplied by helicopter every two weeks. They delivered ration packs which were A,B,C or D etc for a different day of the week. However, some genius sent us C ration pack for weeks. Variety was finally provided by my mother who, under the impression that Borneo was somewhere in the English Channel, sent me Fortnum and Mason hampers with instructions to the RAF to deliver them to me personally. On one occasion they did and the change of diet from tinned Australian Irish stew to foie gras did nothing to improve our productivity.

'My group's ten-week task was first to clear an area in the jungle to enable the Wessex helicopters to land and disembark Gurkha soldiers, should the need arise. Secondly we were given an area of some 10 square miles along the Indonesian border to map and patrol. My group consisted of three other soldiers – a Parachute Regiment medic, an Irish Guards wireless operator and a Coldstream linguist.

'My most dramatic adventure involved my group, with Lord Patrick Beresford, of the Royal Horse Guards, being required to reconnoitre a fort across the border in Indonesia, to see whether it was occupied and what equipment was there. The expedition was called 'Operation Annabel' after the Berkeley Square night club. The patrol involved a tough five-day march to a mountainous range which marked the border. After nearly drowning, I led my party into enemy territory. It now consisted of the wireless operator, and three nomadic Punans. The remainder stayed on the border. The Punans are virtually stone age hunters who are usually extremely timid; they are superstitious of sunlight, and have a bizarre appearance with blue dye on their legs, white bodies and jet-black hair.

'We found the fort which, to my relief, had recently been vacated. I discovered that the Indonesians were using the services of Americans who had carved their names near their wooden beds. They also left American equipment behind. It was quite an experience, feeling so completely alone and isolated. Most signals were by morse in code. On return to Borneo I wondered if the patrol was an act of invasion since normally we were forbidden to cross the border. It was scarcely spying because we were in uniform, or rather military rags after three weeks wading through rivers.'

CSM Woodfield also had many memories of this exceptional tour: 'I interpreted the sketch maps sent back by patrols and put them into map form. Some helped on hearts and minds by treating sick natives. Water with a sterilizing tablet, known as the white man's pill, satisfied the bare-breasted Punan women and seemed to cure them. In one longhouse I saw baskets of skulls, some possibly Japanese. We were entertained by pretty dancing girls and intoxicating recently-fermented rice wine. Despite a hangover, I sang 'Roll out the Barrel' to much applause. We dropped grenades or plastic explosives into the river and then picked up the enormous concussed pike; so we were very popular in providing food.'

The Company's return to Elizabeth Barracks, Pirbright, on 9 November 1964, was delayed somewhat as three RAF Beverley aircraft broke down one after the other.

Thirteen months later most of the Company was back in Borneo in the SAS role once more. It relieved the 1st Australian SAS Squadron. The Company Headquarters was in a famous haunted house overlooking the Sultan's Palace in Brunei. Capt JCF Magnay had now joined as the Second-in-Command to Maj J Head, Irish Guards, the Company Commander. Capt Magnay led small patrols into Indonesia to report on the enemy positions. The second tour, like the first, meant months of no human company, other than their own, and a daily routine of sweating eight hours through the unrelenting rain-forests, of blasting landing zones, of siting

241

ambush positions, of mapping the unknown and of overcoming the diffi-culties of radio contact at distances of up to 250 miles.

On one occasion two patrols together, commanded by Sgt McGill, Scots Guards, had a spectacular success. They set an ambush for 36 hours just in front of a river bank. The cut-off groups were in position when L/Sgt Mitchell, Irish Guards, saw an Indonesian and opened fire with his light machine gun. Four to five Indonesians were killed. The patrols withdrew very quickly, not searching the dead to save time. This was just as well as the Indonesian regulars took immediate follow-up action, with their mortars and machine guns. Both patrols returned safely, but it was worrying as some were missing for a time. One of them was Gdsm Sheppard of the Coldstream. He shot two Indonesians who were pursuing him. 1/10th Gurkha Regiment was mobilized and saturated the area which they were thrilled to do, hoping to avenge the deaths of their men the previous year.

On 6 November 1965 the Company returned to England. Sukarno was overthrown in Indonesia and the emergency was over. His army had suffered severely.

In view of the success of the Company in the SAS role, it was decided that a Guards Squadron should be formed in 22 SAS, based at Hereford. Men were allowed to choose whether to try for the Company or for G Squadron. By late 1966 the Squadron had achieved over half its strength and formed four Troops.

Some pre-training for the selection course was carried out by the Company at the Royal Marine training establishment at Bickleigh. Capt Magnay and Sgt McGill MM were among those instructing.

Two years later the Parachute Company, now commanded by Maj DV Fanshawe, converted to parachuting from the Hercules aircraft. Training also continued in freefall parachuting which enabled soldiers to drop from 22,000 feet by day or night. By this means they could steer across borders, infiltrating into enemy territory without being picked up by enemy radar. In June 1968 the Company, in which Capt PML Smith was also now serving, moved with the 2nd Battalion The Parachute Regiment to Hong Kong's New Territories. Patrolling the remote areas with the civilian police was a foretaste of the Grenadier 2nd Battalion's tour seven years later (Chapter 18).

Maj DV Fanshawe said at the time in a letter to a friend: 'I have the immense privilege of serving in the Independent Company. It is quite small – just 92 Guardsmen and Troopers including the five officers – but its members are superb. Each man is an expert Airborne Soldier, immensely fit and skilled not only in his profession but also in additional specialist prowess such as medicine and surgery, foreign languages, foreign weaponry and, of course vitally, communications, on a worldwide basis. There is in the Company a wonderful, almost relaxed, feeling of total reliability – and of course endless excitement and indeed merriment. We are all very lucky to be part of such an organization.'

G Squadron SAS also deployed frequently far afield. In September 1968, for example, it was in the Rocky Mountains acting as a guerrilla force operating against Canadian units. Each troop attacked targets, observing wireless silence, living off the land and avoiding capture.

On return to the UK some members of the Squadron did a month's medical refresher training at hospitals throughout the country. One Guards officer, a future Chief of the General Staff, gave treatment to an elderly lady. She later returned to the hospital treatment room where potential doctors were watching procedures. Brushing off the doctor on duty, she insisted on being treated by 'a proper doctor' who was no other than the officer dressed in a white coat. One G Squadron officer did not deny putting stitches into a Nigerian's swollen lips after a fight, unfortunately stitching them together.

The following year every member of the Squadron trained abroad at least four times. In 1970 it spent four months in Sharjah where a Grenadier Company was then serving with the Scots Guards (Chapter 15). The Land Rover Troop drove hundreds of miles across every type of desert ending with a three-week patrol in the Liwa. The Mountain Troop climbed in the Jebel Akhdar area of Oman before joining the Boat Troop at their beautiful camp near Kho Fakkan. The Free Fall Troop amassed a most impressive number of descents. The Guards Parachute Company, meanwhile, was surveying the 'Peace Line' in Belfast, based on a dirty and draughty bottling factory.

By the spring of 1971 G Squadron received its first non-Guardsmen. It was sad to see the wholly Household Division aspect disappearing, but the Squadron's strength could not be allowed to fall below an operational level. After service in the Squadron, virtually no NCO or Guardsman chose to return to his Regiment.

The Guards battalions, in turn, were reluctant to give up their best to the Squadron; but only the best could meet the SAS entry standards. On the other hand, if some Guardsmen had not joined the SAS, with its very special roles, they would have been lost to the Army. The regular officers, however, had to return to their Regiment, otherwise they would have no career prospects. All Grenadiers who serve in the SAS get an immense amount of experience out of it and benefit greatly accordingly. As stated in Chapter 27, it is not the Army's policy to reveal the names of those who serve in the SAS unless they are killed in action, Capt HR Westmacott MC being an example (Chapter 23). Suffice it to say, Grenadiers played a full part in the Squadron having found the Squadron Commander, Squadron Sergeant Major and many others at various times.

To revert to 1 Guards Independent Parachute Company, by June 1972 it had completed another tour in Northern Ireland. Capt CWJP Langdon-Wilkins joined the Company, and went on an interesting exchange visit to the Moroccan Parachute Brigade. CSM DJ Webster, meanwhile, had served in the Company as both CQMS and CSM.

The Company's role, by now, had reverted to pathfinding for 16 Parachute Brigade, rather than the more recent role of deep penetration patrols on foot. Working with the Special Forces' crews of 36 Squadron RAF, excellent liaison was established. However, the days of the Company were numbered due to cuts in the Defence budget, and improved technology meant that it was considered that pathfinding was no longer necessary.

After a final exercise in Italy, the Company, now commanded by Maj RJS Corbett, Irish Guards, was disbanded. A memorable last parade took place on 24 October 1975 at which Field Marshal Sir Gerald Templer took the salute. 'The secret of the Company's outstanding success lies, of course, in the fact that all of its members brought with them an intense pride in their parent regiment and in their profession as the Household Troops of the Sovereign,' he wrote. 'The Company quickly became greatly respected by both commanders and other units; it proved a most successful example of a happy composite unit.' After the Company was disbanded some transferred to G Squadron. The Company Colour was laid up in the Guards Chapel.

Fortunately the links between the Guards and the Parachute Regiment have remained exceptionally close. Parachute Regiment junior soldiers were trained at the Guards Depot until the formation of the Army Training Regiment at Pirbright. By 1993 they were training together in the Infantry Training Battalion, Catterick.

The Guards Independent Parachute Company, in which Grenadiers had played such a notable part, combined all that is best in the Household Division and the Airborne Forces. Even twenty years after its disbandment, some 180 members of the Company meet annually, such is their *esprit de corps*. Since then Grenadiers continue to serve in G Squadron, participating in operations all over the world, living up to the SAS motto 'Who Dares Wins'.

CHAPTER 20

SOLDIERING AWAY FROM THE REGIMENT

As the post-war years wore on, Grenadier officers and an increasingly large number of men served on the staff or at duty in appointments away from the Regiment.

By the time a regular officer reached his mid-thirties, he may well have spent two tours, some five years, filling staff appointments at Grade Three and Grade Two level, in the ranks of Captain and Major respectively. And beyond his early forties, by which time he will be considered too old to serve in a battalion, the rest of his career will be in command, or, more usually, staff appointments outside the Regiment.

The officer will only get promoted to Lieutenant Colonel if he has received excellent reports both at Regimental Duty and on the staff. The more demanding the appointments he holds, the greater the chances of promotion if he proves himself to be successful in his job. The same can be said for the Warrant and Non-Commissioned Officers.

* * *

This chapter gives some examples of the great variety of appointments filled by Grenadiers over the years. Lack of space sadly prevents more than a few reminiscences being quoted. An entire book could be written about those Grenadiers who served away from the Regiment.

THE ADCs

Between 1945 and the 1970s distinguished Britons continued to be sent as Governor Generals and Governors to Commonwealth countries. They often asked the Major General Commanding the Household Brigade for young Guards officers to serve on their staffs as Aides de Camp.

The Governors, the personal representatives of King George VI or Queen Elizabeth II, were dedicated and highly respected men who usually had a good war record. They often left their friends and families in Britain, serving their country far afield at the request of their Sovereign.

Gradually the Commonwealth countries chose their own nationals to represent The Queen. The new Governors naturally wanted local ADCs and so by 1970 British officers ceased to fill such appointments.

Serving as an ADC overseas made a deep impression on those lucky enough to be chosen, usually totally unexpectedly, for such employment. The reminiscences which follow give a brief glimpse of a different world.

'I was in Haifa in January 1946,' wrote Lt JW Scott (later Sir James Scott, Lord Lieutenant of Hampshire). 'I received a signal stating that I had been selected as ADC to the Viceroy of India – Field Marshal Earl Wavell whom I had never met. On arrival at New Delhi, I replaced another Grenadier, Hugh Euston, now the Duke of Grafton.

'I had to introduce up to sixty guests with tongue-twisting names before the large official dinners which were eaten off gold plates: Viceroy's House was a large Lutyens palace; the grounds were equipped as a country club with a golf course, swimming pool, stables, squash court, tennis court and gardens.

'There was much travelling (by Dakota aircraft rather than the Viceroy's train). Visits were paid for by Provincial Governors and might include a tiger hunt or duck shoot. We were expected to hack, hunt, race, and pigstick despite no previous experience of a charging pig: it was certainly exhilarating.

'Lord Wavell was replaced by Mountbatten who had a totally different regime. Two other Grenadiers arrived – John Lascelles, PA to Lord Ismay, the Chief of Staff, and Ralph Selby on the High Commissioner's staff.

'Appalling riots followed as India was granted independence. Leave was rare but I took a chance to enjoy a holiday with another Grenadier, Hew Hamilton-Dalrymple, who was ADC to Gen Sir Frank Messervy, the C-in-C Pakistan Army. Visits to Kashmir were not encouraged because of communal tension, but if anything happened one of us would be on the right side.

'After enjoyable but unsuccessful days searching for bears, we learnt that tribesmen were on the move. It seemed timely to return by river (with challenges during the night from soldiers on bridges). We found that artillery was firing on the airfield as we finally, after much persuasion, took off for Delhi. Eventually Lascelles and I returned to England with Mountbatten.

'After four years abroad we were finally granted two months' leave – all August and September. Plans were made for Scotland. This all seemed too good to be true. It was. On 15 August Alan Breitmeyer, Adjutant of the 3rd Battalion, ordered us to report immediately to Windsor. We soon found ourselves sharing a tent in Malaya – a very different life to those exotic Indian days.'

In 1956 Capt NA Tunnicliffe became ADC to the Governor of Cyprus at the height of the Emergency.

Between 1959 and 1964 Maj Gen Sir Julian Gascoigne was the Governor of Bermuda. Capt DHC Gordon Lennox was his first ADC. 'We arrived in an ocean liner flying Sir Julian's personal flag. His remit was to break down racial prejudice in Bermuda; the politics were basically black versus whites: the schools and social lives were segregated. We lived life to the full and gradually the social barriers were removed.

'One day at 3 pm the phone rang: it was Prime Minister Harold Macmillan's private secretary saying that the PM would like to spend that night staying with the Governor. He was returning from New York where he had been addressing the United Nations. I found the Governor on the golf course and told him that Mr Macmillan was coming. The RAF said they were landing at 6.30 pm. The Governor went to the airport where the PM greeted him: "Julian, I don't think you should be meeting me." The Governor replied, "I'm meeting you as an old friend and fellow Grenadier, rather than as the Queen's representative, but, as a Grenadier, I wasn't expecting you to be an hour late!" The RAF had forgotten the time difference and had landed at 7.30.

'At the very relaxed private dinner party which followed, the Prime Minister drank half a bottle of brandy before announcing that he wanted to swim. So I alerted the police and we went to the private beach. Macmillan sat in a deck chair from midnight until 2 am, working with two private secretaries.

'On another occasion I met Truman, the former American President. I introduced him incorrectly as Mr Truman and was told in no uncertain manner, "Once a President, always a President. I'm Mr President Truman".'

The Governor's next ADC was Capt JP Smiley: 'On about 10 December 1961 Mr Macmillan telephoned to ask His Excellency if he could invite President Kennedy to a Summit meeting to be held in Government House. Within 48 hours of the invitation being accepted, the American Secret Service arrived and tried to take over Government House. Since the Prime Minister particularly wanted the atmosphere to be like a country weekend house party, the Governor told them all to go away, except for one he would allow outside the President's study and one outside his bedroom. They had clearly never been spoken to like that before! When Mr Macmillan flew in on 20 December we heard that the President's father, Joseph Kennedy, had had a stroke and that it was doubtful whether the President could come,' remembers Capt Smiley. 'The Prime Minister offered to meet the President elsewhere, but confided in us, "The President owes his allegiance to his country, not to his father."' Fortunately Joseph Kennedy recovered and the President flew in to Bermuda on the 21st.

'The Governor very properly received him in full dress and with all appropriate ceremonies,' wrote Mr Macmillan in his memoirs *At The End of the Day*. 'Kennedy was at first rather amused by the cocked hat, feathers, sword and gold spurs, but he soon began to realize the real quality of our host.'

Capt Smiley remembers: 'The first morning the hot water in the President's room was cold. We discovered that, while checking the bedroom for listening devices, the American Secret Servicemen had turned off the immersion heater and forgot to turn it on again. I had to arrange for hot water to be carried up to him. When I told the President what had happened he replied, "The bloody Secret Service are always mucking up my life."

'At the dinner party and afterwards,' continues Capt Smiley, 'I saw history briefly being made. The Prime Minister and President, with Lord Home and Dean Rusk, thrashed out the Anglo-American policy on negotiations with Russia and the problem of Berlin. They also showed each other their secret correspondence with Khruschev. While sitting on a sofa they discussed the bloodshed in the former Belgian Congo. Macmillan had earlier remarked, "I don't know why they don't eat each other up. It would solve all our problems there." I never stopped offering them cigars and whisky to catch up on their conversation. The following day Kennedy flew off by helicopter – all the lovely plants put nearby blew away in a whirl of dust. The Prime Minister sat down and I offered him a drink. "Yes please," he replied. "I feel just like a father who has seen his daughter off after her wedding."'

In 1961 Viscount De L'Isle VC became the last Englishman to be appointed Governor General of Australia. Capt CR Acland accompanied him as his ADC but 'it was a traumatic time,' he recalls, because Lady De L'Isle developed cancer and died. 'I arranged with the Regimental Adjutant for her to be buried at Penshurst alongside her father, Viscount Gort, who had won a Victoria Cross as a Grenadier in the First World War. A Grenadier bearer party carried the coffin.'

Capt PGC Maxwell later served as an ADC in Western Australia.

Capt DH B-H-Blundell was appointed ADC to Lord Cobham, the Governor General of New Zealand and, like Lord De L'Isle, a former Grenadier. 'I haven't forgotten the pomposity of it all,' he noted. 'The footmen wore knee breeches; all the house staff were English; the band played "God Save the Queen" everywhere. White tie and tails were worn for some twenty dinner parties, and dinner jackets for the remainder. Iain Erskine was the Comptroller.

'The King and Queen of Thailand visited us. I was captivated by the beauty of one of the Ladies-in-Waiting, and so I put a teddy bear in her bed with a message saying how much I admired her. Unfortunately Lady Cobham checked the rooms first. I got a rocket.

'The Governor General did an annual trip round such islands as Fiji and Tonga. He couldn't land on one island due to the reef, and so two people had to carry him through the water: he was wearing his gold sash, sword and everything else.'

The last British ADC in New Zealand was Capt AC Roupell.

One Grenadier, an ADC to a Governor General in Africa in 1962, recalls being summoned to arrange a tour far afield. 'I want to fly to Nyasaland in early November,' he was told by the Governor General, 'to visit the Chitedge Agricultural Research Station, a couple of schools, and several hospitals, before flying to Chileka to attend the Inter-Services Dinner and stay at Government Lodge. Her Excellency's programme is to include a visit to Poor Clare's Convent and she will go to the Ex-Servicewomen's Dinner. We will give a Sundowner party for about 150. Arrange it all.' Half a dozen other such tours followed over the next ten months, most of the work for coordinating the programme falling on the Provincial Commissioners. A Lady-in-Waiting was also on the Governor General's staff to look after his wife. It can be seen that being an ADC involved rather more than pouring out the whisky and sodas and adjusting the height of the tennis court net.

It was not unknown for a Governor General to be boycotted because the colonialists whom he was trying to serve mistakenly assumed that he represented an unpopular British Government rather than The Queen.

Following a party on the night of an election in which the Colony had smugly voted in an extremist party, one Grenadier accompanied his Governor General on a forlorn walk at 2 am to commiserate with the defeated Prime Minister. The Union Jack was lowered for the last time at Government House a few months later; the ADC disposed of the Governor General's wine cellar and the five Government House cars. Within a few years a long and bloody civil war had broken out.

Two ugly murders occurred at one Government House. Near midnight on 10 March 1973 the Governor of Bermuda, Sir Richard Sharples, and his Welsh Guards ADC, Capt HRL Sayers, were both shot dead in the gardens of Government House while giving the dog a walk.

ADCs to Generals filling military appointments in the UK, Germany and elsewhere learned useful lessons, but few could have enjoyed themselves as much as their colleagues who served Governors in the Commonwealth.

*   *   *

Hitherto this History has been told in a chronological order – starting in 1945 and ending in 1995. However, it may be simpler to tell of the experiences of a few Grenadiers serving away from their battalions by referring to the countries in which they served, starting in America, before turning to Europe and the Middle East, finishing off with the Far East.

Those quoted below were chosen because most of them contributed to this History. They provide an interesting cross-section of the sort of appointment which Grenadiers filled, and illustrate where many others once served.

Maj Gen Gascoigne was in Washington 1946–1949, commanding the British Army Staff. He had many meetings with his American opposite numbers on the standardization of weapons in the two armies, 'but they were a pure waste of time,' he wrote later, 'since none of the officers I met had any powers of decision.'

Brig EJB Nelson was also posted to Washington, to join the staff of the Standing Group which acted as the agent of NATOs Military Committee. He found that work in the Pentagon was frustrating due to the tedious processing of all plans through the time-consuming channels of so many national Governments. In 1956 he saw the damage done when Prime Minister Eden 'lost all confidence in Dulles and refused to tell him his intentions' over the Anglo-French invasion of Suez. Brig Nelson was visiting the Canadian Guards when news reached him of the British ultimatum to Egypt. He was encouraged to hear the views in the Canadian Sergeants' Mess: 'My hosts declared their sincere hope that the predicted hostilities would continue long enough for the Canadian Army to be able to take part,' he wrote later. But it was not to be.

At least four Grenadier officers served with the Canadian forces on exchange. One of them was Maj ET Hudson: 'I was commanding a company in 3rd Battalion the Royal Canadian Regiment in 1971 when one of the staff of the British High Commission was kidnapped. I was regarded as something of an expert on internal security having come from Northern Ireland's troubles.

'On one occasion we flew down by helicopter from Petawawa at short notice to a high security prison and remand centre. Riots had led to inmates barricading themselves in, molesting the teenagers. They hung bloody blankets and mattresses outside the prison windows. It lasted three days and was all quite horrific.'

The responsibilities of Maj OJM Lindsay, serving 'on exchange' on the staff in Ottawa, included that of being secretary to the Canadian Forces Airborne Advisory Board which had been set up due to the large number of fatal accidents amongst Canadian parachutists. On his first morning's visit to the Canadian Airborne Regiment he was persuaded to parachute several times from a Twin Otter which was so small that the parachutes had to be put on in the plane. Fortunately the landing zone was very soft, being covered in snow.

Turning to Grenadiers' staff appointments in England, reputations were quickly made or lost in the Ministry of Defence in Whitehall. Some officers, on leaving the Staff College in their early thirties, were immediately plunged into very intense and testing staff appointments. For example Maj PR Holcroft, in the Military Operations Branch of the MOD, regularly had to brief the Chief of the General Staff, Gen Sir John Stanier, on highly complex operational or long-term planning issues. They often involved inter-Service rivalries such as the pros and cons of the fourth Trident submarine or whether the theatre nuclear capability should be air-delivered by the RAF or missile delivered by the Army.

'General, now Field Marshal, Stanier, was renowned for his deep probing, and his intolerance of ill-prepared briefers,' noted Maj Holcroft. 'Officers who lacked self-confidence or knowledge didn't last long.' The RN and RAF were less inclined to rely upon young briefers, thereby denying the potentially outstanding the opportunity to gain such experience.

In 1973 Lt Gen Sir David Fraser became Vice Chief of the General Staff. During his stewardship there was a change of government and the 1975 Defence Review directed at a massive reduction of Defence Expenditure which led to the 'restructuring' of the Army in following years. General Fraser's primary function was to direct this Defence Review as far as the Army was concerned. His principal preoccupation was to ensure that regiments were not destroyed by the 'scissors of the ignorant' and that manpower savings were achieved by reorganizational measures rather than by disbandments. Pressure was considerable but success was to a large extent achieved.

Three officers served with the British Military Mission at Potsdam in East Germany. They were the Hon MF Fitzalan Howard MC (later the Duke of Norfolk), Brig LAD Harrod OBE who had transferred to the Welch Regiment, and Maj N Boggis-Rolfe. 'We toured round the twenty divisions of Group Soviet Forces Germany, photographing them from cars and getting intelligence,' Brig Fitzalan Howard wrote later. 'We were very conscious that our Headquarters at Potsdam was totally bugged by listening devices but our offices in Berlin were safe. Nevertheless all our journeys were always greeted at the Glienicke Bridge with Volkspolizei cars which trailed us throughout. We had many subterfuges for blocking them from following us when we wanted to take a photograph of some sensitive target behind signs which forbade Allied missions to pass. I was arrested ten times – on one occasion being kept shivering with cold for 10 hours in a bugged room with my driver. We always had a hot thermos for such eventualities.

'We had got in the middle of the Russian manoeuvres and took invaluable photographs including one of the secret infra-red equipment. We were very conscious of the bugging and never talked during these detentions. But I established very good relations with the Russians.

'The reason why we were followed by the Volkspolizei was that a copy of our programme was being passed on by a spy to the Russians. While in Berlin the Russians sent up their first "sputnik" which really alarmed the West and was no surprise to us who felt their armaments were far ahead of ours.'

Brig Harrod knew that 'there was no point in having our house in Potsdam swept for listening devices as the Russians would put in new ones. One of our team in the house said that Russians were all hooligans. A week later a Russian interpreter, a young serving officer, laughingly assured us that they were not all hooligans.'

Maj JRS Besly served two tours with the French Army, one being at St Cyr, the equivalent to Sandhurst, in 1968 to 1970. 'The whole question of loyalty arose because the Army was split over de Gaulle's Algerian policy,' he recalls. 'There were student riots in Paris: all troops were put on alert including our cadets with their divided loyalties. After a frightful row St Cyr had to be pulled out of the order of battle. The worried St Cyr Commandant asked me my views of the cadets' loyalty. "Ask their platoon commanders, rather than me", I said. He replied, "You seem to be closer to your cadets than the French officers, so I value your views."'

Grenadiers served in NATO appointments in Brussels. In 1975 General Sir David Fraser became the United Kingdom Military Representative to the NATO Military Committee. Here he represented and spoke with the authority of the British Chiefs of Staff.

Brig DH B-H-Blundell also served there: 'In 1988 I became the Secretary to the Chiefs of Staff at Supreme Headquarters Allied Powers Europe in Mons, Belgium. The job entailed running a multi-national secretariat of 130 officers, men and women. Demands for information descended from the Supreme Allied Commander and others, in curt notes such as: "What is the exercise pattern of the Soviet air defence systems in East Germany? Reply by 1600 hours Wednesday". There were endless conferences and committees. In the event of nuclear warfare, it could have been my responsibility with another officer to press the button for atomic weapons to be used. The two black American Staff Sergeants in the war bunker were both pregnant.'

During the period of this History Berlin was often the centre of confrontation between the West and the Soviet Union. As with the American and French, the British sector of Berlin has been subject to a Military Government, the responsibilities and powers of which dated back to the Potsdam agreements and then to the time of the Berlin Blockade.

The British Commandant was the Head of this British Military Government in Berlin, and also the GOC of the British Forces in Berlin. Of the last twelve British Military Commandants, eight have been Guardsmen, and of these three have been Grenadiers, namely Major Generals EJB Nelson, JFC Bowes-Lyon and BC Gordon Lennox. Together with them many Guardsmen and Grenadiers, such as Lieutenant Colonels AJC Woodrow and ACMcC Mather have served on the staff.

As well as their military and security duties, these Commandants were the senior diplomatic representatives of the British Government in Berlin. In this last capacity they were the servants of the British Foreign Office and of the British Ambassador in Berlin, and in this role they were assisted by a Foreign Office Minister, and by a sizeable diplomatic staff.

The purpose of retaining British and Allied Military Governments was to maintain a situation which allowed the Tripartite Powers to negotiate about Berlin with the Soviet Union rather than with the East Germans.

In practical terms Major Generals Nelson, Bowes-Lyon and Gordon Lennox had not only to train and to administer the Servicemen under their command, but also to be responsible for the large Berlin budget, much of which was paid for by the West Germans, to act as the Senior British Diplomatic Representative in West Berlin, to act as the sheet anchor for the British intelligence agencies based in Berlin, and to ensure that Great Britain and its protecting role was understood and supported by the Berlin Senate and Business Community in West Berlin. For all three it was a challenging but fascinating and ever changing task. Together with their diplomatic colleagues they were successful, for in the end West Berlin was not challenged by Soviet military power – and the Wall came down.

Majors N Hales Pakenham Mahon, HML Smith and HA Baillie meanwhile served on the staff in Cyprus.

Numerous Grenadiers have also served in Africa including Maj SEH Baillie in Ethiopia in 1950. Col DV Fanshawe OBE was in the Sudan: 'It was a difficult tour as the whole structure of the country was fast crumbling and there was little one could do to support our old friends the Sudanese people. Her Britannic Majesty's Embassy occupied the top floors of a derelict Shell building. Huge fluorescent letters proclaimed this as "Shell House". After a while the "S" was carried away in a sandstorm. Of course there was no suggestion of replacing it.'

At least three officers, much earlier, saw active service with the King's

African Rifles fighting the Mau Mau terrorists. Among them were JCA Russell-Parsons, PM Lambert, and ER Johnstone who commanded a 7 KAR platoon. Johnstone recalls one patrol in the forests near Mount Kenya: 'I was with my platoon who were from half a dozen different tribes, up by a track, well hidden. I took a section of six men forward down a slope. We suddenly saw nine Mau Mau in a diamond formation coming towards us on a path below. They were going to pass within 30 yards of us; we had better weapons but I was uncertain of the outcome.

'They probably smelt me as they were like animals. They suddenly turned away. We gave chase, running downhill; then we got down and opened fire. A man on my left hit a Mau Mau 400 yards away with his fifth shot. Another KAR patrol opened fire over our heads and dropped three of them. I found a rifle alongside our dead Mau Mau. Beads were all over his beard as his necklace had broken. The Askaris got out their bayonets and stuck them into the body which was their tribal custom. We left the body on the ground – there were lots of vultures and hyenas. The dead man was carrying papers on him and was later identified as a Captain, perhaps the Quartermaster. A Minister later authorized the release of the terrorist's gun. It's now in my local pub at Ickham.'

Maj A Dobson MBE had more than a rifle pointing at him in unusual circumstances: 'I was the Quartermaster of 1st Battalion Tanganyika Rifles. In January 1964 we did a tour of barracks. I pointed out that the Guardroom's only cell, eight foot by five, was much too small and should be enlarged. Three days later we heard that a Battalion had mutinied 600 miles away. Orders were given to sort it out and ammunition was to be distributed accordingly. I said that they were all blood brothers and so it was unwise to issue the rounds, but I was told to obey orders.

'We were up at 5am the following morning when we heard a big explosion and bullets were flying everywhere. I saw the new British Commanding Officer with his hands up. We were quickly surrounded by 300 very hostile men in uniform. The whole of our Battalion had mutinied, including the Mons Officer Cadet School officers, except for a Sandhurst-trained Captain who ended up as a General. We were marched to the tiny cell where the CO, 2IC, Adjutant, ORQMS, a Staff Sergeant, the MT Sergeant and finally the Paymaster were all imprisoned with me. A Bren gun faced us through the window. Meanwhile my wife and daughter in our quarter heard the firing and hid in the wardrobe, but were found.

'At 2pm we were taken outside and put on three ton vehicles – two of us on each, surrounded by blacks; to be paraded round the camp. Some of us were hit by rifle butts and badly bruised. We were driven to Tibora airfield where I met my wife and daughter, and the British residents who had not been arrested. They got picnics together for us and really helped.

254

At 9.30pm an East African Airways aircraft arrived, but it was overloaded so we had to leave our suitcases behind. The residents said they would get them to us. On landing at Nairobi we were greeted by Middle East HQ personnel who had already gathered everything that we could need. The 2nd Battalion Scots Guards was also there. They took all the families to buy warm clothing for the UK, and all the suitcases turned up. We never got to the bottom of the mutiny which had been planned over several months by small-minded young African officers. I was next posted to Northern Ireland, and then to Aden where there were plenty of rebels about.'

Maj PFL Koch de Gooreynd and Capt PZM Krasinski served in Mozambique, while Brig J Baskervyle-Glegg MBE was one of many selected to train the Zimbabwe Army. He was also responsible for advising Prime Minister Mugabe and found him intelligent, well read and a good listener. Brig Baskervyle-Glegg later went, on promotion, to be the Senior British Loan Service Officer in Oman. Dozens of Grenadiers served on the staff or in command appointments in the Middle East.

Lt Cols GEV Rochfort-Rae and PT Thwaites commanded battalions of the Abu Dhabi Defence Force and the Muscat Regiment at Dhofar respectively. Maj TN Bromage served in Jordan, while Capt PH Cordle was sent in 1967 to Aden with a Foot Guards team of thirteen at a fortnight's notice to train four battalions of the Federal Guards prior to their becoming part of the Federal Regular Army.

Moving to the Far East, numerous Grenadiers served in Malaya including Lt Gen Sir Rodney Moore who was appointed Chief of the Armed Forces Staff and Commander in Chief of the Malayan Army in 1959.

Maj DMA Wedderburn was serving in the Singapore Regiment in February 1960 when he was ambushed by criminals while returning to the Battalion with the pay. He managed to carry on to the camp gates before he collapsed. He died in the ambulance on the way to hospital.

Lt Col ATW Duncan MVO later commanded the Royal Brunei Malay Regiment, and Brig PGA Prescott MC commanded a brigade in Hong Kong. Capt FS Acton served with the United Nations in Cambodia, as did Capt MGP Lamb who survived a Khmer Rouge ambush when travelling in a vehicle convoy. In Korea Maj DJ Beaumont-Nesbitt MBE was employed on the staff, as was Capt CWJP Langdon-Wilkins. Captains I Reid and GCR Gottlieb served with 1st and 3rd Battalions the Royal Australian Regiment respectively.

*　　　*　　　*

For every officer serving away from the Regiment, there were probably several Grenadier Warrant and Non-Commissioned Officers and Guardsmen doing likewise.

Let us take the year of 1964, for example. The Chief Clerk of Headquarters, Household Brigade was RSM JS Bird MBE who held the appointment for over 16 years. At Sandhurst RSM T Taylor still had 12 years service in front of him. He was to play a major part in the planning and execution of such unique events as the granting of the Freedom of Windsor to the Foot Guards in 1968, the Service participation in the Investiture of the Prince of Wales in 1969, the Lying in State and Funeral of the Duke of Windsor in 1972, the Wedding of Princess Anne in 1973, the funerals of the Duke of Gloucester and three Field Marshals, the Silver Jubilee in 1977 and much else. He retired in 1977 after 35 years service, with the MVO and MBE.

Also at Sandhurst in 1964 was RSM GR Whitehead who was to reach the rank of Lieutenant Colonel. RSM S Lowe was at the Royal Military School of Music, Kneller Hall; B Owen was RSM at the Army School of Education, RQMS L Braine at the Staff College, Camberley, while others were scattered around the world – RSM J Murrant with the Kuwait Liaison Team, RSM L Price in Bahrain, RSM R Jordan with the Jamaica Regiment and RQMS J Walmsley with the Malaysian Rangers.

Almost 20 years later we had RSMs PT Dunkerley in Hong Kong, CE Kitchen in Sudan, D/Sgt RD Hobbs in Abu Dhabi and CSM SC Allen in Gambia. Ten years later the 1994 *Grenadier Gazette* (so brilliantly edited each year by Lt Col HS Hanning and Capt BD Double) shows 114 Grenadier Warrant Officers, NCOs and Guardsmen serving 'extra-regimentally'. RSM AJ Green was now in Zimbabwe and ML Pearson in Hong Kong. The influence of these Warrant and Non-Commissioned Officers and the others in the Household Division was very considerable and, indeed, worldwide.

In 1970 RSM RP Huggins achieved his great ambition when he was appointed Academy Sergeant Major at Sandhurst, which is regarded as the senior Warrant Officer post in the British Army. Grenadier RSMs have held this distinguished appointment for some 50 of the 95 years of this century. Huggins had served in the Regiment in most parts of the world and represented his Battalion in half a dozen sports. He retired from the Army in 1980 with the MBE, Meritorious Service Medal and Cross of Recognition by the French Army to become Deputy Administrator at Blenheim Palace.

No officer cadet at Sandhurst will ever forget the name of his instructors such were their personalities. In 1994 the Grenadiers there were the Academy Sergeant Major (DL Cox), two CSMs (SP Milsom BEM and PM Ladd BEM), and no less than nine Colour Sergeants, (AM Harding,

P Parker, J Allen, M Thompson, WT Orton, BF Broad, SW Dodd, PA Cartwright, and DA Harrison). I Kime and J Finch were also on the Long Service List there. A further five NCOs were in Northern Ireland.

Finally, a reference must be made to the eighteen NCOs and Guardsmen serving on long tours with 1st Battalion Coldstream Guards in Germany in the 1990s. They provided vital continuity in the highly complex mechanized battalion. They deserve special mention because they greatly distinguished themselves when with the Coldstream Battalion in Bosnia (part of the former republic of Yugoslavia) between October 1993 and April 1994. The Coldstream and Grenadiers were in Bosnia and Hercegovina exercising 'the skills of traffic policemen, diplomat, nanny, soldier, mechanic and mediator' in often appalling weather conditions. L/Sgt PW Murray made relentless attempts to evacuate casualties to hospitals; the gallantry of L/Cpl SJ May also received official recognition.

Maj RG Adams, Capts AA Pollock and JA McDermott also wore the blue berets of the United Nations in Bosnia. They endured exciting tours amid some danger. Perhaps this can be said of many others who served 'extra-regimentally' throughout the world.

CHAPTER 21

ASSISTANCE TO CIVIL AUTHORITIES

Troops have been used by the Government in Great Britain in some forty industrial disputes between 1945 and 1995. Grenadiers stationed in or close to London have played their part in ensuring that, where possible, essential industries did not collapse due to industrial action.

The soldiers' role was to take over the work of the strikers; any public order problems were handled by the police. Troops have been used to intervene in strikes since at least the 19th Century. Since the General Strike of 1926, the army's role on work of urgent or national importance, to maintain essential services and supplies, most usually (but not uniquely) when they are disrupted by industrial dispute, has been that of an unarmed, reserve labour force, providing what is called 'Military Aid to Civil Ministries'.

The Trade Unions' opposition to military interventions in strikes became much less violent. This was because the striking workers in 1945–1951 regarded the National Servicemen waiting to be demobilized as 'workers in uniform'. Moreover, the strikers were placed in a difficult ethical and social position; if they were to prevent troops maintaining essential services they might have placed the lives of the public in jeopardy – an action for which they would have had to take the blame. An opinion poll in February 1979 at the height of the 'Winter of Discontent' showed that 78 per cent of the public were in favour of the use of troops in key industries, with 71 per cent of Trade Unionists supporting this as well.

Between 1945 and 1950 dock strikes were frequent due to protests over wages and working conditions. Food supplies were affected. Troops intervened in two strikes in 1946, one of them being at Smithfield meat market. CSM FJ Clutton MM recalls seeing 'the 1st Battalion's lorries returning to Chelsea Barracks with sides of bacon and joints which should have been delivered to the butchers. An extra joint was usually left on the trucks for the Grenadiers. It was put in the cookhouse stores; strict rationing was in force and so steaks were a great treat. Acting as meat porters, shifting

258

carcasses and delivering them to the butchers was very hard work, but the Guardsmen did very well. They were very popular with the butchers from whom they received tips.'

On 6 July 1949 8,500 dockers and ninety-two ships were idle due to the dock strike. 'We spent the next morning organizing ourselves into gangs to work the first three shifts – collecting meat from liners from the Argentine and Australia,' wrote Lt CCW Hammick. 'Reveille was at 5.30 am and we were rarely back before 6.30 pm. Twenty Grenadiers worked in the hold and thirty on the quay. On the meat ships the temperature was well below freezing.'

On 11 July a State of Emergency was declared – a very rare event in Great Britain. The 1926 General Strike files in London district were dusted off. (The reminiscences of Maj Gen Sir Allan Adair, *A Guards' General*, contain a unique account of the General Strike in London.) An Emergency Committee met daily to control the Port of London.

The Grenadier Battalion had under command 100 RN, 200 RAF, 80 from the Life Guards and 100 from other regiments. They handled well over 10,000 tons, either loading or unloading. In the Royal Group of Docks nobody was seriously injured thanks to the conscientious work of the Royal Navy who operated the winches and cranes.

On 21 July the strikers voted to go back to work, having obtained minor concessions. The Government's *Review* concluded that the armed forces had done a 'magnificent job and the rate of output which they achieved gave great satisfaction'.

Less than five months later manual workers at three London power stations started unofficial strikes over possible loss of pay at Enfield, Dartford and Willesden. Over 1,000 workers struck and were replaced by troops on 12 December 1949. 'Our cherished vision of a Christmas at home after Malaya began to fade,' wrote a Grenadier in *The Guards Magazine*. 'No 3 Company, 3rd Battalion, well bolstered up from other companies, was dispatched to Enfield to look after Brimsdown Power Station, while another small detachment was sent off to Willesden. After five days' shovelling coal, during which the Guardsmen looked more like black minstrels than soldiers, the strike was finally called off and, much to our relief, the detachments returned to Chelsea.'

Nearly the whole Battalion therefore enjoyed Christmas at home for the first time for several years. The Communist newspaper, *The Daily Worker* recorded that: 'The Ministry of Labour has the troops in the docks, the meat market or the power stations in the twinkling of an eye. The Trade Union leaders who remain silent in such a situation are acquiescing in a vile, anti-Trade Union practice which cuts at the very foundations of the movement.'

The unofficial strike of oil tanker drivers in London started in October 1953. By the 23rd 3,000 workers were on strike. That evening 2,000 troops were moved into London. 'Nobody relishes strike-breaking but it was an interesting exercise,' wrote Lt Col John Nelson who was responsible for operations in HQ London District. 'The Royal Navy manned the container pumps and the Army and RAF found the tanker drivers. The Union pickets were surprisingly cheerful and did not attempt to interfere. After a weekend of blissful traffic-free streets, the petrol stations were replenished and the cars came back on the roads. I always believed that one of the reasons why the strikers went so quickly back to work was because of their reluctance to see their "invisible" perquisites going into the pockets of their Service understudies. One per cent of every 1500 gallons of petrol transported was written off against "evaporation". In fact, there was virtually no such loss and garage proprietors used to buy the balance direct from the tanker driver. The soldiers and airmen shared the spoils with their sailor colleagues who were confined to the pumps.'

The oil tankers drivers' strike was the last occasion for over twenty years on which a complete labour force was replaced by troops.

The Government's efforts to enforce strict incomes policies led to the intensification of labour disputes and industrial strife in the 1970s. All this was made worse by the rapidly rising level of unemployment.

Some strikes, such as that of the local authority manual workers, were minor affairs. At Tower Hamlets the borough's Medical Officer of Health said that the mounting piles of uncollected refuse on the streets were a health hazard. The Labour-controlled council appealed to the Government for help. Their request for troops was passed on to the Ministry of Defence: 'It was, as always, over a weekend,' wrote Capt HMP de Lisle. 'I was Assistant Adjutant of the 1st Battalion at Chelsea. Peter Ratcliffe at HQ London District called us in for a briefing at the Horse Guards.' At 2 am on Saturday 24 October 1970 a Grenadier convoy of eighteen vehicles, including two mechanical shovels and a dozen tipper trucks with almost forty Grenadiers and Royal Engineers, moved into Tower Hamlets to clear away the worst five of the eighteen major rubbish dumps. As the rotting rubbish included slaughterhouse refuse, the task was extremely unpleasant.

The rubbish was loaded into trucks which were driven to a dump 12 miles away where another party of Guardsmen unloaded it. One pile alone took twenty-five three-ton loads to remove. With the usual friendly co-operation of the Metropolitan Police, the operation was completed without incident. 'I was interviewed on television,' recalls Capt de Lisle. 'Giles in *The Daily Express* did a large cartoon on me, and so did *The Daily Telegraph*; I gave a short briefing at the Staff College, Camberley, as this was reckoned to be the first time the Army had been used in aid of the Civil Power in England for some years!'

In 1972 Prime Minister Heath ordered a fundamental review of state contingency planning. It was decided that the old emergencies organization should be removed from Home Office control and be based in the Cabinet Office.

From 1972 to 1974 there was serious industrial and social conflict within Great Britain. The operations branch of HQ London District had contingency plans for using the military in the event of the breakdown of a number of essential services. The 1974 miners' strike contributed to the fall of the Heath Government. But there was no use of troops to break the strike probably because the Services were not capable of taking over the very complex mining industry.

From September 1977 the Labour Government's increasingly tough pay policies for the public sector aroused militant action from the Unions. Over the next twenty months, the military intervened in five disputes.

The 1977–1978 Fire Brigade's strike was a turning point in the development of Military Aid to Civil Ministries: almost the entire 30,000 strong Fire Service was replaced for two months by 20,750 military personnel. On 8 November 1977 the Chief of the London Fire Brigade told HQ London District that the strike was imminent. He passed on the Home Secretary's decision that neither the London Fire Stations nor their communications would be made available to the Army.

Five days later HQ London District, responsible for Greater London, set up and exercised its command structure. Each Foot Guards Regimental Headquarters had been deployed with responsibility for a number of service fire stations in various barracks. Col GW Tufnell and the Grenadier RHQ staff were based on Chelsea Barracks with fire stations at Knightsbridge, St John's Wood and Holloway. The drivers of the Army's fire engines, called Green Goddesses, came from the 1st Battalion. All had to be Heavy Goods Vehicle trained. The Guards Depot's staff and recruits largely made up the balance of the crews in the Grenadier sector.

A joint control HQ was set up in the Greater London Council Flood Emergency Headquarters. It was situated in the former underground tram centre in Kingsway and manned principally by RHQ Household Cavalry. On the following day the strike began. The 11,000 troops already on standby were mobilized under the authority of the Emergency Powers Act 1964 which permitted them to be used on 'urgent work of national importance'.

Every Serviceman received one half-day's training in basic fire fighting and in the Green Goddesses' equipment. There was initially a general air

261

of pessimism among the London Fire Brigade authorities; they felt that the Servicemen would be unable to cope.

Those on strike still had access to their own fire service radios and there was apprehension that they would interfere with the Army's communications. To overcome this possibility, the Home Office re-crystalled all the Kenilworth radios using a sixth channel on an exclusive frequency. This led to some delay because the crystals had to be obtained from Berlin. When all the communications were coordinated and operated by the Chief Royal Signals Officer, London District, rather than by the Home Office, there were considerably fewer problems.

Gdsm S Milsom was waiting to go to the 1st Battalion from the Guards Depot and was based on St John's Wood. 'I worked a six-day cycle with two day shifts, two night shifts and two days off duty,' he remembers. 'I fought ten fires all started by troublemakers. We had one Green Goddess and just gushed water everywhere. Although we had a driver from the Royal Corps of Transport, who was a CSM, we were all commanded by a Scots Guardsman from the Depot.'

Gdsm A Thomson was based on Hounslow and combined fire-fighting with Public Duties at Buckingham Palace. 'I attended half a dozen fires,' he recalls. 'We were fairly amateurish. I rushed into one home with a vast hose after a frying pan had led to a very small fire. I'm afraid that the damage we caused cost nearly £45,000. My fiancée and I were later shopping in Woking. Firemen on strike stuck their collection boxes under her nose demanding money. She gave them an ear full in her Cockney accent as I'd been on fire duty over Christmas. We had to drive past the strikers in our Green Goddesses. They were grouped around their braziers by their fire station at Hounslow. There were sarcastic comments to start with but after several weeks it was quite chummy.'

Maj PAJ Wright felt that: 'visiting a fire was a bit like hunting. I remember setting off with four Green Goddesses; another eight joined us en route. Seeing tall buildings in flames collapsing was a spectacular sight. I was very apprehensive when two Guardsmen were overcome with fumes: oxygen was vital. Overall, in London District, there were fifty-seven Service casualties, none being serious. Servicemen's pay became a major issue with the media. Their courage and organization were greatly praised by the press and public. Winston Churchill MP launched a Christmas appeal for the troops. Some units spent the donations on the children and families of those who had been on duty; they went to pantomimes, for example.'

The strike was almost total. Nevertheless, faced with a reluctant Union leadership, little financial backing and an intractable Government, the strikers had to compromise, accepting an immediate ten per cent offer. A key factor in their decision to resume work on 16 January 1978 was the relative success of the military fire-fighting capability.

In the Grenadier sector alone 526 fires had been put out. In London District some 12,000 telephone calls reporting fires were received of which 4,000 were hoaxes. One penalty of the military involvement was that Servicemen at the Guards Depot lost about nine weeks' training. However, the military benefited from its experience during the strike and the reputation of the Servicemen was further enhanced.

In 1979 Grenadiers were also involved in the strike by Ambulance crews who were claiming a two-thirds pay increase. In London fifty Army vehicles and eighty-five police vans were brought into use. Union officials were angry at the decision to use troops, although, as usual, there was little hostility directed at the troops themselves.

During the railway workers' strike Grenadiers from the 2nd Battalion had to construct car parks in Hyde Park. And so it went on – prison officers were on strike in 1980 and ambulance crews again in 1981 and 1982, among others. (An account of the 1989–90 Ambulance dispute is contained in *The Guards Magazine* of Autumn 1990.) It is doubtful if the Services will be able to make a major contribution to helping Civil Ministries in the future due to the dramatic reductions in the size of the Armed Forces and the loss of key specialists.

Reference should be made to assistance provided in two other very different areas which might have led to the saving of many lives.

Until the construction of the Thames Barrier, near Woolwich, there was a serious threat of flooding in Central London. In 1972 much of the London District annual Study Day was taken up with planning against such a disaster. Boats were available to Guards Battalions; duty officers were occasionally woken at night with alarming reports of surge tides approaching the Thames estuary. Fortunately floods did not materialize and troops did not deploy in London in the 1970s. The operation was appropriately named 'Giraffe'. During the first ten years of operation, the Thames Barrier has been closed eleven times to protect London.

A much larger operation which has frequently involved all Guards Regiments and the Household Cavalry in turn relates to the terrorist threat of hijacking aircraft, largely at Heathrow Airport.

In 1972 the Ministry of Defence instructed Maj Gen James Bowes-Lyon, then commanding London District, to liaise with the police in New Scotland Yard and at Heathrow to prepare the appropriate military plans.

Maj OJM Lindsay had regular meetings with Assistant Commissioner John Gerrard over the following two years; a roster was produced with Guards Companies on standby, and discreet liaison was established at Heathrow. The principal plan anticipated the use of the Guards and other

Servicemen after a hijacking had taken place. The military had a greater capability against terrorists than the police. An alternative plan ensured that troops would deploy to Heathrow to protect the runways as a general deterrent to potential hijackers of whom there were a number, particularly in the Middle East.

Towards the end of his tour, late one evening, Maj Lindsay was telephoned by Assistant Commissioner Gerrard and told that some terrorists were believed to have anti-aircraft weapons; an attack at Heathrow could be imminent. 'Could the principal Commanding Officers on the roster move there immediately?' the Assistant Commissioner enquired.

The chain of command for such requests was that a Minister had to ask the Ministry of Defence to deploy the force. The MOD, in turn, had to order London District to do so. However, the MOD duty officer was scarcely able to comply with the Minister's request because, most unusually, he was not aware of the combination number of the cabinet which contained the plan. Fortunately Maj Lindsay knew the duty officer well. It was therefore arranged for Lt Col JACG Eyre, responsible for operations in London District, to go to the Ministry of Defence and order Maj Lindsay to implement the plan. Maj Lindsay instructed the acting Commanding Officers of the 1st Battalion Irish Guards (Maj HC Blosse-Lynch) and of the Blues and Royals to report discreetly to the police station adjacent to Heathrow Airport.

Unfortunately the press quickly picked up that a unique operation was imminent. Majors RAG Courage, the Army Public Relations Officer in London District, and Lindsay spent the night dealing with media enquiries.

Fortunately everyone involved in implementing the operation had known each other extremely well over a number of years and so there was no possibility of misunderstandings or hoaxes.

This deployment of armoured vehicles of the Blues and Royals, Grenadiers and Irish Guardsmen in early January 1974, assisting the police in the terrorist alert, was unprecedented in recent years. This form of 'Military Aid to the Civil Power' had not been seen in Britain since the General Strike. The Government stated that Palestinian guerrillas were intending to shoot down an airliner; *The Guardian* said it was 'basically a public relations exercise to accustom the public to the reality of troops deploying through the high street' – a view which was disputed by the military and police. According to Steve Peak's book *Troops in Strikes* the Heathrow incident was seen by many as tantamount to military intervention in the prevailing industrial and political conflict.

However, the threat to Heathrow was very real and the public became increasingly reassured to see Guardsmen deployed there fairly frequently thereafter. The Guards' cooperation with the Metropolitan and Thames

Valley Police Forces and the British Airports Authority Police was outstandingly successful and the commitment was generally welcomed.

By the early 1990s the police presumably felt that they could cope themselves without calling upon the military so frequently. Military deployment to Heathrow became increasingly rare, despite terrorists mortaring the Airport on consecutive days from a hotel car park in 1994.

CHAPTER 22

SCOPE FOR THE ADVENTUROUS

Few Regiments can have attached as much importance to adventure training as the Grenadiers. As this chapter indicates, every effort has been made to get officers and men away from routine duties in order to develop their leadership and, literally, to broaden their horizons. Lack of space prevents more than a brief summary of a cross-section of such activities. Some notable expeditions have already been covered in earlier chapters.

Those who enjoyed big game hunting found much to shoot in Africa in the post-war years. Maj SE Bolitho MC, for example, took, a month's leave in 1948 to shoot in Kenya two elephant, a rhino, leopard, three buffalo and much else.

Others followed his example. Lt OJM Lindsay with 2Lt GN Tait of the Royal Horse Guards drove 8,000 miles across country through Nigeria, Chad, the Cameroons, Sudan, Uganda and Kenya, living off the land by what they shot, and sleeping in the bush by their Land Rover. They were accompanied by two Nigerian servants, one of whom prayed ten times a day facing Mecca in the east which was useful when they were all lost. In a night club in Nairobi they met a girl who asked them to stay with 'Daddy'. He turned out to be the Governor General of Tanganyika. Their frayed, smelly bush clothes were not ideal for smart parties at Government House, Dar Es Salaam.

Revolutions, widespread famine and the establishment of game reserves led to few such expeditions after the early 1960s.

When time permitted, opportunity was taken to join a Battalion far afield, or to return to England from the overseas posting, in a more adventurous style than taking a train, ship or plane. Maj IM Erskine, in the 3rd Battalion in Cyprus in 1959, encouraged eight junior NCOs and Guardsmen to hitchhike in small groups to England via Egypt, Greece, Italy and France – an exciting journey, particularly in those days.

A rather quicker journey was undertaken by Capt DJC Davenport three years later. He participated in the 1962 Monte Carlo rally. Despite having to drive occasionally at over 100 mph on a road no wider than a cart track, he reached Monte Carlo sufficiently quickly to qualify for the speed trials on the Grand Prix circuit. His Vauxhall by now possessed 'no power and little brakes, but it was great fun,' he wrote later.

In 1963 Capt the Hon TRV Dixon took part in a race that was to make him famous. Three years previously he had started to work in earnest to win a place in the Winter Olympics. Despite a most spectacular crash, both he and Tony Nash were nominated to represent Great Britain in the two-man bobsleigh event.

On arriving at Innsbruck 'everyone had mixed feelings', he recalls. 'We knew it would certainly not be fun and we had fears that it might be hell. On the first night we paid a last tribute to our friend and team-mate K Strepetzki who had tragically lost his life that day on the toboggan run. The Olympic flag and Union Jack were lowered to half-mast for 48 hours. They had hardly been raised when the flags had to be lowered again for a young Austrian skier who was killed, as was a Liechtensteiner. Our race turned out to be one of the most exciting ever.' The result was not known until the last boblet had made its descent.

'Tony and I did not really feel the full excitement despite leading at the end of the first day. We knew that the sure way of not winning was to get excited or emotional in any way. It was obvious that the second day would be a battle of nerves and, as it turned out, the Italian Sergio Zardini threw the race away by cracking under the strain in the last lap.

'He left us the winners of an Olympic gold medal by 0.12 of a second, approximately six feet after four miles and fifty-two corners of racing over two days. Standing in the middle of the stadium watching the Union Jack go up to the music of the National Anthem was like a dream come true.'

In April 1968 the *Daily Mail* announced that 'the greatest Air Race of all time' would be held to celebrate the crossing of the Atlantic fifty years earlier. The two check points were the Post Office Tower in London and the Empire State Building in New York. Capt ADM Clark had been invalided out of the Grenadiers with polio in 1943 while serving in Italy. Despite being unable to stand or walk unaided, he decided to enter the race. Col AN Breitmeyer in Regimental Headquarters, and Maj DV Fanshawe, commanding the Guards Independent Parachute Company, obtained volunteers on both sides of the Atlantic to form a small but efficient team. Richard Hall had served with The Queen's Company and was now with the British Airways Authority. He organized all manner of short cuts.

Sunday 4 May 1969 dawned a lovely bright day. Within 29 minutes of leaving the GPO Tower at 10.15 am they were at Heathrow where Capt Clark was transferred to another wheel chair. Guardsmen linked arms to clear a path from other passengers. The pilot waited fruitlessly for two other competitors before taking off 16 minutes late. With the help of three Grenadiers in America, (Edward Parry, Piers Dixon and Gerry Hacquebard), Alan Clark reached the 86th floor of the Empire State Building 8 hours, 18 minutes and 30 seconds after leaving London. 'I had to use five wheelchairs, three cars, a helicopter and airliner,' he wrote later. 'We beat a lot of others, including the XIth Hussars and the Sandhurst cadets; our time was very creditable, and was entirely due to the great efficiency of the serving and past members of the Regimental Association. It all added up to an experience that I wouldn't have missed for anything.'

Those commanding London District Battalions tried to get their platoons well away when possible. 'One Queen's Company platoon enjoyed a kaleidoscopic caricature of Scottish life,' wrote a Grenadier in October 1970. 'They learned something about grouse driving and the habits of the red deer. No 3 Company sent one platoon to Norfolk on army bikes to find out that Norfolk was not as flat as it looked. Another platoon commander read the book *Three Men in a Boat* and sailed with his men down the Thames from Oxford to London in canoes. A third looked ahead to Kenya and used a certain peer's Wiltshire estate as an aperitif training area complete with lions.' And so it went on.

During the massive build-up of the forces of the Third Reich in the 1930s Goering believed that young officers would benefit by sailing large yachts which had no engine in the Baltic. *Gladeye*, which was then called *Reiher*, was the prototype of a number of yachts. After the war these yachts were taken as 'prizes of war'. Some were towed to England by the Royal Navy and distributed to Service yacht clubs.

The Household Brigade's yacht was named *Gladeye* after the famous ever-open eye of the Guards Armoured Division. '*Gladeye* served the Household Division for 32 years,' wrote Col DW Hargreaves in 1980. 'During that time thousands of Guardsmen of all ranks have sailed in her. Many liaisons have been made in her, some leading to marriages. There are several instances of father and son having been skippers. Many Guardsmen remember her with affection for what they have learnt about sailing.' Sgt DA Roberts was among the numerous Grenadiers who had many weeks' enjoyable sailing in *Gladeye*: 'She covered many thousands of miles a year in which she had good, successful sailing without incident. With her tall, raked masts and low black hull, she looked a real thoroughbred.'

In a typical season there were two 'skippers' courses', and some 360 officers and men and 240 Junior Guardsmen sailed in her. She cruised to many of the channel ports. In 1994 the Household Division took delivery of yet another *Gladeye*.

Another exciting development was the purchase in 1968 by the

Household Division of the Guards House 'Folda' at Glenisla in Perthshire as a skiing and adventure training centre. In 1972 Guards' families could also use it in the summer months. Folda was put to good use. For example in February 1974 2Lt CAH Wills oversaw sixty Guardsmen of the 2nd Battalion staying there and skiing above the Devil's Elbow at Glenshee. No photograph exists of the expedition, perhaps because nobody could stay on their feet long enough to be photographed!

L/Cpl Appleby was among the Grenadiers who participated in the 1974 Zaire River expedition which marked the centenary of Stanley's epic journey. L/Cpl Appleby received considerable praise for getting the 140-strong expedition's Land Rovers across Africa from Livingstone Falls to the Atlantic.

Meanwhile Captains CH Walters, CRJ Wiggin and the Hon JA Forbes were driving from Germany to Algiers, and thereafter across the Sahara, Nigeria, the Cameroons and the Central Africa Republic to Kenya.

Two years later Maj A Heroys was told by his Commanding Officer, Lt Col BC Gordon Lennox, on joining the Battalion, 'Now, Alexander, as you are a mountaineer, I should like you to go up Mount Kenya first to find out the form for the Companies and check their route, and then, no doubt, you would like to take a summit party up to the very top, and finally you had better go up with your own Company.' In January 1976 the 1st Battalion started a six-week training exercise in Kenya (where Lt Col DHC Gordon Lennox took over command from his brother).

The mountain is the remains of an extinct volcano. The summit party consisted of Maj Heroys, Capt the Hon James Hogg, C/Sgt Higgins, L/Sgt RG Garmory and Gdsm Robinson, Vince and Bexton. It was decided that Hogg and Garmory were the two strongest and had adequate rock experience to be the lead pair. All was going well until Hogg fell off a blank wall, bouncing off the rock on the way down; he had been held by the rope after 75 feet. He was badly bruised but they went on to reach the summit of Nelion (17,022 feet). The following day a blizzard was blowing: conditions were Arctic and there was danger of frostbite.

Fortunately the weather was better when four Companies attempted a mass climb of Point Lenana. Each set off with up to seventy-five men with slightly different aims and routes, but with the object of giving as many as possible the experience of high-altitude mountaineering with its attendant feeling of achievement. Each Company spent between three to five nights on the mountain. Breathing above 11,000 feet was an effort and some found that sudden movement caused dizziness. Most took the route of the Sirimon Track across to the Liri and Mackinder Valleys before ascending steeply to Simbra Col and round to the summit of Point Lenana.

The plan of attack went well. Each days' climb was accomplished leaving sufficient time to pitch camp and cook the evening meal of compo before the early dusk of 6 pm and twelve hours of darkness. 'Sleep was not always possible due to the shortage of oxygen,' wrote Maj Heroys. 'By the time the summit of Point Lenana was reached clouds were already drifting in. What then had we achieved during our three-week assault on Mount Kenya? We had climbed Nelion and put 150 Guardsmen on Point Lenana which, at 16,355 feet, surmounts any Alpine peak. We had given a further 150 Guardsmen experience of moving and living at high altitude in the mountains and on average we had all walked at least 30 miles at over 12,000 feet. Above all, the Guardsmen had faced a challenge and achieved success.'

Also in 1976 Capt AGH Ogden and Lt AJS Scott-Hopkins, on an exchange with the French Army, climbed the north face of Mont Blanc du Tacul, a steep snow route through eerie crevasses and ice caves. The view from the top covered Courmayer in Italy, the Dents du Midi in Switzerland and the Jura in the north.

Grenadiers volunteered and were then chosen to be responsible for transporting overland all the equipment and food for an assault on Mount Nanda Devi (25,645 feet) in the Indian Himalaya. This, the highest mountain in India, was first climbed in 1936 and for 16 years was to remain the highest peak in the world conquered by man. The Grenadiers involved were Captains RCM Wynn-Pope, the Hon J Forbes, Lt NP Sandford, Sgt Nesbitt, L/Sgt Fielding and Gdsm Laird. They travelled via Belgium, Austria, Greece, Yugoslavia, Turkey, Teheran, Lahore and Jessimath where they met the climbers. A roadhead was established near Lata (7,300 feet). Between May and June 1977 various peaks were climbed. Returning to Elizabeth Barracks, Pirbright, and Queen's Guard two days later, was rather an anti-climax.

There was certainly no anti-climax about Maj Heroys's most successful expedition to climb Mount Columbia, one of the highest peaks in Alberta in the Canadian Rockies. On 2 July 1978, due to his leadership, ten completely novice Grenadiers reached the summit. 'During the exercise all of us learnt to remain alert in spite of fatigue and fear; we also learnt the importance of self-reliance and team work,' he recalls. 'Above all, we had a thoroughly exhilarating time.'

Some Canadian Grenadier Guards participated in the expedition. One of the more frightening moments occurred when CSM Lukasic of that Regiment fell down a crevasse and out of sight. His rope held, but he could not be pulled out as his foot was stuck. L/Sgt Fielding quickly had himself lowered down to dig him out. On another occasion Gdsm Horner went up to his waist through a weak snow bridge, feeling free air beneath his feet, but was safely held on the rope.

38. The Colonel visits the 1st Battalion at Hounslow, 1985.

39. The Colonel presents a silver model of a Warrior in 1989 to Lieutenant Colonel E H Houstoun, commanding the 1st Battalion, to commemorate the entry of Warrior into service in the British Army.

40. The Colonel at a reception for long-serving Branch Secretaries of the Association with their wives, at Buckingham Palace 19th November 1986, together with the President of the Association and the Lieutenant Colonel.

41. The Garden Party at Buckingham Palace in 1988 to celebrate the seventieth anniversary of the Grenadier Guards Regimental Association. Colonel N Hales Pakenham Mahon accompanies The Queen, followed by Major General C M F Deakin, Lieutenant Colonel A Heroys and the Colonel.

There were opportunities for adventure training for battalions stationed in Germany when time permitted. Before the 2nd Battalion's vehicles had even returned to Munster from Exercise Crusader in 1981, Headquarter Company's expedition had departed for Norway. Capt DP Ratcliffe and CSgt D Hardman arranged climbing and canoeing in the Harz Mountains while The Inkerman Company went canoeing down the Danube.

No 1 Company had two weeks in Bavaria; the signals platoon went to Cyprus; Lt A St JP Hamilton took some of No 2 Company on a cycling ride round Holland while others toured Morocco in Land Rovers with Lt OP Bartrum.

The Household Division battalion stationed in Munster had its own ski hut called Alpen Gasthof at Grasgehren, about 15 km south-west of Sonthofen. The skiing in the area was enjoyed by many units and was called Exercise Snow Queen. The hut was first used by the 2nd Battalion in 1981. At 1500m it was one of the highest and was ideal for both novice and intermediate skiers, with three ski lifts nearby. Most units used farm houses or barns and fed on compo, whereas the Grenadiers lived in a small hotel whose staff provided three full meals a day. During the season over 200 Guardsmen learnt to ski. Some won the Army Downhill Bronze Medal and everyone benefited from learning a new sport.

Sometimes adventure training expeditions were linked to visits to Grenadier Second World War battlefields. Also in 1981, Capt ET Bolitho set out from Berlin with the eccentric desire to be the first to drive across 300 miles of totally uninhabited area of sand dunes and rocky desert known as the Grand Erg Occidental in Central Algeria. Accompanied by nine Grenadiers, he arrived in Tunis in January, just missing the ambassador, Sir John Lambert who had served in the Regiment.

'For four days we travelled up to 200 miles away from the nearest habitation, navigating by day using the sun compasses, and by night checking our position by astral navigation, using the stars, a theodolite and immensely complex charts understood by our navigational guru, Tim Breitmeyer,' noted Capt Bolitho. 'In an area where the annual rainfall is negligible, we endured a 16-hour rainstorm, causing L/Sgt Goodenough to dash off to the nearest hotel dressed in a huge polythene bag! We later got arrested three times for espionage. L/Cpl BW Lawson in charge of our Clansman 320 radios talked daily in morse and voice with CSgt Franklin in Berlin.'

The expedition successfully accomplished, the group stopped at Medjez el Bab to study the 5th Battalion's battle for Grenadier Hill. The signs of battle were still visible. They even found an unexploded shell on the summit. They camped for the night in the shadow of the Bou, as did the

5th Battalion when they hung on so grimly for ten days under continual fire in 1943.

On return to Europe they visited Monte Cassino and listened to a peaceful evensong in the Abbey there. Post-exercise euphoria was short-lived for within eight hours of returning to Berlin everyone deployed on an alert exercise. The expedition team also included CQMS Ridley, Sgt Gormany, L/Cpl Watson, Gdsm Wallis and Dmr Westwood.

Capt Bolitho took another twenty Guardsmen to Morocco in 1985. 'After the blizzards of the Atlas Mountains they found the sun and sand of the Sahara and the seas of the Atlantic. More potential "Desert Rats" were trained,' noted *The Guards Magazine*. 'A new and exhilarating sport was attempted, that of dune canoeing, and many received a good bruising canoeing in the surf of Agadir.'

In 1986 Capt JPJ Garratt was selected for the Joint Services Everest Expedition. Sadly the weather prevented them reaching the summit.

Three years later Lt GV Inglis-Jones set out to swim the English Channel: 'The weather was perfect,' he wrote. 'Within a few minutes of starting at 8.0 am I was attacked by fish which nibbled my leg. I looked forward to each hour's feeding time but at the second stop I was badly sick. After about 12 hours I could see Cap Griz Nez quite clearly and I knew I had three miles to go. Successive hours reduced this to one mile but I was then told that the tide had swept me beyond the Cap which meant another three hours to go: morale dropped. I had not anticipated swimming in darkness.' Eventually he waded to the shore having swum about 24 miles altogether in 16 hours, raising £2,000 for charity – a tremendous achievement.

Even these brave endeavours were eclipsed by the 1994–1995 Roof of the Americas Expedition in which Grenadiers played a full part. Five teams of Household Division Servicemen ventured through Alaska to the most northern and western points in the Americas, down through the entire land mass of North and Central America, climbing Mount McKinley (20,320 feet), and kayaking in the white waters of the Colorado River in the Grand Canyon on their way. The jungles of Guyana offered the Phase 4 Team a challenge of finding the source of the Mazaruni River. The next team climbed Mount Aconagua (22,850 feet), the highest mountain in the Americas. The expedition finished at Cape Horn, Lt RAJ Phasey BEM having played a particularly notable part. A dozen Grenadiers participated, including Capt SW Gammell, Lt JGN Geddes, L/Sgts W Scully and AK Bissett.

Regimental Headquarters is to be thanked for helping to subsidize the cost of Grenadier participation in almost all the more recent adventurous training described in this chapter.

Moreover, funds for such activities were specifically donated in 1979 and 1987 in memory of Capt Robert Nairac and Lt Christopher Fagan respectively. Robert's murder by the IRA is recorded in Chapter 23.

Christopher died in a motor accident aged 23. A financial award is made annually to young Grenadiers who have made an outstanding contribution in the field of adventure training or sport. In 1992, for example, Christopher's father, who served in the Regiment in the 1960s, presented the award to Sgt Westwood for sub aqua diving and L/Cpl K Wilkinson for bobsleighing. The income from Robert Nairac's fund is used to subsidize the personal expenses of Grenadiers who participate in worthy adventurous training exercises. Robert and Christopher's zest for life and love of adventurous and outdoor training have been shared by many Grenadiers over the generations, as this chapter indicates.

CHAPTER 23

NORTHERN IRELAND: THE MIDDLE YEARS

1976 – 1988

1ST BATTALION ARMAGH OCTOBER 1976 – JANUARY 1977

On 26 October 1976, a cold, foggy day, the 1st Battalion, commanded by Lt Col DHC Gordon Lennox, took over responsibility for the rural area of North Armagh. Battalion Headquarters, The Queen's Company (Maj AA Denison-Smith MBE) and Support Company (Maj A Heroys) were in Armagh City while No 2 Company (Maj EH Houstoun) operated from Newtown Hamilton and No 3 Company (Maj JVEF O'Connell) from Middletown. All the Company Commanders were to command Grenadier Battalions. The map on page 171 refers.

Within minutes of taking over, an Ulster Defence Regiment Lieutenant under command was murdered in his shop in Armagh City. However, compared to their last tour in Northern Ireland (Chapter 17), the operational incidents were relatively few. Apart from a well-organized lunatic fringe of hard-core terrorists, most people in the area were quiet and simply wanted to be left alone. There was no contact with the local community for there was no middle ground. It seemed futile to persevere over hearts and minds for the Republicans regarded the Grenadiers as enemies.

The tour consisted largely of a continuous round of patrolling, guarding, observing, making arrests and finds, and dealing with bomb scares and hoaxes, all in the grip of an Irish winter.

On 28 December the Battalion took over responsibility for a much bigger area including Dungannon, previously controlled by the Royal Hussars. Lt Col Gordon Lennox had under command some 1,800 soldiers, including two battalions of the Ulster Defence Regiment and elements of the Black Watch and 3 Royal Tank Regiment. The Quartermaster, Capt

GR Whitehead, was involved in handing over or taking over eleven camps. The Battalion's frequently changing boundaries led to some confusion.

Few were sorry when the Battalion's tour was cut short by one month. It was good to return to the families on 26 January 1977 without a casualty.

CAPTAIN RL NAIRAC

During the emergency in Northern Ireland, some Grenadiers have served there with the Guards Parachute Company, the Special Air Service, in surveillance units and on the staff. In 1977 Capt RL Nairac was on his fourth tour in Ulster. One of them is described in Chapter 17. He had worked with the SAS on occasions, according to his Brigade Commander, Brig DM Woodford CBE, but was employed as a liaison officer at HQ 3 Infantry Brigade responsible for surveillance operations.

On the night of Saturday 14 May 1977 Capt Nairac was on an undercover operation in plain clothes carrying a concealed 9 mm pistol. He went to the Three Steps public house at Drumintee in South Armagh, three miles from the Border. At least seven men abducted him from outside the pub. There was a fierce struggle; bloodstains were found in his specially equipped car; the windows were smashed and the car dented. Capt Nairac was taken across the Border into the Republic where he was subjected to a succession of exceptionally savage assaults in an attempt to extract information which would have put other lives and future operations at serious risk. These efforts to break his will failed entirely. Weakened as he was in strength, though not in spirit, by the brutality, he yet made repeated and spirited attempts to escape, but on each occasion he was eventually overpowered by the weight of numbers against him. Capt Nairac knew that he was to be executed and asked for a priest. After several hours in the hands of his captors, he was callously murdered by a gunman of the Provisional IRA who had been summoned to the scene. His assassin subsequently said, 'He never told us anything.'

Cardinal Hume, Archbishop of Westminster and former Abbot of Ampleforth, where Capt Nairac went to school, made an appeal calling for his safe return before it was known that he was dead.

LP Townson aged 24, an unemployed joiner from Meigh, Co Armagh, was arrested on 28 May and convicted of the murder. He had taken policemen to a field where two guns were recovered.

Brig Woodford said at the trial that 'Robert Nairac was very well known by the Army, the RUC and the local people. He is one man, a very brave man, out of a number of brave men, who have been trying to deal with terrorism.' The self-styled 1st Battalion of the Provisional IRA in South Armagh admitted murdering him after recognizing him from a number of their photographs. In all six terrorists were convicted and jailed for his kidnapping and murder. Two of them received life sentences, one having been convicted in the Irish Republic.

Capt Nairac was posthumously awarded the George Cross. The citation read: 'His exceptional courage and acts of the greatest heroism in circumstances of extreme peril showed devotion to duty and personal courage second to none.' The award, equivalent to the Victoria Cross, filled all Grenadiers with the deepest sense of pride and gratitude. We salute his shining example. Tragically, his body has yet to be recovered. Even now the IRA will not say where it is hidden.

The BBC later made a television programme. Maj PR Holcroft, then serving in Northern Ireland, was cross-examined by the interviewer who wanted to adopt the IRA line that Capt Nairac had taken excessive risks. Maj Holcroft felt that this was not the case. On being asked repeatedly if Nairac had taken numerous risks, Maj Holcroft replied, 'Yes, but calculated ones only.' He was bitterly disappointed, on seeing the programme; his last four words had been cut out – an object lesson in dealing with the press.

LONDONDERRY 2ND BATTALION NOVEMBER 1977 – MARCH 1978

On 9 November 1977 the 2nd Battalion moved to Ulster under command of Lt Col DH B-H-Blundell. His aim was to minimize violence, to give a greater sense of security to the innocent, and to assist the RUC in upholding the law. Like all Commanding Officers, he was very conscious of the need to avoid risking his men's lives for so little reward. When he was last in Londonderry in 1974 there were more than three major units west of the River Foyle, whereas now the Battalion was responsible for the entire area.

No 1 Company (Maj CSS Lindsay) lived in a camp set up in the car park of the Masonic Hall in the city walls with the responsibility for the Bogside. No 2 Company (Maj JM Craster) was based on the Creggan Camp to look after the Creggan Estate and the Rosemount areas. Support and The Inkerman Companies, commanded by Maj the Hon Samuel Coleridge and Maj PR Holcroft respectively, were in Bridge Camp by Craigavon Bridge on the Foyle. They oversaw Brandywell with the large commitment for manning the vehicle and pedestrian checkpoints which enclosed the whole area of the walled city and city centre. Also under command was A Squadron of 2 RTR, a company of the 2nd Battalion Coldstream Guards and, from the New Year, C Squadron of the Life Guards commanded by Maj CN Haworth-Booth. The rest of the Coldstream Battalion was on a residential long tour east of the Foyle.

Maj CJE Seymour was the Community Relations Officer. Each Sunday twenty-five local children visited Creggan Camp to play football, use the trampoline and enjoy an enormous tea – all very different from three years before.

There were a number of incidents. They included incendiary devices, the stoning of a mobile patrol, claymore mines, numerous armed robberies and several shooting attempts. The Battalion was very busy, but, operationally, the tour was comparatively quiet. 'Part of the reason that life was quieter,' wrote the Commanding Officer, 'was because we had learnt from our mistakes in the Creggan on the 1974 tour. We did not go out of our way to provoke. By 1977 we were much more professional and in some respects this was a much more important tour.'

The last six weeks were more lively: more weapons were found and arrests made. The Battalion returned to Chelsea Barracks on 13 February 1978 without a casualty, by now an important aim, and with a feeling that much had been accomplished.

1ST BATTALION SOUTH ARMAGH
NOVEMBER 1978 – MARCH 1979

The tour of the 1st Battalion in South Armagh from 15 November 1978 promised to be very different. Seventy soldiers had been murdered there over the preceding ten years.

South Armagh is a 200-square-mile pocket of Northern Ireland protruding down into the Irish Republic. It is mainly open, rolling country with small farms, and solidly Catholic. Minor roads cross the border at forty-seven different places along its 30-mile length. Six IRA had been killed in the area and many convicted of serious crimes. Their hold over the local population was considered to be diminishing but isolated communities did not dare to resist them actively.

The harsh realities of the situation were made clear to the advance party prior to the take-over from 42 Commando, Royal Marines. A mixed patrol of Grenadiers and Marines was subject to a bomb attack in Crossmaglen. Marine G Wheddon was fatally wounded. L/Sgt K Regan and L/Cpl K Kinton of No 2 Company were both slightly hurt. They had been wearing Marine berets to prevent the IRA knowing of the imminent change-over.

The Commanding Officer, Lt Col MF Hobbs MBE, put his Headquarters with The Queen's Company (Maj DM Braddell) in half of a flax mill at Bessbrook. To the south-west was Crossmaglen where No 2 Company (Maj AJC Woodrow QGM) lived in rough and muddy conditions. The Royal Engineers were constructing mortar-proof bunkers for their accommodation, adjacent to the RUC station. No 3 Company (Maj JS Scott-Clarke) lived in rather less unpleasant quarters to the east, close to the Dublin/Belfast railway. No 4 Company (Maj TJ Tedder), consisting of the Anti-Tank Platoon and Drums, commanded by Lt NP Sandford and D/Maj T Dove BEM respectively, were at Newtown Hamilton to the north. A company of the 2nd Battalion the Light Infantry looked after Newry. (See map on page 171).

The Battalion relied entirely upon air support from RN, RAF and AAC light helicopters. The helicopter landing site at Bessbrook was reported to be the busiest heliport in Western Europe, handling well over 100 movements a day. Lt DT Ashworth and his able assistants, CSMs J Morris and TA Rolfe, often controlled the activities of eight aircraft at once. The pilots gave the most excellent service, often flying to the very limits of weather and machines.

The Battalion's Senior Major was N Boggis-Rolfe, with other key appointments being Maj the Hon Philip Sidney MBE (HQ Company Commander), Captains GF Lesinski (Adjutant), DCL De Burgh Milne (Signals), REH Aubrey-Fletcher (Close Observation Platoon), RSM P Richardson and D/Sgt D Rossi.

Crossmaglen and Forkill were so close to the Border that it was easy for terrorists to execute a well-planned attack and return to the south within minutes, although the effectiveness of the Garda (the Irish Police) to impede them in the Republic was thought to be improving.

The Battalion was extremely busy throughout the tour. On 20 November a train en route from Dublin to Belfast was hijacked. It was left at Newtown Bridge where eight more terrorists loaded the locomotive with 100 lbs of explosives. The area was cordoned by The Queen's Company and the device eventually removed. The whole operation took five days in bitterly cold conditions; the threat of booby traps was constant.

No 2 Company was then involved in the search for and recovery of a Scout helicopter which crashed into Lough Ross, west of Crossmaglen, killing the pilot, Capt Stirling, and navigator, Cpl Adcock.

On 6 December an explosion damaged a road near Newtown Hamilton just missing an approaching civilian lorry. The RUC started placing barriers across the road when a second explosion knocked them off their feet.

After another incident, ten days later, a headless body was found near Forkill. On 18 December a patrol came under high-velocity rifle fire. Three men were seen to be running across the fields carrying firearms. In the nick of time the Section Commander prevented his men from firing; the terrorists from concealed positions had 'flushed' three sportsmen who were shooting game. The terrorists beyond them escaped.

Sgt RG Garmory was serving with No 2 Company at Crossmaglen. His recollections are worth quoting in full: 'Crossmaglen was rather like the Somme that December with duckboards and mud everywhere. We were all living on top of each other; morale was very good – the strange thing with Guardsmen is that the worse the conditions, the better they do. No Army vehicle was about – all movement was on foot or by helicopter. The food was adequate with vast egg sandwiches at all hours. The RUC policemen were very good. They didn't come on to the streets unless they were well protected, and then only to serve a summons for example. Every time we stepped out of the camp we knew we were an available target to an IRA sniper. It was real soldiering and so I enjoyed the danger as did most of the others. All the locals were unfriendly: nothing could be done to break down their hostility.

'The four-month tour consisted of three-day periods. In the village of Crossmaglen itself we moved to the surrounding countryside where the patrols varied from several hours up to fourteen. And finally we went on camp guard when it was largely two hours on duty followed by four off.

'We usually patrolled in Crossmaglen with about twelve men, divided into three groups of four. We were always in touch by radio with one group who may be in the north, the other perhaps in the centre near the village square, with the third in the south – all moving fast, ready to support each other in a crisis. We were patrolling in such a manner on 21 December. We heard a shot. It came from the camp, so my "brick" ran back towards the camp, but we learnt on the radio that it was an accidental discharge. I decided to lead my "brick" into Newry Road which led away from the border from the square. The road had a gradual bend to the right before straightening out. We usually didn't patrol much beyond the church because we would become too vulnerable. I was leading the four-man "brick" in a staggered file formation with Duggan carrying the LMG behind me, Johnson across the road and Ling to his rear. Although it was market day there was scarcely anyone around.

'On coming round the bend near the Rio Bar I saw, forty yards away, what looked like a British Rail parcel delivery van parked partly on the pavement on the left facing away from us. It had an eighteen-inch tail board with a roll shutter that could be pulled down.

'The van immediately struck me as highly suspicious because I saw what looked like cardboard boxes piled to the top in the back of the van, all flush with the tailboard – so they would fall out if the van moved off fast. I instantaneously put my magnifying sight to my eye and saw four firing slits, two above another two, among the boxes. I immediately opened fire. A lot of things then happened at once. In a fraction of a second I was under very heavy fire from, I knew by experience after numerous tours in Ireland, three Armalites and a Kalashnikov. Two terrorists were probably firing standing, and two lying. I feel I killed the man detailed to take me out. Even so, bullets passed through the sleeves of my smock. I could see them hitting the pavement in front of me and also the wall just to my left. Having fired off my whole magazine I whipped back behind a wall, gave a quick contact report and saw my three Grenadiers had been hit. I ran back to Duggan who was still conscious although an Armalite round had penetrated his liver. I told him to hang on. I was still under fire and the van was still stationary which made me think I'd hit the driver too. I fired off Duggan's LMG's magazine from the hip in one quick burst which quietened them down a bit. I could see the van's rear must have armoured plating behind the "boxes" because every third of my rounds was tracer and I could see them ricochetting upwards off the plating. When the magazine was empty I ran across the road to Johnson, moved his body off his rifle and fired more shots at the van as it careered off, partly hitting a telegraph pole just in front of it.

'Just then one of the other patrol's "bricks" doubled up from behind me. I detailed them to each casualty. The Ferret scout car with its Browning machine gun roared up. It should have been in the village square as a back-up but had gone back to camp to refuel. Had it been in the right place I'd have chased the terrorists anywhere. I sent it up the road to give cover. Maj Woodrow was quickly on the scene and a doctor and nurse from the adjacent health centre were doing all they could for the dying.

'I went with Maj Woodrow and commandeered a civilian wagon to carry the casualties back. Nobody would give me the wagon's key so I knocked the handle off the vehicle's door with my rifle butt, after which they produced it. A Wessex helicopter very quickly arrived with our own Doctor, but too late to save the three lad's lives, two probably being killed instantaneously.

'It was very strange later having three new men in my brick. Although my nine years' service expired shortly afterwards, I signed on for the duration of the tour so I could remain in command rather than a young L/Sgt. Looking back on it, as I've done so often since, the terrorists were highly professional. Their supporters' signalling presumably indicated to them that our armoured car was being refuelled. They had been waiting for us, as before perhaps. They'd expected us to turn our backs on the van, anticipating our returning to the square. If there were IRA casualties then, I think they'd have been killed. I heard later that the van had been taken to the south and sunk in a bog. The Irish may have been marginally more sympathetic over the following week, particularly as it was Christmas, but they daren't show it: they'd have been sorted out by the IRA if they had. Months later I heard I'd got the Military Medal. The Queen presented it to me at Buckingham Palace in front of my mother, father and fiancée. It was all very grand.'

Gdsm K Johnson, G Duggan and G Ling were buried with full military honours in their home towns of Blackpool, Lymm and Saxmundham respectively. The 2nd Battalion from Chelsea Barracks provided the bearer and firing parties.

In early January heavy snow and bitter cold made the rugged South Armagh countryside even more difficult to patrol. The Quartermaster, Capt GR Whitehead, succeeded in obtaining the only snowsuits in the whole of the Army.

Two electrical pylons were blown up near Crossmaglen. No 2 Company discovered an unexploded device on a third pylon which provided valuable information on its composition and origin.

Guards Battalions serving overseas are visited so frequently by the Lieutenant Colonel Commanding the Regiment that such visits are seldom mentioned in this History. However, reference should be made to Col DV Fanshawe's visit. He stayed six days and covered more area on foot than most did in a month, for he spent 24 hours in each Company location, seeing everything and giving everyone an opportunity to chat to him.

During his visit there was a major operation to deliver 2,000 tons of engineer stores into Forkhill. All normal movement was by helicopter because of the effectiveness of terrorist bombs. However twenty-one convoys moved the stores by road after the Battalion was reinforced by companies of the 2nd Battalions of the Light Infantry and the Royal Anglian Regiment, thirteen Royal Engineer search teams, eight more search dogs and six more helicopters. An equally successful operation moved building stores and equipment into Crossmaglen when a company of Green Howards was under command.

South Armagh, nicknamed 'Bandit Country', lived up to its reputation. The danger was constant. A Queen's Company patrol of No 2 Platoon (Lt JG Heywood), led by C/Sgt S Allen, found some wires. A full-scale clearing operation resulted in the successful disarming of a terrorist bomb of 200 lbs in two milk churns. The patrol had originally come across the command wires leading from the bomb to a firing point some 400 metres away on high ground where batteries and an initiating switch were positioned.

Eight months later at Warrenpoint, some eight miles to south-east, the IRA placed 500 lbs of explosives in milk churns which were detonated in the Republic by remote control as men of the 2nd Battalion Parachute Regiment drove past. A further 1000 lb device nearby exploded after the

Queen's Own Highlanders had arrived on the scene. Eighteen soldiers were killed including the Highlanders' Commanding Officer, whose remains were never found. Lord Mountbatten had been killed on holiday in Southern Ireland the same day. If a bomb went off, there was instant concern on where a secondary or even a third bomb may be located. One bomb in South Armagh had five further devices to catch following-up soldiers. Remotely-detonated bombs caused nearly 30% of the security forces' deaths in 1979.

In the case of The Queen's Company's patrol in January 1979, referred to above, the IRA were probably planning on killing an RUC vehicle patrol until the Grenadiers disturbed them. Soldiers were very vigilant. A sudden flight of pigeons alerted one Grenadier to imminent danger. A large bomb was found in a culvert nearby.

On 3 February the deceptive calm of Crossmaglen was shattered by a fire bomb in the back of a car. It was detonated by remote control as Gdsm W Lundy was passing. He was enveloped in a sheet of flame, but, showing amazing presence of mind, he ran, still burning, into a house where he jumped into a bath while Gdsm R Davis turned on the cold tap, saving his life. He gradually recovered at Woolwich from the 33% burns. The Battalion had earlier that day reported a suspicious car. The Garda subsequently saw it and recovered a cocked Armalite, ammunition and bomb-making equipment.

A major attack on Crossmaglen was thwarted on 22 February. Maj Woodrow, accompanied by C/Sgt D Wilmot in an Army Air Corps helicopter, saw two vehicles parked just north of the border near Coolderry. Maj Woodrow told the pilot to drop him and Wilmot off so they could investigate. The pilot did so, regained height and saw, to his amazement, a lorry and van further south with what appeared to be twenty armed men approaching the two Grenadiers. In the back of the lorry there was a construction covered by a tarpaulin; it was to all appearances the classic IRA mortar vehicle. The pilot picked up Maj Woodrow and C/Sgt Wilmot who, despite the odds against them, landed beside the road in an attempt to stop the lorry. They opened fire, hitting it, but the lorry and terrorists fled across the Border, abandoning the van which contained disguises and the welding equipment used to set up the mortars. Lt Gen Sir Timothy Creasey, the GOC Northern Ireland, turned up, as reinforcements flew in from Crossmaglen, to get a first-hand impression of the incident. (Most senior officers visited Crossmaglen since it is the most notorious spot in the Province.)

During the afternoon on 3 March there were reports of vehicle hijackings near Cullaville, south of Crossmaglen. Maj Woodrow had a premonition that another mortar incident was a possibility. He flew in a

Scout helicopter and saw a tipper lorry and another vehicle south of Glassdrumman. The helicopter made a low pass to investigate and came under heavy automatic fire. Over 100 empty cases were later found in three camouflaged fire positions. The aircraft was hit. Maj Woodrow received gunshot wounds to the right leg. The pilot temporarily lost control of his aircraft when his face was cut by flying perspex, but he flew it back to Crossmaglen. This almost unique incident brought the Battalion's tour to a close. It had been a momentous four months. Had it not been for the vigilance and professionalism of all ranks – for there were other shooting incidents not recorded here – it is inevitable that casualties would have been greater.

Looking back on the tour, the Commanding Officer noted that 'in South Armagh, more than anywhere else, the task of the Battalion was largely reactive. The proximity of the border, the complete hostility of the population and the professionalism of the terrorists meant that there was little intelligence available to regular soldiers on the ground.'

On 7 March 1979 the Battalion handed over to 3rd Battalion the Queen's Regiment and returned to Pirbright for a little extra leave because Lt Col Hobbs had decided against the usual Rest and Recuperation short break during the tour. Everyone disappeared to the far corners of the world: officers and men were reported to be in India, Australia, America, Canada, Europe and Malaya. However exotic the leave, South Armagh remained in peoples' thoughts.

Nobody will forget the incessant clatter of helicopters, the bitter cold, the long hours on radio watch or the impenetrable blackthorn hedges. They will remember the very exceptional skills of the pilots of the Royal Navy, Royal Air Force and Army Air Corps, the Royal Engineer Search Teams and the Ammunition Technical Officers who dealt with suspect bombs. Without them the difficult job for the Grenadiers would have been impossible.

2ND BATTALION FERMANAGH
MARCH – JULY 1980

The tour of the 2nd Battalion in Fermanagh March to July 1980 was a strange contrast. At one time the Battalion was housed in ten separate locations in three different counties, working with four different police divisions, temporarily under command of four different Commanding Officers and two Brigadiers.

Lt Col HML Smith found he had little to command other than the 4th Battalion Ulster Defence Regiment which was placed under command to mitigate against the loss of Grenadiers to other areas. His Battalion Headquarters was based on a disused airfield in the marshes of Fermanagh, two miles north of Enniskillen. Platoons of No 1 Company (Maj HA Baillie) redeployed as necessary as did those of The Inkerman Company (Maj HMP de Lisle). No 2 Company (Maj EJ Webb-Carter) was based on Fort George in Londonderry.

The Corps of Drums, commanded by D/Maj J Evans, was based on Rosemount Police station in Londonderry looking after the once infamous Creggan Estate. Six years before an entire Grenadier Battalion had been responsible for the area. Three years before a Company was required; and now, in March 1980, a platoon sufficed. This indicates how the situation has improved over the years.

On 9 May 1974 the Most Reverend Dr Daly had told Grenadiers in the Creggan, which he knew better than anyone, that the then prevalent tactics of continuous presence and indiscriminate operations were highly unpopular, counter-productive and, combined with the IRA pressures, had reduced cooperation to zero. He said that a scaling-down and modification of general Army operations would win the confidence of the people and make his own appeals for peace more credible. 'Popular influence will be mobilized that will eliminate the gunmen,' he said. The reduction in violence in Londonderry since 1974 is due to a number of complex factors; but clearly Dr Daly had a point worth recording here.

The best tribute to those Grenadiers who served in Fermanagh and elsewhere for four months in 1980 was the fact that only three major incidents occurred, whereas normally the level of violence was higher. It was hoped that the closure of various border crossing points during the tour would be a lasting hindrance to the terrorists. The Battalion returned to Germany on 15 May 1980 after their quietest tour on record.

CAPTAIN HR WESTMACOTT

Capt HR Westmacott had been posted from the Grenadiers to be employed as a Troop Commander with the SAS since February 1980. In April he personally led an assault which captured a senior terrorist who was responsible for up to twenty murders. Forensic evidence subsequently proved that the terrorist's Armalite rifle had been used in six murders and eight attempted murders.

On 2 May 1980 Capt Westmacott and his troop were tasked for an operation against an IRA gun team, believed to be planning a shooting attack using an M60 machine gun and a number of assorted weapons. At 1130 am the firing position was located in one of two houses in the Antrim Road district of North Belfast. Speed was of the essence. Capt Westmacott's plan was to assault the front of the building while placing a blocking position in the rear.

The speed, aggression and courage of Capt Westmacott and his troop in their assault so stunned the four-man gun team that, within a few minutes, they surrendered. Unfortunately he was killed, while coolly returning the fire of the gun team to give cover to the other members of his assault group.

This particular gun team had been responsible for a large number of shooting attacks, many of which resulted in death or injury to the Army and Police. Previously they had always eluded capture after their ambushes, and evidence with which to convict them was not obtainable. The loss of the gun team and its weapons had a profound effect on IRA morale in Belfast.

'The outstanding leadership and courage shown by Capt Westmacott will have results far beyond those achieved in the action itself,' reads the citation on him. 'There is no doubt at all that it will save the lives of many other members of the Security Forces. Capt Westmacott, in paying the supreme sacrifice, displayed selfless disregard for his own safety. His conduct represents the highest possible devotion to duty in the best tradition of his Regiment and the British Army. He deserves very special recognition for his valiant actions.'

Richard Westmacott received the first posthumous Military Cross ever to be awarded. A Commandant at the Royal Military Academy, Sandhurst had earlier decided that portraits would be painted and hung at Sandhurst of the first two RMAS University graduates killed in action. As fate would have it, they were both Grenadiers – Captains Robert Nairac GC and Richard Westmacott MC. Richard's young German wife, Victoria, named their newly-born daughter Honor. As in war, the bravest sometimes die young.

The terrorists involved in the above incident escaped from the Crumlin Road jail five weeks after their capture. Among the convicted killers was one who fled to America, but was deported to Northern Ireland in 1992 after a nine-year legal battle.

✻   ✻   ✻

Due to the security forces' successes there was a significant drop in violence in 1980 throughout Northern Ireland. Eight regular soldiers were killed compared to thirty-eight in 1979 and 103 in 1972. The IRA had switched to the softer targets of UDR and RUC, murdering eighteen of them in 1980. 671 Loyalists and Republicans were charged with terrorist offences.

According to a book, *Big Boys Rules*, by Mark Urban, the security forces decided in the late 1970s not to confront the IRA at their arms caches.

Killing terrorists simply added to the number of martyrs in the Republican community. Urban states that miniature transmitters were instead planted inside weapons found in such dumps; the devices were then activated when the terrorists collected the weapon. He claims that this led to some spectacular successes, but by 1984 the IRA apparently discovered that their weapons were being tampered with.

In March 1980 the IRA sought wider publicity by attacking British soldiers in Europe. This aroused more hostility than support. In January 1981 twelve members of the IRA murdered Sir Norman Stronge, the 86-year-old former Speaker of the Stormont Assembly, and his son James, MP for Mid-Armagh, in their house. The latter had served in the Grenadiers in 1951/52.

The IRA encouragement of hunger-strikers at the Maze Prison led to more funds reaching them from America. On 1 March 1981 Bobby Sands began a new hunger-strike campaign. He was elected MP for Fermanagh and South Tyrone on 9 April and died on 5 May. Widespread rioting gripped Londonderry, Belfast and Dublin. Nine more on hunger strikes died over the following fourteen weeks.

Meanwhile the British Government continued to seek a political solution without which there will be no peace in Northern Ireland. Constitutional talks on power-sharing had ended a year earlier with inconclusive results, but Prime Minister Thatcher and Mr Charles Haughey, the Irish Prime Minister, agreed on close cooperation.

IRA attacks on the Irish Guards at Chelsea Barracks in October 1981 and on the Parachute Regiment at Aldershot murdered civilians. The IRA decided on avoiding armed soldiers altogether when they killed seven musicians in the band of the 1st Battalion Royal Green Jackets in Regent's Park. A bomb had exploded beneath the band stand. On the same day men of the Household Cavalry Mounted Regiment had been cut down by a car bomb as they rode through Hyde Park to ceremonial duties in Whitehall. Regiments returning from the victorious campaigns in the Falklands were the first to send their commiserations.

In May 1983 the British Government extended Direct Rule over Northern Ireland for a further 12 months. On 25 September there was a bitter set-back. Twenty-two terrorists successfully escaped from the Maze Prison. Among them were some of the most ruthless. One returned to South Fermanagh to bring death to the countryside. Another was later convicted of attempting to assassinate the Prime Minister with a bomb which killed five people at the Grand Hotel in Brighton in 1984.

1ST BATTALION SOUTH ARMAGH
OCTOBER 1983 – FEBRUARY 1984

After completing a two-month package of excellent training, the 1st Battalion, commanded by Lt Col A Heroys, moved to County Armagh in October 1983. Battalion Headquarters and No 3 Company (Maj JP Hargreaves) were at Drumadd Barracks in Armagh City with No 2 Company (Maj ET Hudson) at Bessbrook.

The Queen's Company (Maj EJ Webb-Carter) was at Crossmaglen. One of the platoon commanders there was Lt GK Bibby: 'The atmosphere in our base was claustrophobic and oppressive. Being surrounded by a 40-foot wall, one could only see the sky. Jeremy Wills was the Second Captain; Mark Fisher and David George were the other platoon commanders. We felt we had everyone in the Army supporting us, including up to ten helicopter lifts a day. They even delivered fresh milk. We had videos and a sauna: money was no object to ensure we performed to the best of our ability. We never had more than five hours in bed – often only three. We had to dominate the area, although Crossmaglen is small. We now had excellent, fairly light bullet-proof vests.

'On one patrol I saw a Ford Escort which was brand new and therefore suspicious. I knelt one foot away from it at 2 pm, checking it out on my radio. Within 20 seconds I was told it wasn't stolen and so my patrol – three half-sections each of four men – moved on. Nine days later, on 26 October, at exactly the same time, L/Cpl Taverner, of the Devon and Dorsets under our command, stopped in exactly the same place. There was an explosion. He was mortally injured. We put nine cartridges of morphine into him. Three civilians were hurt too. We still don't know how the bomb was detonated. We had a cordon round the area for three days while scientists looked for clues.'

The RUC in Crossmaglen were very gallant. Nine of them shared the base on two-year tours. Each patrol was usually accompanied by one of them.

Six years earlier a new policy had been introduced to re-establish the primacy of the RUC. By 1983 this had been largely accomplished. The Commanding Officer was called upon by the RUC to deploy Grenadiers anywhere within police division H. Support was given to all the RUC forward stations. The RUC was regarded as very dedicated, but not, of course, trained for war.

In December Prime Minister Margaret Thatcher visited the Battalion. 'She was quite wonderful,' recalls one Grenadier. 'She was charismatic, relaxed, warm-hearted, honest and down to earth.'

Another frequent visitor was Brig MF Hobbs CBE who now commanded

39 Brigade. In his typical, imaginative way, he ensured that soldiers were used where they were most effective. 'In a sense my approach was to try to avoid regular soldiers being tied to static defensive tasks – in essence to remove the targets from the IRA's sights. It could only work if the policy was a joint RUC/Army one,' wrote Brig Hobbs much later. 'The major change over the 1975–1985 period was the change to RUC primacy. What also changed was better intelligence which led to more covert operations; "better" terrorists led to fewer but more fatal incidents; RUC primacy led to a more passive (and frustrating) role for the Army, and the rise of the UDR led to the misuse of regular soldiers.'

To increase mobility a helicopter force was built up which, after a successful trial, ensured that No 3 Company was air mobile. They could be quickly deployed for several days. They acquired the nickname 'Martini Company' ('anytime, any place, anywhere . . . it's the right one').

There were dangerous moments. Sgt KM Gibbens with seven men were in two OPs in Scott Street for two days. Six hours after leaving, there was a massive explosion causing over £1m damage. They had expected to be shot at by the IRA but not bombed.

January 1984 was tense but fairly quiet, apart from two tragic RUC incidents. On the 10th Constable Fullerton was shot dead on his way home from Warrenpoint and on 31 January Sergeant Savage and Constable Ringham were blown up on a landmine on their way from Forkhill to Newry.

On 3 February the Battalion was visited by The Colonel. The visit gained more publicity than any of the terrorist incidents throughout the tour. Prince Philip met Grenadiers at Armagh and Bessbrook, but the weather prevented him going to Crossmaglen. His visit, as always, was a very cheerful and proud occasion and enabled the tour to finish on a successful note. Two weeks later the Battalion handed over to 1st Battalion the Staffords and returned to England.

1ST BATTALION ARMAGH
SEPTEMBER 1986 – JANUARY 1987

On 28 September 1986 the 1st Battalion started their eighth tour in Northern Ireland. Much to the frustration of the Commanding Officer (Lt Col EJ Webb-Carter) the rifle companies were deployed all over the Province. Indeed, it was a matter of conjecture who had the most Guardsmen – Lt Col Webb-Carter or Lt Col ET Hudson MBE, commanding 8 Ulster Defence Regiment, who was responsible for so many Grenadiers at various times that 8 UDR was nicknamed the 4th Battalion.

The Queen's Company (Maj GF Lesinski) was based at Ballykinler under 3 UDR, No 2 Company (Maj Sir Hervey Bruce Bt) at Middletown under 2 UDR, Support Company (Maj ARK Bagnall and then Maj CTG Bolton) under 1st Battalion Scots Guards, and No 3 Company (Maj MGA Drage and then Maj ARK Bagnall) under 6 UDR.

The RUC had identified an increased threat to the security of their police stations throughout the Province; this partly accounted for the wide deployment.

On 20 November The Colonel visited the Battalion. Although his visit was kept secret some of the Irish Army deployed close by across the Border.

There were similar incidents to those experienced by Grenadiers on earlier tours. However, that described by Gdsm SM Faithfull was exceptional: 'On 23 November I was on stag in a sanger at Middletown looking down the High Street when I heard a muffled explosion. For a split second I hesitated, but then I thumped the mortar alarm bell. A second later I saw three bombs tumbling down on the base. One hit our buildings with a deafening roar. Two bursts of automatic fire were directed at the base. Gdsm CP Sanders in the operations room below came up on the intercom and just kept saying "Don't panic". After the attack I said to L/Cpl PA Gabb, "I don't give a . . . about the orders, I am lighting a fag now." There was a pause before he replied, "Yes, so am I!"' Gdsm Faithfull's quick reaction gave people just sufficient time to take cover. Six out of eight mortar bombs had been fired from a vehicle 300 metres away, south of the Border. Three Grenadiers were slightly hurt.

Mortar attacks were highly dangerous which is why Maj Woodrow at Crossmaglen in 1979, and others, were so vigilant. In 1985 nine RUC police officers were killed when one bomb at Newry crashed through the roof of their canteen. Since then most bases have been protected by blast-deflecting walls and a reinforced roof.

Gdsm Faithfull's adventures were not over: 'Two weeks later, on a town patrol, there was a sharp crack of a high velocity shot. We hit the deck and cocked our rifles. I saw a puff of smoke. L/Cpl Gabb gave a fire order and we all fired a few rounds into the hedge to our front. It didn't seem to be anything more than an exercise; there was no time to be scared. Our training took over. The quick reaction force arrived in seconds. We wanted to go after the gunmen across the Border but permission was refused as it may have been a come-on.' It later transpired that the IRA had several well-prepared firing positions to kill anyone they hoped would follow up. Shortly afterwards morale was boosted by a visit from the Prime Minister.

The Battalion's tour ended uneventfully. Younger officers on returning to Germany were sad at losing the friendships that had been formed with the kind and hospitable Irish.

THE ULSTER DEFENCE REGIMENT 1969 – 1992

When the police B Specials were disbanded the Ulster Defence Regiment was formed in 1970 to guard key points, man vehicle check points and mount mobile patrols. By the mid-1980s their strength reached 6,000. 10% were female and nicknamed Greenfinches. Lt Col Michael Dewar's excellent book, *The British Army in Northern Ireland*, has a useful chapter on the UDR.

Over 40% of the UDR soldiers were regular cadre with an average age of 32. At one stage the UDR amounted to eleven Battalions, commanded by regular Army officers. 8 (County Tyrone) UDR was affiliated to the Household Division. Lieutenant Colonels RS Corkran OBE and ET Hudson MBE were among its distinguished commanders. Many other Grenadiers served in 8 UDR as Training Major, Quartermaster, RSM, or in other leading appointments. Within their tactical area of responsibility, as has been seen, they frequently assumed operational control of Regular Army companies. By the mid-1980s the UDR was responsible for providing military support to the RUC in some 80% of the Province. Being locally recruited, and largely Protestant, they were particularly hated by the IRA. 190 UDR soldiers were killed in the Regiment's first fifteen years.

The Grenadiers and their families who served with them were, of course, under equal threat. 'It wasn't easy adjusting to living in Northern Ireland or returning to school in England,' recalls one serviceman's daughter. 'When in Ireland I could never tell anyone my father was in the Army. And back at Tudor Hall School I amazed everyone by always looking under cars for bombs. Moreover, I would never throw anything such as letters away, because we had been taught in Ireland that they could only be shredded to preserve security.'

Those who have been privileged to visit 8 UDR would have been amazed by the amount of blue, red, blue. They loved their affiliation with the Regiment and Household Division. In 1992 the UDR was re-capbadged, forming the Royal Irish Regiment on amalgamation with the Royal Irish Rangers.

2ND BATTALION BALLYKELLY
JANUARY 1986 – MARCH 1988

Hitherto all the Grenadier tours in Northern Ireland had been of four months' duration. However, in January 1986 the 2nd Battalion was stationed for 2¼ years with their families at a deserted air base at Ballykelly on the southern shores of Lough Foyle in the north-west of the Province.

Apart from the Commanding Officer (Lt Col AJC Woodrow MC QGM) all but two officers changed over as the tour wore on. No 1 Company (Maj RG Cartwright, then REH Aubrey-Fletcher) were the first to be dispatched to the Border. No 2 Company (Maj AD Hutchison, then Lord Valentine Cecil) were responsible for the Ballykelly base with its 14-mile perimeter fence. Support Company (Maj EH Houstoun MBE, then OP Bartrum) departed to assist Lt Col Hudson's 8 UDR in the protection of RUC stations in the Dungannon area and The Inkerman Company (Maj CRJ Wiggin, then DCL De Burgh Milne) became the brigade reserve. The Senior Major was TTR Lort-Phillips, then JS Scott-Clarke, the Adjutant was Capt AJ Fraser, followed by GCR Gottlieb and EC Gordon Lennox, the Quartermaster was Capt DR Rossi MBE, then MB Holland and the RSM was DA Moore, then BMP Inglis.

The operational cycle, once established, seldom changed. Every five to six weeks the companies changed roles between the border tour, brigade reserve, barrack protection and leave/training. The Battalion also manned a permanent position at Auchnacloy.

When the weather is fine Ballykelly is a most beautiful place. The area is overlooked by a spectacular escarpment to the east, while the hills of Donegal lie across Lough Foyle in a view that is forever changing in colour, light and shade. The early months were marred by the tragic deaths of Gdsm B Hughes and P Macdonald in an accident when their landrover was hit by a train.

One highlight of the year was the weekend of 12 July. No 2 and Support Companies deployed to Portadown to keep the peace among the Loyalist marches. Due to the Anglo-Irish Agreement, the Loyalist factions became much more difficult. Grenadiers were totally neutral. This led to complaints from Protestants when soldiers were seen to be searching both sides.

To avoid repetition, these pages will briefly dwell on matters other than the various operational incidents which occurred.

The Battalion acquired a boat, a successor to the *Sir Allan Adair* of Hong Kong days. She was regularly chartered for fishing by Grenadiers and their families. Capt Rossi and Lt CE Kitchen (the Transport Officer) soon had the Messes' deep freezes full of cod, mackerel, flat fish and plaice. Some Grenadiers and two wives completed free-fall parachuting courses, while CQMS A Bradley and two other Grenadiers attended gliding courses.

On the sporting side, the Battalion won the UKLF Skiing Championships, the Northern Ireland Athletics and Basketball. They were runners-up in the Cricket league, and reached the Army finals in the Rugby 7-a-side. It is remarkable how the exceptional sporting ability of the 2nd Battalion has endured over the generations. Capt Gordon Lennox played for the Combined Services at cricket and rugby, and Gdsm Armstrong represented Great Britain in the European Championships in the Four-Man Bobsleigh, carrying on a Grenadier tradition.

The turmoil and tensions of the troubles in Ireland tended to pass by the Grenadier families in Ballykelly. The Wives' Club was very active. The pinnacle of their achievements in 1986 was arranging a families' 'Grenadier Day'. The families' quality of life in Ballykelly was not as bad as many had feared thanks to the hard work of the families office led by Lt DW Ling (followed by Lt Kitchen) and WO2 D Bradley BEM. They were helped by wives running the thriving Youth Club organized by Mrs Konarski, the crèche with Mrs O'Keefe and the horses with Mary Wiggin. It was not easy for the wives to get used to security restrictions but they remained cheerful. There were cinemas, restaurants, beaches and golf courses where they could go with their unarmed husbands. Most people probably enjoyed the tour.

'When our husbands were in Northern Ireland,' remembers Mrs Sue Daniels, 'we felt they were serving their country, keeping the peace and, hopefully, keeping their heads down too! I felt they should have been paid more as they were on call 24 hours a day. The wives became very close in Ballykelly: there was always another wife one could turn to. Most worried about their husbands' safety. There was a lot of separation. Husbands couldn't always get leave, for example when a first baby was due, and so relatives were important. Most Irish were friendly, but we were very wary and kept our distance.'

Unfortunately it was a dreadful winter with freezing conditions and over 100 burst pipes. Fourteen families had to be evacuated. However, it was still a merry Christmas with a visit by the Regimental Band and many parties.

During April and June UDR Battalions often trained in England and so the Grenadiers took over their responsibilities in Ulster.

Ten Grenadier wives at Ballykelly joined the UDR, becoming Greenfinches. They learnt drill, signals, map-reading, first aid and field-craft before being committed both on the ground and in operations rooms. 'The UDR has been the most demanding job both physically and mentally that I ever had, but certainly the most enjoyable,' wrote Mrs Jean Lannon.

The Mortar platoon and Milan anti-tank platoons trained in England, as did L/Cpl M Boulton of the Battalion's dog section which won the UK Championships. The Battalion also found most of the staff for a youth camp at Magilligan Point.

The conversion to the excellent new SA 80 rifle, to replace the old and trusted self-loading rifle, was a notable event. The long tour in Ireland enabled many to attend worthwhile driving, signalling and external courses. In addition, all the platoons went away on adventure training in the UK, France or Germany.

The need for vigilance at Ballykelly was constant. Five years earlier a bomb at a bar there had killed eleven soldiers of the Cheshires, injuring sixty.

However, all continued to go well for the Grenadiers. In March 1988 they returned to Caterham after a most successful tour. The Commanding Officer, Lt Col AJC Woodrow OBE MC QGM, said his farewells to the whole Battalion in the cinema. There were not many dry eyes by the end of his speech. Everyone was greatly moved to see him leave after over two difficult years in command in Northern Ireland during which the Battalion received no casualty from terrorist action.

CHAPTER 24

MORE BLUE BERETS IN CYPRUS

1982 – 1983

In March 1982 the 2nd Battalion Grenadier Guards returned from leave to Chelsea Barracks with the prospects of a quiet summer with much adventurous training planned. The Falklands War changed all that. The Battalion was overstrength; two-thirds had undergone prolonged battle-group training in Canada from Germany; the anti-tank and mortar platoons were at high pitch. As mentioned in Chapter 27, it is therefore both remarkable and disappointing that other Guards Battalions were committed to the Falklands on the grounds that they had been in London longer.

By July the 2nd Battalion was the only Battalion in London. They mounted twenty-seven Queen's Guards, twenty-three of which were consecutive, while No 2 Company provided the administrative backing for the Royal Tournament.

Quite regardless of this heavy strain of Public Duties, it was a difficult time to be in London: the IRA had attacked the Household Cavalry Mounted Regiment and Royal Green Jackets Band in the Royal Parks, tragic incidents which heightened security in London. A disturbed man had climbed into Buckingham Palace through a window left open by a Palace official. He sat on The Queen's bed; the police did not respond despite The Queen pressing her alarm bell. Furthermore, an ambulance and railway strike added to the commitments. Grenadiers had to construct car parks in Hyde Park.

Lt Col AA Denison-Smith MBE, commanding the Battalion, maintained morale by talking to all ranks frequently and by inviting senior officers to visit. The Army Commander was among those who came.

The prospects of the Scots Guards returning to Chelsea Barracks after so distinguishing themselves in the victorious Falklands War was unlikely to make some things easier.

It was, therefore, thrilling news when Grenadiers were warned for a six-month tour with the United Nations in Cyprus, starting in November.

*   *   *

The 1st Battalion had left Cyprus after filling a similar role seventeen years before (Chapter 13). Since then the situation had become more serious. A lasting peace had become further away than ever. This was because the Turkish Army had invaded northern Cyprus in 1974, due, they said, to the sufferings inflicted on the Turkish Cypriots by the Greek Cypriots. Thousands of Greek Cypriot families had fled from their houses in the north. The Turks sliced the island in two, shattering the Island's economy, convulsing both Greek and Turkish Cypriot communities and creating two Cypruses, north and south.

By 1982 the Republic of Cyprus meant the Greek south. The north lay in a zone controlled by the Turkish military, whose Unilateral Declaration of Independence in 1983 was recognized only by North Korea and the Turkish mother state. A buffer zone had been established between the two communities to be patrolled by United Nations troops, cutting across Cyprus like a scar. The zone ran through Nicosia, leaving it the only divided capital in the world, apart from Berlin and Beirut, both of which have since been reunified.

The aim of the United Nations in Cyprus continued to be, as previously, to pacify, observe, report, restore confidence and act as a quick-reaction force to prevent incidents developing into more bloodshed.

Lt Col Denison-Smith deployed No 1 Company (Maj AJC Woodrow MC QGM) and No 2 Company (Maj RCM Wynn-Pope) to the Buffer Zone to the west of Nicosia, to keep the peace between the Turkish Army and Greek Cypriot National Guard. Battalion Headquarters was at St David's Camp near Nicosia. The Quartermaster, Capt DJ Webster, said that, beyond a minefield, was the site of the old Tunisia Camp where the 3rd Battalion was stationed on internal security duties between 1956 and 1959 (Chapter 8.)

A second Headquarters, commanded by Maj CJE Seymour with the acting rank of Lt Col, was responsible for the security of the Eastern Sovereign Base Area based on Alexander Barracks, Dhekelia. With him were No 3 Company (formerly Support Company), commanded by Maj Sir Hervey Bruce Bt and The Inkerman Company (Maj DP Ratcliffe).

Thus the two halves of the Battalion were separated by a two-hour drive along the then rather inadequate Cypriot roads. The Adjutant was Capt

ET Bolitho, the RSM MJ Joyce and the two Drill Sergeants AA Ferneyhough MBE and J Morris.

Six platoons deployed along the Buffer Zone. They were based on observation posts spaced about four kilometres apart along the ceasefire line. Each post differed enormously. Some were isolated in orange groves, some in villages and some in disused factories. Some were in primitive windswept huts, while others were in luxurious houses with hammocks slung across garden trees. Great effort went into improving the general standard of the posts.

In some of No 1 Company's OPs inexplicable shooting could occasionally be heard at night. This was put down to the Turkish Army disciplinary problems. On one occasion fields caught fire. Four Guardsmen went to extinguish the flames. It was afterwards realized that the area was heavily mined. Sgt CK Cox, then a platoon sergeant, regarded his independent command in Cyprus as the best six months of his life, partly due to the job satisfaction achieved by such a role. 2Lt RHG Mills commanded a platoon on the extreme left, south-east of Morphou. His men saw the Turks throw one of their soldiers off a high water tower OP for allegedly being asleep. 2Lt Mills responded by staging a mock execution. After a theatrical inspection, he pretended to stab with his sword the Grenadier "miscreant" in full view of the Turks. Their reaction was never discovered. At night one could sometimes hear the Turkish sentries calling to each other to check that they were awake.

Gdsm JA Hinds manned a No 2 Company two-man OP near Nicosia Airport, with the platoon base a mile away. 'We had to be awake throughout our twelve-hour stag,' he recalls. 'We did a five-day cycle with adventure training and rest thrown in. Lots of interesting things happened. Very friendly Turkish soldiers came to the OP to borrow tea, sugar and the electric kettle which they put on a fire as they couldn't plug it in. At 3am our OP caught fire due to a paraffin heater. Two rifles melted and ammunition went off. We had to sit in a tin hut thereafter. It snowed later.'

DL Cox, the CQMS in No 2 Company and the future Academy Sergeant Major at Sandhurst, spent most days delivering food and supplies to the far-flung positions, occasionally giving cigarettes to the Greek and Turk sentries. He recalls one most unusual competition thought up by 2Lt DMC Fisher. It consisted of his Guardsmen having to catch, kill, pluck, dress and bake a chicken. The CQMS judged the winner. The platoon commanders in No 2 Company were Second Lieutenants JLJ Levine, JPW Gatehouse and HVL Smith.

To raise money for the Commanding Officer's charity for disabled children at Lord Mayor Treloar College in Hampshire, a platoon arranged a darts marathon for thirty-four consecutive hours, enabling them to achieve a score of one million. When Christmas came, the Corps of Drums played carols around the UN married quarters, raising a significant sum for the charity.

As Christmas approached 6 Platoon asked a local priest to get them a Christmas tree. He sent tractor-loads of green foliage which must have left the local gardens very bare. This illustrates the friendliness of most Cypriots.

Capt EF Hobbs, the Battalion Operations Officer, attended frequent meetings with the Greek or Turk military in turns, accompanied by a Grenadier company commander. The Turks always arranged an elaborate lunch which followed speeches. They ate lamb, goat and sliced apples and pears during which business could not be discussed. If a plate was finished, it was automatically replaced by a full one. The UN officers were busy and so the prolonged meals were rather frustrating, particularly as not much was concluded. The Turkish officers did not seem to have much to do. The Greek National Guard's hospitality was similar but less ostentatious.

The fields in the Buffer Zone were fertile and a great many were un-cultivated. Lt Col Denison-Smith decided to improve the situation by negotiating the opening of fields for Cypriots to farm. By lengthy and patient negotiations conducted with great skill by Capt TEM Done farming arrangements were dramatically improved over six months to both sides' advantage. One major advance was when the Greek Cypriots allowed the Turkish Cypriots to take water to irrigate their land. These developments led to an easing of tension and a tentative feeling of trust between farming communities which everyone wanted. The Austrian commander of the UN in Cyprus, Maj Gen Gunther Greindl, was very supportive of the Battalion but there was great pressure not to upset the delicate status quo.

In January 1983 a Grenadier visited Varosha, an area of Famagusta, which had contained the banks, the smartest hotels and night clubs. It had become a ghost city for it had been captured by the Turkish Army in 1974 and kept in a time vacuum thereafter because no civilian has been allowed back there. He found empty streets and buildings, abandoned in panic nine years previously. Apart from the dust, the evacuation could have taken place the previous day, for nothing had been touched. In deserted restaurants, plates and cutlery were still on the tables – only the food had gone, eaten by rats. In one luxury hotel bedroom, he saw a Reader's Digest open by a bed. It was dated, predictably, August 1974. Today weeds have now grown into small trees, and pushed through the tarmac of abandoned streets. Overlays of sprayed graffiti adorn each crumbling, ransacked ruin as well as most of the north's churches, which lost their treasures to professional art thieves after the Turkish Army had damaged and desecrated some of the graveyards.

To the south-west of Varosha, in the Dhekelia Sovereign Base Area, the two other Grenadier Companies were training hard and doing all they could to enjoy themselves and all that Cyprus had to offer. One of the Companies was always deployed on OP and patrol duties along the Sovereign Base border for the SBA contained 9 Signal Regiment which played a crucial role. There were, therefore, numerous guard duties and fatigues.

Maj Sir Hervey Bruce formed a mounted patrol on horses, just as Lt RL Fanshawe was to do later in the Falklands. There were plenty of volunteers to ride because it was said that there were female instructors. The mounted patrol saved much time and effort.

The Inkerman Company played an enjoyable football match against a Turkish Special Forces unit. Sir Hervey inspected the Turkish team while their officer inspected the Company's.

All sports in Cyprus were exceptional. When the opportunities arose, Grenadiers participated in sailing, canoeing, skiing, windsurfing, hang-gliding, parachuting, riding, gliding and climbing. Most Guardsmen were introduced to at least two new pursuits. This was undoubtedly one of the most successful aspects of the Cyprus tour.

The United Nations Military Skills competition took place in Dhekelia and was organized by Capt GVA Baker. The Battalion took this UN challenge very seriously; the previous one had the humiliation of being beaten by a team from a cavalry squadron. Danes, Canadians, Swedes and Austrians were competing. The Battalion entered two teams, one from No 1 Company, the other from Support Company. The latter consisted of Lt EC Gordon Lennox, Sgt K Smith, L/Cpl SD Ashley, Gdsm P Godby and SB Higgins. They triumphed over all and won the arduous and complex competition convincingly. It was the first time a British infantry battalion had won it for some time. The champagne flowed. There were many other successes including the Battalion skiing team beating those from such notable skiing nations as Canada, Sweden and Austria.

The end of February marked the halfway stage in the Battalion's tour in Cyprus. The Grenadiers therefore swapped roles – No 3 and The Inkerman Companies deployed to the UN Buffer Zone, while Nos 1 and 2 Companies defended the SBA. On the operational front, despite Greek-Cypriot elections and Turkish-Cypriot alerts about Armenian terrorists, all remained encouragingly quiet, although rumours occasionally led to increased tension. L/Sgt SP Milson was serving with The Inkerman Company in an isolated OP: 'There were lots of complaints that Turks were raping Greek Cypriots who were tending the olive groves. There was no evidence that this occurred, although the Turks were definitely coming across the border to steal from shops. We could watch the Turks and their Russian T55 tanks through massive binoculars. I had a great command of ten men although I was scarcely aged 22.'

The spring provided an ideal time to use the excellent training areas around Episkopi. No 2 Company provided the enemy for Victory College, RMA Sandhurst, on their final internal security exercise. The Guardsmen enjoyed the opportunity to exercise against potential officers. Maj Woodrow busied himself turning No 1 Company into Special Forces while the Corps of Drums (D/Major JB Evans) went off Assault Pioneer training, learning how to overturn assault boats and blow up Ferret scout cars. The Regimental Band paid an enormously successful visit to Cyprus, drawing large crowds wherever they played. Rugby and skiing made way for cricket, polo, athletics, volleyball and more waterskiing.

Meanwhile in London Maj JFM Rodwell and Capt DD Horn, followed by Capt TS Nolan, did all they could for the families. Some of them visited Cyprus during the Grenadier two weeks' 'rest and recuperation', much enjoyed by everyone.

Lt Col Denison-Smith arranged for a remarkable parade and march past to be held to receive the United Nations medals on Nicosia airport coinciding with the visit of the Regimental Band. Many compliments were paid to the Battalion for its high standards. Maj WR Clarke, the Battalion's former Quartermaster and now the QM to the UN Support Regiment, chose to be presented with his MBE by the British High Commissioner amid the Battalion rather than at Buckingham Palace.

Gradually thoughts turned more and more to a damp, cold but welcoming England. By the beginning of June 1983 the Battalion was back in Chelsea Barracks with forecasts of a relatively quiet summer in London – unlike the previous summer, they hoped.

The tour in Cyprus was an unqualified success. The Battalion won a great reputation with the British and other contingents alike. Much was achieved in helping the community and keeping the peace. It was a memorable and rewarding tour.

CHAPTER 25

BELIZE: A THREAT AVERTED

1984 – 1985

The 2nd Battalion had served in Belize, or British Honduras as it was still then widely known, in 1971–1972. The threat of a Guatemalan invasion had led to the Grenadier six-week exercise being extended to a seven-month tour. This was described in Chapter 16.

Five years later the Guatemalans deployed towards Belize's border. Reinforcements were immediately despatched. They consisted of a second infantry battalion, a close reconnaissance troop of armoured cars, one field battery of 105 mm light guns, several Blowpipe low-level surface-to-air missile detachments and a squadron of engineers. The RAF flew six Harrier vertical take-off fighters into Belize. They were protected by Rapier missiles. A Royal Naval frigate provided cover against air attack for the Battalion Group in the south. This show of force had the desired effect: the Guatemalan threat temporarily subsided. Even so, the border was patrolled by the British forces and permanent observation posts were set up overlooking Guatemala.

Although Guatemalan engineers constructed a main road within 50 yards of the Belizian border, it was possible to reduce the force level to a four-company battalion backed up by the RN and RAF.

In 1981 Belize became an independent state within the British Commonwealth. The following year, during the Falklands War, the possibility arose of Guatemala siding with Argentina, but that invasion scare passed.

The British Government announced that the Army would remain in Belize 'for an appropriate period' to safeguard the country's territorial integrity. Accordingly in August 1984, the 2nd Battalion, commanded by Lt Col JVEF O'Connell, having just won the Army Athletic Championships for the first and last time, put away tunics and bearskins at Chelsea Barracks. After training at Thetford and embarkation leave, the Battalion deployed to Belize. Much had changed since the Battalion was there twelve years before.

300

Meanwhile, the diverse Belizian population of Mayan Indians, Creoles, Spanish, Caribs, Lebanese and a few Europeans had seen a succession of equally diverse British units deploy in turn to defend them. After the 2/2 Gurkhas, Black Watch, Cheshire Regiment and the Royal Irish Rangers, amongst others, it was the turn of the Grenadiers once more.

The threat was more evident, perhaps because Guatemala had significant internal problems. External aggression might win the sort of local popularity which President Galtieri of Argentina had sought when he invaded the Falklands.

To meet this threat, the Battalion had to create two Battle Groups each of two rifle companies with integral support weapons. To create such a force London District and HQ Household Division arranged for a troop of Life Guards, in an infantry role, and a platoon from 2nd Battalion Coldstream Guards to be attached to the Battalion. This not only provided extra troops to meet the operational role, but also, equally important as far as the Commanding Officer was concerned, gave the Battalion the flexibility, if needed, to make maximum use of the valuable training and recreational opportunities offered by Belize. The two Groups were known as Battle Group North and Battle Group South.

Battle Group North consisted of No 2 Company (Maj TEM Done), B Troop of F (Sphinx) Battery 7 Royal Horse Artillery, and a close recce troop of the Queen's Own Hussars. Their area of responsibility stretched about 125 miles from the Mexican border in the north to the Maya Mountains in the south. (See map page 190). They were based near to the Western Highway close to the border. Also in the north was No 3 Company at Airport Camp close to Belize City and the international airport was within the camp's perimeter. No 3 Company (Maj CRJ Wiggin) had four platoons, including the Corps of Drums and a Coldstream platoon. (When it came to an inter-platoon patrol competition the Coldstream, under command of Lt M Giles, won it!)

Battalion Headquarters, under the watchful eyes of the Senior Major (HA Baillie and then TTR Lort-Phillips), the Adjutant (GVA Baker) and RSM TA Rolfe were also based on Airport Camp.

Battle Group South, 80 miles away, consisted of The Inkerman Company (Maj DP Ratcliffe) and a small tactical Headquarters, commanded by Maj AD Hutchinson, at Rideau Camp which lay three miles north-west of Punta Gorda on the Southern Highway, the only road linking the north and south.

Maj Ratcliffe had under command a troop of F (Sphinx) Battery and a Royal Engineers Field Troop.

Less than an hour's journey north-west of Rideau Camp was another base called Salamanca where No 1 Company (Maj AH Drummond, then

Maj RG Cartwright) was stationed, with a troop of the Life Guards under command.

All Companies had the tasks of manning observation posts on the border, carrying out regular patrols daily to prevent Guatemalan encroachment, showing the flag and jungle training. Four years previously a Gurkha patrol stopped the Guatemalan Army building a road into Belize.

The Battle Groups in the north and south each needed their own anti-tank weapons and 81mm mortars. Support Company was therefore split into two platoons, each equipped with these weapons. Captains EC Gordon Lennox and JPW Goodman were therefore under command of The Inkerman Company and No 2 Company respectively.

The platoon commanders in the south were Lt ORP Chipperfield, Lt ERW Stanley, 2Lt PDR Landale (No 1 Company), and Lt MCJ Hutchings, Lt FA Wauchope and 2Lt RM Allsopp (The Inkerman Company). Those in the north were Lt JLJ Levine, Lt DJH Maddan, 2Lt JPW Gatehouse, 2Lt RD Winstanley (No 2 Company) and Lt RHG Mills (Corps of Drums).

The total strength of the British forces in Belize was 2,000 Army and RAF. In addition the West Indies Royal Naval Guardship occasionally visited.

Operational command was split, with the Senior Major commanding Battle Group North and the Commanding Officer commanding Battle Group South. Nevertheless, for training and administration the Commanding Officer continued to command both the Battle Groups; he and the Sergeant Major therefore spent the tour on a continuous circuit of visits to each company, isolated platoons and the various outposts.

In spite of these frequent visits, the Company Groups were largely autonomous and, as so often happens in such circumstances, each developed its own character and style.

The Battalion's mission was to act as a deterrent to aggression from Guatemala. They were there to support the small Belize Defence Force. The United States were anxious that the British forces remained in Belize; the Central American states were largely unstable, and there was the continual fear, particularly on the part of the US, that Belize would become a Communist foothold in Central America.

Communication between Battle Group South and Airport Camp relied upon an hour's flight in a helicopter, twelve hours by road, or ten hours in a flat-bottomed landing craft run by the maritime wing of the Royal Corps of Transport. The craft were called RPLs standing for Ramp Powered Lighter.

Telephone wires were usually inaudible and the teleprinter was unreliable. The southern Companies therefore, despite the efforts of the

42. A patrol of the 2nd Battalion leaves Creggan Camp, 1977. (A watercolour commissioned by the Regiment and painted by Joan Wanklyn.)

43. A patrol of The Queen's Company near Crossmaglen, November 1983. Back left to right, Guardsmen Meak and Rushton and in front Guardsman Elliott and Lance Sergeant Watson.

44. The Quartermasters' lunch at The Barracks, Caterham 20th April 1990.
    Fourth Row: Lieutenant Colonel C A Atkins RADC, Major F J Clutton, Lieutenant
    M C Hutchison, Captain P F Richardson.
    Third Row: Second Lieutenant J A Heroys, Second Lieutenant S W Gammell, Captain
    W Williams, Captain S Tuck, Captain D Mason, Captain T S Nolan.
    Second Row: Major Sir Hervey Bruce Bt, Captain P Harris, Lieutenant J T F Fagan,
    Captain B Everest, Lieutenant W Coulson, Captain G P R Norton.
    Front Row: Major L C Drouet, Major A Dobson, Captain L E Burrell, Captain A
    Holloway, Captain D W Ling, Captain B D Double, Major P A Lewis, Lieutenant
    Colonel R B Bashford, Major D R Rossi, Captain C E Kitchen, Major B T Eastwood,
    Captain J A Sandison, Captain P T Dunkerley, Lieutenant The Hon E Brassey.
    Seated: Major A G Everett, Lieutenant Colonel A M H Joscelyne(Commanding Officer),
    Major G C Hackett.

45. The Queen with her Grenadier Knights of the Most Noble Order of the Garter 1990.
    Left to right: Major The Lord Carrington, Lieutenant Colonel The Earl of Cromer,
    Colonel The Viscount De L'Isle, The Queen, the Colonel, Major The Duke of Grafton,
    Major General The Duke of Norfolk.

Commanding Officer and Regimental Sergeant Major, felt gloriously free and independent.

The observation posts in the south were very different in character. To the extreme south-west was one of The Inkerman Company's OPs, reached only by helicopter. It was perched high on a pinnacle of rock overlooking a Guatemalan military camp. Traffic could be watched along three Guatemalan roads. The Union Jack fluttered defiantly from the top of the OP.

The other one was 10 metres from the border and faced the end of the road built by Guatemala with a view to providing a link to the Caribbean. It was an eerie post to man, particularly at night when every sound might be a precursor to invasion.

Belize has the second longest barrier reef in the world after Australia's. The Inkerman Company's area of responsibility included the southern half of the reef amid some 300 tropical islands. The southernmost island is Hunting Cay, one of the Sapodilla Cays. It is within 35 miles of the coast of both Guatemala and Belize. A section of two NCOs and six Guardsmen spent a week at a time on the island. Their main role was to oppose any seizure of the island by the Guatemalans which would have caused an international incident. A subsidiary role was watching shipping without being diverted by the many Guatemalan tourists who visited this attractive 'Robinson Crusoe' island.

These interesting and worthwhile patrols usually consisted of eight Grenadiers, visiting remote areas. After being dropped off by boat, road or helicopter, they spent up to a week moving from village to village along tracks and paths. The reception by the Mayan Indians was always friendly, a refreshing change from Northern Ireland. The children, overawed by the white giants, huddled together in timid groups, or clutched their mother's skirt, often the only item of clothing in their mother's wardrobe. Boat patrols in Royal Engineer raider powerboats were fun. The patrols were good experience for young officers and NCOs.

As Grenadiers had learned many times before in similar surroundings, the training benefits of jungle patrolling were enormous. Officers and NCOs discovered what command really means when alone with five or six Guardsmen far from support or assistance. They developed self-reliance and adaptability and learned the importance of a comrade's support.

For example L/Sgt TA Sentance took a section on one ten-day patrol to check that border cairns were still in position and to make a map of the tracks. On another of his patrols, the section supported eight American archaeologists who were examining Indian graves. L/Sgt Sentance was responsible for providing them with water, radio back-up, first aid and, in an emergency, evacuating them by air. The oppressive atmosphere in the jungle was not easy to get accustomed to. Those with vivid imagination wondered if the rustling amid the vegetation was an enemy, some wild animal or a slithering snake.

One Inkerman Company patrol was out during torrential rains. The Sarstoon River started to rise rapidly and the patrol had to evacuate their position very quickly as the water was already up to the Grenadiers' waists. As they made their escape they were approached by an alligator upon which they fired. The Armalite rounds were not as heavy as the 7.62 and just bounced off the alligator's thick skin, but they were enough to persuade it to find a meal elsewhere.

In late November large-scale exercises began in torrential rain during which the Grenadiers were visited by the new Force Commander, Brig DBW Webb-Carter OBE, MC, formerly Irish Guards.

The Commanding Officer ran test exercises for each Battle Group, in each of which the other Group acted as enemy. The enemy faced gruelling infiltration marches through some of the hardest jungle in the world. Belize suffers from regular hurricanes and, as a result, the 'deadfall', as it is known, creates extremely inhospitable secondary jungle growth. In January 1985 the Force test exercise, which was set by the Force Commander, took place. This was designed to test Battalion flexibility and stamina. Battle Group South Companies were tasked to carry out search and destroy operations against an aggressive enemy led by the Senior Major and found largely by No 2 Company. Capt Gordon Lennox and the stalwarts of the anti-tank platoon were given a stay-behind role, being left to observe and report enemy movements, and, thanks to their success, the enemy HQ was located and destroyed just as the exercise reached its climax.

In March 1985 a company was placed on short notice for impending operations in the Caribbean. The Commanding Officer and a small tactical headquarters was despatched shortly afterwards on HMS *Arrow* to the Turks & Caicos Islands: the Prime Minister of this small colony was about to be arrested in Miami for dealing in drugs. Sadly, from the Commanding Officer's viewpoint, although he was appointed Commander Joint Forces Turks & Caicos, this force never landed. Instead, it learned about life on a naval frigate and sailing in a "box" – going round and round the same bit of sea for days without seeing land – though knowing it was just over the horizon.

2Lt RD Winstanley was told to accompany the British Ambassador from Jamaica on a two-day visit to Haiti. He met the notorious son of the former dictator 'Papa Doc'. Winstanley saw the misery of the locals living in the police state. They were so backward that some were frightened of being photographed, believing that they would lose their spirit to the camera.

When planning the tour, the Commanding Officer had the benefit of having been a Company Commander in Belize in 1971–72. He therefore knew that the opportunities for recreation and adventure in Belize were unrivalled and set up two teams to ensure that everyone was able to benefit from this. The first was spearheaded by SQMS Bojtler, the Battalion Physical Training expert. He set up an adventure training camp on one of the Cays – islands on the reef – where Guardsmen were taught to wind-surf, scuba dive, canoe and fish. Each Platoon went there for a week at a time. The second team was masterminded by CSM Bradley, who became Messrs Thomas Cook & Co for the Belize tour. He arranged R & R for every member of the Battalion, booking accommodation and holidays in such far-flung spots as Orlando, Miami, Cancun in Mexico, New Orleans, Panama and Honduras. In addition, he was instrumental in helping a number of wives to visit Belize – a trip which those who managed it will never forget.

Some envied one Guardsman in the Transport Platoon who caught an indulgence flight to Miami for £10. He went to the local military base there, claiming to be a Sergeant and lived in luxury for a fortnight for a dollar a day, wined and dined every night by hospitable, if gullible, Americans.

Grenadiers who had been unsuccessful in their application to run in the 1984 London Marathon decided to raise money for health workers in Belize by running a Marathon there instead. Chris Brasher, the organizer of the London Marathon, sent his encouragement and guaranteed places in the next London Marathon. Forty-seven runners set off in Belize as dawn was breaking. Half of them were Grenadiers; the other runners included four Coldstream and some Royal Horse Artillery, Garrison troops and Belize Defence Force. At the 20-mile point everyone felt ghastly. By 8.0 am the temperature was in the nineties and some were near collapse. But only eight failed to finish at Punta Gorda which, by a coincidence, was being visited by the Prime Minister of Belize. He was interviewed for Radio Belize with the Commanding Officer. The marathon became the one news item of the day. £1,300 was raised. Some Grenadiers later ran in the 1985 Marathon in London amid 19,000 others and a different atmosphere.

The visit of the Regimental Band was a great success. They gave a very good performance at Punta Gorda followed by an evening concert in Rideau for Battle Group South. At Airport Camp the Sergeants' Mess members proudly watched them perform to representatives of many different Regiments and Corps. It is always a great moment after their invariably excellent music to hear them march off to 'British Grenadiers'.

D/Sgt JA Sandison QGM and CSM A Myles took Band Sergeant Major T Dove BEM on a day-long expedition for lobsters, but the boat broke down and they returned the following day so scorched by the sun that they were the colour of the lobsters.

The Battalion was not encouraged to become involved in internal issues but this proved impossible since the greatest threat to the stability of the country was not from Guatemala but from involvement with the drug industry, particularly the growing of marijuana. Much of this was well-organized and Mafia-controlled. A crop of marijuana could be sold for ten times more than any other crop, so it was not surprising that farmers were attracted by such profits.

Although, therefore, the Guatemalan Army was the top priority, preventing drug-smuggling became important, although this was primarily the responsibility of the Belizian Government. Belize was a source of the world's highest grade marijuana; the American Drug Enforcement Agency estimated that 54 tons were produced there in 1992. Grenadier patrols occasionally came across drug smugglers who dropped their loads and ran.

One typical operation, which lasted six days, involved a Grenadier platoon working with the Belize Defence Force. Half a platoon was positioned by helicopter in a cut-off role while the remainder manned road blocks. The Defence Force then swept the area, arresting some twenty-five locals on suspicion of carrying drugs.

Maj Cartwright was in a Gazelle helicopter when he saw an abandoned, brand new fixed-wing aircraft which had crashed. On investigating, he found all the evidence of a drug smuggler – burnt documents, a coca-cola tin made in Bogata, extra white plastic fuel storage tanks to increase the aircraft's range, and very sophisticated avionics.

Lt RHG Mills was told to help seven armed, uniformed Special Investigation Branch Belizians who were to search for weapons used by drug smugglers guarding marijuana fields. Ten Grenadiers and SIB were dropped off by helicopter near the 15-acre field. The SIB burnt abandoned huts but no weapons were found. On leaving, the SIB stuffed their pockets with so much marijuana that they could easily be followed, like a paper chase, by 2Lt Allsopp who accompanied them.

The impoverished farmers had difficulty making a living and so they turned to the drug industry: cannabis grew like weeds everywhere. Their marijuana crops were destroyed without compensation by the Belize Defence Force. The locals in such areas could be hostile and were quite capable of attacking outsiders with machetes. Despite this Grenadiers had no special powers of arrest.

The opportunities for Guardsmen to meet Guatemalan soldiers were limited. L/Cpl J Hinds saw a few wandering aimlessly beyond the border. L/Cpl D Groom chatted to several friendly private soldiers; there was no sight of their officers or NCOs. Sgt S Murphy misread his map and led his section across the border but they were released after two hours.

Maj Cartwright took some of the Brigade staff by Puma helicopter to meet, near Treetops, their Guatemalan opposite numbers all of whom carried pistols. They sat together in comfortable chairs, sipping soft drinks. Maj Cartwright found himself next to a sophisticated Guatemalan Colonel who had served in Rome and well remembered Maj DCL De Burgh Milne, a Grenadier who had also served there. 'Give my love to Anna,' (Maj De Burgh Milne's wife) he said. It was difficult imagining going to war against such friendly people. Indeed, the Guatemalans seemed much more concerned about their own internal terrorist threat. On one occasion The Inkerman Company Guardsmen from an observation post played volley ball against the Guatemalan Officers.

The Battalion's tour gradually drew to a close. Much had been learned. 'For me Belize offered a glimpse into the world of the old Empire,' wrote one Grenadier afterwards, 'where duty called and good men took up the White Man's burden: the foot patrols, moving from village to village among people whose prized possession was a machete or a football. With barebreasted women and barefoot children, we entered the prehistoric world of wooden huts thatched with palm leaves. It was a far cry from the unyielding gravel of Horse Guards and a rare chance to put the clock back a hundred years when a Briton was a demi-god with his pale face and high technology.'

The benefits of the tour are best summed up by the Commanding Officer in his final address to the Battalion: 'I hope everyone who has been in Belize will be able to look back on an enjoyable and worthwhile tour. We are all fitter than on arrival; everyone has become more self-reliant and we are able to survive and fight in what is a very tough, demanding environment. We have walked hundreds of miles, visited remote corners of the world and met some very different people! Above all we have earned a high reputation in the local community for being smart, courteous and professional.'

By 9 April 1985 the 2nd Battalion was back at Chelsea Barracks once more, preparing for Public Duties and their tour in Northern Ireland which was due to start in January 1986, another example of the varied life led by Grenadiers at this time.

In due course the relationship between the British Government and Guatemala over Belize further improved. Successful negotiations were held over territorial disputes. Both Belize and Guatemala became interested in benefiting from the tourist potential and frightening off illegal immigrants and drug smugglers. Over the years the British Army and the Grenadiers had, by deterring Guatemalan aggression successfully, 'bought time', enabling peaceful solutions to be found.

CHAPTER 26

THE END OF AN ERA – BAOR

1986–1991

Of the eleven post-war tours of Grenadier Battalions in Germany, the last, in 1986–1991, for the 1st Battalion, was probably the most traumatic. It had everything from the privilege of being the first to receive the new infantry fighting vehicle, Warrior, to the Gulf War; an operational tour in Northern Ireland, and battlegroup training in Canada, to the final, sudden collapse of Communism which led to the unification of Germany, the disintegration of the Warsaw Pact and, over the following five years, withdrawal of most British soldiers from Germany. It was the thankful end to the longest period of tension and confrontation in modern history.

* * *

In May 1985, the Commanding Officer, (Lt Col EJ Webb-Carter) issued his first Training Directive at Cavalry Barracks, Hounslow: 'There are, I believe, two things to be done,' he wrote. 'First we need to sell Germany and the way of life in BAOR. There are many soldiers wavering towards termination who need some encouragement to remain in the Army. Secondly we need to prepare professionally for the demanding role in 4 Armoured Brigade. Standards are high.'

That month each Company was made responsible for a Study Morning. No 3 Company (Maj MGA Drage) started with an introduction to BAOR. The Queen's Company (Maj REH Aubrey-Fletcher, later GF Lesinski) led on the threat; Support Company (Maj JS Scott-Clarke) and No 2 Company (Maj ET Bolitho, later Sir Hervey Bruce) covered battle drills, while Headquarter Company (Maj DNW Sewell, later GJS Hayhoe) presented on logistics.

The Study Mornings were followed in August by Tactical Exercises Without Troops and Combat Decisions Games on a cloth model. In September an Infantry Tactical Training Wing from Sennelager visited the Battalion. This was followed by Command Post Exercises and two periods on Battle Group trainers to try out tactical development. The following month, during a Soviet Studies week run by the Intelligence Officer (Lt CC Allen), everyone had the opportunity of firing Soviet weapons, eating Soviet food and attending lectures by the Soviet Studies Team from Sandhurst. More exercises followed with the emphasis on shooting, tactics, Nuclear Biological and Chemical Warfare, and battle fitness, including a march of 25 miles carrying equipment weighing no less than 56 lbs in under 10 hours. Courses were started on learning German. Some wives attended.

The above pre-BAOR training can be compared with the inadequate preliminary pre-BAOR exercises in England in the 1960s and '70s. But by the late 1980s the equipment in BAOR was extremely sophisticated and peacetime soldiering in Germany had never been taken more seriously. Nevertheless there were dangers in this increase of activity. Soldiers and officers alike could be disenchanted with BAOR soldiering unless the hard work was leavened by a good deal of 'play'.

Over Christmas the Battalion found six Windsor Castle Guards, including one on Christmas Day which was commanded by the Adjutant (Capt EF Hobbs). The Battalion was also responsible for the security commitment at Heathrow Airport. It moved to Oxford Barracks, Münster, in January 1986.

The manpower levels for a battalion in BAOR were significantly higher than in England, and so the Grenadiers were reinforced by a platoon of Irish Guardsmen for six months. (The Platoon Sergeant was a piper, which led to some fear of noisy reveilles.) The outgoing Battalion's rear party of experts remained for three months in Germany. The Army later introduced an Armoured Infantry Manning Increment (AIMI) which, for Foot Guards Battalions, consisted of a mixed or 'multi-regiment' group. It remained in BAOR, bridging the gap between the two different establishments. The Guards Division has operated on a system of 'composite' battalions for over 300 years, and so took such a change in its stride. Grenadiers therefore continue to serve in Germany with the AIMI of whichever Foot Guards battalion is stationed there.

Germany created a sense of purpose. There was a 'call out' exercise, called Active Edge, which normally entailed early morning recall and deployment of all battalion vehicles and personnel. These were frequent occurrences and not always popular.

The wives had been briefed that an invasion from the East could, in principle, happen at any time. 'We were told that, should such an event occur, we would have to put our pets down or give them away,' recalls Mrs Nancy Joyce, the wife of the Quartermaster. 'We could take only one suitcase each and we must have a torch for when the lights failed. We took it all fairly seriously, but by 1991 all such briefings had stopped.' Her husband, Capt MJ Joyce, takes up the story: 'All Companies had to plan to evacuate the barracks; the recce group had to be out in four hours and the whole Battalion including all vehicles had to be at the harbour area within 24 hours, having emptied the ammunition bunkers and taken with them three days' rations and fuel. When I was in Berlin earlier, some wives felt that they would never get out in time, whereas the feeling in Münster was that they would make it to Holland and Belgium.'

Opposite 1st British Corps were East German Army units and behind them was the formidable Warsaw Pact 3rd Shock Army, with its Headquarters only 35 kms beyond the border in the (East) German Democratic Republic. This Army had three tank divisions and one motorized infantry division.

Each tank division had three tank regiments equipped with T-64s or T-54/55s. The 3rd Shock Army had its own integral 27th Aviation Regiment with eighty Mi-24 Hind helicopter gunships. It contained 60,000 troops, 1,240 tanks and 480 guns. It boasted its own missile regiment, an air defence regiment, an artillery brigade and amphibious troops. The 3rd Shock Army was said to be fully trained and maintained at a constant state of readiness. As new equipment became available, such as the new 152 mm self-propelled gun with nuclear capability, this Army was among the first to receive it. The potential enemy was therefore a menacing foe, but the men were largely conscripts. Whether their morale would be affected should they invade West Germany, fighting on unfamiliar ground amidst a hostile population, was unknown.

The organization of the British Corps in Germany changed every seven or eight years, usually as a result of a defence review. But during this era the Corps comprised two forward armoured divisions and a reserve armoured division. The two forward divisions had to provide their own covering force brigades whose task it was to delay any enemy attack until the two other brigades had prepared their General Deployment Plan positions. These GDP positions were meticulously planned down to the last detail, not unlike the Maginot Line in some respects. This was positional defence. However, the third division, which comprised just two brigades, provided the reserve manoeuvre force for the Corps. These brigades were trained to carry out a series of counter-moves – counter-attack, counter-penetration and the most-quoted counter-stroke. 4 Armoured Brigade, which had replaced the old 4 Guards Brigade in Münster, was one of these brigades and was the specialist for the counter-stroke. This operation was designed to strike the rear of a Russian Army from an unexpected flank depending on a precise timing and the shock action of 114 main battle tanks. A brigade was the smallest formation able to mount such an operation. It would have been desirable to have had another brigade to improve success. These tactics depended very much on surprise.

Although there was no General Deployment Plan for them, the training required for this role relied on perfect coordination, agility of mind and an aggressive spirit. The 1st Battalion was the infantry component of 4 Armoured Brigade and the battlegroup, as they are called, comprised one Grenadier Company (The Queen's Company) and two armoured squadrons of fifteen tanks each. The remaining two Companies were grouped with the tank regiments (17/21 Lancers and the Queen's Royal Irish Hussars, later 14/20 Hussars).

The Battalion arrived in Oxford Barracks, Münster, an old home for those who had served with 2nd Battalion in February 1986, and soon were at work on the old armoured personnel carrier 432s. Activity was intense with two weeks at Soltau and later field firing at Sennelager. The routine of training had not changed greatly but there was a great awareness of the Battalion's dashing role in the Brigade. There was more emphasis on the attack and less on defence. This was welcomed by the Guardsmen who did not enjoy digging!

Despite the level of training the Battalion flung itself into the sport and adventurous training activities of BAOR. The shooting team under Lt HVL Smith, the best shot in the 3rd Division, led the team to 2nd place at Bisley while the swimming team won the Army Finals. Both these achievements in 1988 were largely the fruits of the previous Commanding Officer's (Lt Col A Heroys) enthusiasm and planning.

One, perhaps unexpected, theme of all Lt Col Webb-Carter's training directives was that life must be fun – 'the giggle factor' as he called it: 'Let us "play hard and work hard" and above all be optimistic in approach and enjoy it,' he wrote, concluding one directive: 'Laugh and the world laughs with you; weep and you weep alone'.

A sense of humour was particularly necessary when the Battalion was unexpectedly ordered in early June to train for an emergency tour in Northern Ireland. The Commanding Officer was much dismayed but the Guardsmen responded in typical style; within a second they concluded that operations in Northern Ireland would be more amusing than hours spent on the vehicle park maintaining the ageing AFV432.

No 3 Company went with all its vehicles to Denmark to train with the Guard Hussar Regiment for two weeks. The Danes had rather different priorities: a 24-hour exercise was halted for four hours so that the Danes could return to barracks to watch their football team compete in the World Cup.

The Battalion's tour in Northern Ireland from September 1986 to January 1987 is briefly described in Chapter 23. The rifle companies were all deployed under other battalions to the frustration of the Commanding Officer (a foretaste of what was to happen in the Gulf War). Fortunately

Grenadiers were able to leave Ireland for courses, and Capt TH Breitmeyer with 2Lt PHM Swire competed in the BAOR skiing championship.

After leave the mechanized training started all over again – Exercise Active Edge, Soltau and the vehicle park. Münster in March is cold and often covered in snow. The arrival of VIPs in helicopters invariably turned the square's light snow into a blizzard and the welcoming party into snowmen.

The trend for having more women in the Army at last touched the Battalion. That month the Battalion's Medical Officer, Capt BL James, gave a lecture to the Battalion on AIDS, despite the misgivings of the Commanding Officer. She was unperturbed by the explicit nature of her talk. In April 2Lt DM Beck joined the Battalion, the first member of the Women's Royal Army Corps to do so. She became the Assistant Adjutant. The subaltern officers' only complaint was that the Adjutant refused her request to carry out Picquet Duties. The Senior Major, RMC Wynn-Pope, was to be seen coming into work with a noticeable spring in his step; he shared an office with her. Nevertheless her reception was somewhat guarded. The Captain (Maj GF Lesinski) did not approve until she beat him at backgammon. Thereafter she never looked back.

Training in BAOR resumed its intensity. Serving in Germany had its frustrations due to financial constraints. The local overseas allowance had been cut and track mileage was severely controlled, although the Brigade Commander paid little attention to that. There was insufficient fuel and ammunition. All this led to an erosion of job satisfaction throughout BAOR. The Battalion was exceptionally short of officers, which made the life of the few subalterns in the Battalion very tiresome. A Battalion short of officers changes its image very quickly. It is a lesson the Regiment keeps on re-learning.

The drive for professionalism was continuous. The Commanding Officer, who placed a considerable emphasis on vehicle husbandry, personally carried out inspections of vehicles every Friday, and a permanent equipment management team was formed under the control of the Assistant Quartermaster (Capt PT Dunkerley).

Guardsmen in the Battalion carried the 7.62mm LIAI rifle. It was semi-automatic with a 20-round magazine and a combat range of about 300m, increasing to 600m as required. Its replacement was to be the 5.56mm individual weapon which was shorter and had an optical sight. The general purpose machine gun was the 7.62mm LTAZ with a fire rate of up to 750 rounds per minute. It could be vehicle-mounted and was to be replaced with the 5.56mm light support weapon. Each rifle platoon had the 51mm light mortar for high explosives, smoke and illumination. The Mortar Platoon (commanded by C/Sgt R Newman and then Lt AA Pollock) had the 81mm mortar with a range of 5,800m and rate of fire of twelve rounds a minute.

Platoons' anti-tank weapon was the 84mm L14A1, known as the Carl Gustav, which fired the 2.59 kg HEAT round. Its effective range was little more than 500m against a static target. In the late 1980s the Light Anti-Armour weapon 90 anti-tank rocket launcher was introduced. The anti-tank platoon (Capt GPR Norton and then RD Winstanley) had the MILAN guided missile with a range of 2,000m and sixteen launchers. It was not appreciated in the late 1970s that the infantry had no anti-tank weapon capable of knocking out the latest Russian tanks head-on.

In September the Battalion exercised with the 432 and received the new rifle, the SA 80.

Two days after the exercise two coaches set off for an exceptional five-day battlefield and bottlefield tour. At Waterloo Gen Sir David Fraser gave an account of the battle of 18 June 1815. Of particular interest was the farmhouse at Hougoumont, to which the QM (Capt MJ Joyce) later returned with Pioneers to repaint the inside of the Chapel. That evening everyone attended a reception given by the Ambassador, a former Grenadier, Peter Petrie. The Corps of Drums under Drum Major R Sergeant played. While wives shopped, the Grenadiers visited the battle-field at Villers Cotterets with Lt Col JM Craster who had arranged for the party to visit Moët et Chandon at Epernay. The next reception was given by the Mayoress in the Hotel de Ville at Epernay – the home of champagne – where the Corps of Drums again performed. The evening was spent with 34 Regiment de Genie. On the following day Patrick Forbes, once a Grenadier, gave a talk on champagne. Being the Managing Director of Moët et Chandon, nobody was better qualified. A visit to part of the 19 miles of cellars was followed by prolonged tasting accompanied by French trumpeters. That evening was spent at Château Saran drinking magnums bottled for the Queen's Silver Jubilee. On the following day Alistair Horne talked at Verdun, the final battlefield in the historic tour. Such activities were valuable to balance the overall programme in BAOR.

The exercise season now over, the Battalion could turn to its extramural talents. The Cross Country skiing team, run by Capt GPR Norton, distinguished itself, winning the Infantry Cup. The Downhill Team, trained by 2Lt JMB Cayzer-Colvin, won a similar trophy. Capt HM Everist again ran the Battalion skiing hut in Bavaria, largely for single Guardsmen, who enjoyed themselves enormously. The Commanding Officer and his family spent Christmas and New Year there.

After Christmas leave Exercise Trading Places led to all key personnel swapping their appointments with Guardsmen and Lance Corporals for a day. The Commanding Officer and Quartermaster became the Sergeant and Corporal of the Barrack Guard. RSM D Hardman was locked up several times for idleness on leaf-sweeping fatigues. L/Cpl Wall was rather disconcerted on being given the Assistant Adjutant's WRAC uniform to wear on assuming her appointment.

The whole of 1988 was dominated by the conversion to Warrior. The new infantry fighting vehicle was awaited by the whole Battalion with pride and excitement. The Grenadiers were to be the first Armoured Infantry Battalion to receive the vehicle, 'and quite rightly so in view of our forbears,' reported the *Grenadier Gazette*. 'There was trepidation – everyone, not least the Royal Armoured Corps, would be looking to see how we would manage, and perhaps we were right in sensing some scepticism.'

The Warrior is a tracked 20 tonne, 75km/hour mechanized combat vehicle capable of carrying a section of eight men and all their equipment. Most significantly it was equipped with a 30mm Rarden gun and a 7.62mm Hughes Chain gun in a two-man turret occupied by the commander and his gunner. This produced a dramatic increase in the firepower of the infantry in BAOR. Powered by an 800hp Rolls Royce turbo-charged diesel engine and fully NBC protected, the vehicle was expensive.

By March the first gunnery, driving and maintenance, and tactics courses were over. Next came live firing of the vehicles' Rarden and Hughes Chain (machine) guns at Hohne, followed by the hurdle of tactical training at Soltau. Warrior proved itself to be fast and agile, able to outpace tanks, fulfilling both its roles as a battlefield taxi and as a weapons platform. It transformed the Battalion's capabilities. Finally, in September, Warrior was combined with field firing which ended with an all-arms live firing exercise. The new armoured vehicle was regarded as superb.

Throughout the Battalion's troop trial of Warrior which lasted a year (April 1988 to March 1989) the stream of visitors was constant. They ranged from a Minister to a retired Field Marshal, from the Directors of GKN Sankey who manufactured the vehicle to numerous Generals. 'Compiling the report was a mammoth task,' wrote Capt M Manning, RHG/D afterwards. 'The report is an example to all on how a trial should be conducted. It is a reference book on all things Warrior. The majority of the points were borne out in the Gulf War. The Grenadiers were the experts.'

The challenge of being the first Battalion to convert to this great vehicle had been exhilarating. The conversion had gone very well because of the outstanding support of the Director of Infantry and others. The determination that the vehicle was to prove a success was evident in every quarter of the Army. The Battalion received great credit for the success.

The introduction of Challenger, the new main battle tank, and Warrior necessitated new armoured defensive tactics, since both vehicles gave the battlegroup greatly enhanced firepower, mobility and protection.

The Challenger tank's thermal imagery and Warrior's image intensification optics greatly improved the ability to fight at night and further developed the true armour/infantry concept. The days of staid and inflexible positional defence, seen by the 1st Battalion in the early 1960s, had long disappeared.

The pressure of work was considerable. Despite all the training and sport there were many single soldiers who were bored in the evenings and weekends. Provision was invariably made for the married men but little for the single ones. This frequently led to drunkeness and indebtedness, the two scourges of BAOR life. The Battalion made great efforts to occupy the young men by setting up a plethora of clubs. The cultural visits were one of many ideas. They included Paris, Hamburg, Copenhagen and Berlin, all of which were very popular.

As the tour in BAOR wore on, the key Battalion appointments invariably changed. Lt Col Webb-Carter handed over command to Lt Col EH Houstoun MBE in 1988 and RSM Nesbitt to RSM DJ Hardman in 1987. What were the military backgrounds of this generation of Grenadier Battalion commanders and Regimental Sergeant Majors?

Lt Col Webb-Carter had spent over twelve years in the two Battalions, commanding both the Queen's and Inkerman Companies, being twice mentioned in despatches. He had considerable experience of BAOR as he had commanded both a mechanized company there and been Brigade Major in an armoured brigade headquarters – a really key appointment. Another very testing one was in Army Staff Duties in the Ministry of Defence. Useful earlier experience had been gained when Assistant Military Assistant to Gen Sir David Fraser when the latter was Vice Chief of the General Staff. Lt Col Webb-Carter had married the daughter of Lord Wigram and granddaughter of Gen Sir Andrew Thorne, both, like Sir David, distinguished Grenadiers.

Lt Col Webb-Carter's successor had a different background. After Regimental service in Germany, Cyprus and Sharjah, Lt Col Houstoun was selected for the SAS in 1971. He then saw action in most of the world's trouble spots, including two tours in Oman and the command of an SAS squadron in the Falklands War. His staff appointments were as Exchange Officer with the American Army at Fort Carson and with NATO in Holland. He, too, had been mentioned in despatches. His sporting achievements were remarkable – three Olympic selections in sailing, and he ran the 110m hurdles for the Army and Combined Services.

M Nesbitt, RSM of the Battalion 1986–1987, was a Junior CSM at the Junior Leader's Battalion, Oswestry in 1967. He served in Berlin, numerous tours in Northern Ireland, instructed at the Guards Depot, and also on two tours at RMA Sandhurst where he met and married his wife. He was involved in skiing in Bavaria, Austria and Norway, had been on canoeing expeditions in Norway and taken part in a major climbing expedition in the Himalayas. He was to leave the Battalion prematurely in order to be Academy Sergeant Major at Sandhurst, in the steps of many other distinguished Grenadier Sergeant Majors.

RSM Nesbitt's successor, DJ Hardman, had also joined the Army as a Junior Leader at Oswestry in 1967, and instructed at the Guards Depot. He served in Germany, Hong Kong, Cyprus and Northern Ireland. He also instructed on Adventure Training in Devon and for a year in Norway. His wife, Sue, was the daughter of the former Quartermaster of the 1st Battalion, Maj DJ Webster. The RSM had met her in the Sergeants' Mess and married her in the Guards Chapel in 1987.

Such were the varied backgrounds of some of those distinguished men who had risen to the highest appointments in Grenadier Battalions in the late 1980s.

The last really major exercise ever to take place in BAOR occurred in November 1988. It will be described in detail, partly through the eyes of a Company Commander, because it was unique. Moreover it reveals how environmental considerations now ruled training in Germany.

Exercise Iron Hammer was the Grenadiers' golden opportunity to demonstrate their new-found skills with Warrior. As the first snows settled in northern Germany, 3 Armoured (UK) Division deployed from its concentration area west of the River Leine. The exercise involved in all 25,000 soldiers, 300 tanks, seventy helicopters, thirty fixed wing aircraft, let alone the 6,000 tracked and wheeled vehicles.

The first week of the exercise was a Brigade work-up in preparation for the Divisional exercise to follow. The Battalion was split into two opposing sides with No 3 Company (Maj DCL De Burgh Milne) as part of the 17/21 Lancers battlegroup acting as enemy to the Grenadier battlegroup consisting of The Queen's Company (Maj JP Hargreaves), 17/21 Lancers squadron and 14/20 Hussars squadron. The 14/20 Hussars battlegroup consisted of No 2 Company (Maj JS Lloyd) and two squadrons.

'We deployed by rail to Braunschweig, some 25 miles from the East German border and thence to a company laager in the village of Vechelde five miles to the west,' wrote Maj Lloyd later. 'Vehicles were parked up in the midst of the village much to the bemusement of the locals for the next two days – a weekend – as no exercising was permitted over weekends. We were self-contained for fuel, food and water and used a portaloo. We had negotiated with the bürgermeister three weeks before to make the administrative arrangements.'

During this period three coachloads of Warsaw Pact, Swiss, Dutch, Spanish, Swedish and other international inspectors, under the Helsinki Agreement, questioned everyone about the exercise. 'I spoke German,' recalls Capt MCJ Hutchings who commanded the Recce Platoon, 'and was cautious what I told them, but they probably knew the answers already. The East Germans wore brand new uniform which suggested they were intelligence officers rather than regimental soldiers. They could look into Warrior but not enter the vehicle. Later in the exercise I had, at one stage, ten recce vehicles, including two engineer ones, four MILAN anti-tank vehicles, an artillery forward observation officer and a recovery vehicle. They were all difficult to control at night because of radio silence.'

By late afternoon on that first day enemy Orange Forces were inserted by helicopter 35 kilometers behind the start line to the west.

'On Monday we started a brigade advance to contact – west towards Hildesheim and Hanover,' continues Maj Lloyd. 'Initially 14/20 Hussars moved with two squadrons up. My company consisted of 125 men with 14 Warriors; the CQMS had a landrover and 4 tonner. I also had a 432 ambulance and REME fitter section in two 432 APCs.

'We did a right-flanking attack at midday, dismounting on the objective with lots of smoke. During the rest of the day we attacked and cleared several more minor objectives; the enemy consisted of men from 2 Battalion Light Infantry with MILAN anti-tank weapons. They had been transported originally by RAF Puma and Chinook helicopters. We finished up ten miles further on at another small village called Algermissen north of Hildesheim. We then joined up in the dark with the rest of the Battalion in woods nearby for an assault river crossing of the River Leine which lay some seven miles to the west. A very sophisticated damage repair system followed the military. A British team behind us cleared mud off the road, put there by our vehicles. Six million pounds had been budgeted for damage. Challenger tanks invariably ripped up road verges. Rumour had it that when the six million had been spent the exercise must finish.'

'At a formal orders group in a barn Lt Col Houstoun split the Battalion into two. The first group, under command of the Senior Major, consisted of No 3 and Support Companies. They were responsible for securing the near bank. The second group under the Commanding Officer consisted of The Queen's and No 2 Companies. These Companies were to cross the river before undertaking two concurrent attacks on villages just beyond. At 10pm we deployed by Warrior eight miles further west and then on foot to the RV where our assault boats were delivered at 1.30 am. It was back-breaking work, carrying the three assault boats and the outboard motors, paddles and life jackets for an hour across very heavy ploughed fields.' After a 90-minute assault crossing, The Queen's and 2 Companies attacked their respective villages at 5am and cleared them of enemy by 8am before starting to dig in. The Warriors rejoined them two hours later. The Grenadier battle group could not attack the 17/21 Lancers as a funeral brought the exercise to an abrupt halt. The next two days were spent in defence.

On Friday No 2 Company deployed 29 miles south to a hide in woods near the small village of Wangelnstedt to the south-west of Hildesheim. It started snowing quite heavily and the temperature plummeted. The Company, with the 14/20 Hussars again, was responsible for providing the backstop to the Brigade position, plugging the north-west exit out of the Einbeck Bowl. The Brigade mission was to deflect the enemy into the ambush/killing area. Maj Lloyd deployed two platoons dug in on a reverse slope with observation posts forward to give early warning. There were not enough men for standing patrols forward. These platoons were commanded by 2Lt A Holloway and by an Army Air Corps officer on a six-month attachment. The third platoon, the Corps of Drums, was detached to battlegroup Headquarters as a reserve. Maj Lloyd's other two officers in Company HQ were Captains NW Welch of the Royal Australian Regiment and TDG Morris of 7 Gurkha Rifles. (The Rifle Companies had eight officers of other Regiments serving with them – reflecting the temporary shortage of Grenadier officers. Regimental Headquarters Grenadiers' efforts to recruit officers in the 1983 – 1986 period had proved inadequate. Col DV Fanshawe OBE was brought in to recruit more officers, which he did with great success.)

Warrior showed its advantages. The Grenadier Companies prepared false positions, lightly manned and supported by a few Warriors. They were highly effective. 9 Platoon caused a battlegroup from the German Panzer Brigade acting as enemy to deploy for a formal attack. Warrior was able to scoop up its dismounted Grenadiers and whisk them away before the enemy could come to grips.

No 2 Company found it very difficult digging in as the ground was so hard. There was no Royal Artillery forward observation officer or mortar fire controller with the Company, although artillery batteries were available in support.

Unfortunately the onslaught of winter, with blizzards during the day and temperature lows of -20° at night prematurely brought tracked action to a halt for 21 hours. It was so cold that sentries were rotated every 15 minutes. Even so, three Guardsmen suffered frostbite. Maj Lloyd waved down a passing Land Rover to have them evacuated. Fortunately the Brigade Commander, Brig JR Cordy-Simpson, was in it and saw the problem at first hand.

To No 2 Company's frustration the enemy, 23 SAS (TA), failed to reach them. There was no worthwhile activity for three days, whereas the other Grenadier Companies saw action. Maj Lloyd's 'enemy' had become the cold. It was one of the few exercises when Guardsmen needed no encouragement to wear their NBC suits. No 2 Company's next phase, with the Battalion once more, involved a divisional night move. Two minutes before crossing the start line for a 30-mile westward night deployment one of the lenses of Maj Lloyd's glasses fell out; he was leading his Company vehicles.

As weather conditions improved, the Division was re-armed, re-supplied and moved to new assembly areas for the counter-attack. The re-supply phase included moving all the Divisional artillery and support elements across the River Weser by floating bridges formed by M2 Alligators.

'We crossed the river amidst three brigades and moved into Company hides for over 24 hours,' noted Maj Lloyd. 'At a Battalion orders group there was great excitement when we were told that 3 Armoured Division would practise the counter-stroke. It would be used for real in the event of a Russian attack. We moved up to the start line, observing radio silence. Suddenly we were told it was the end of the exercise. The ground was now thawing, there was lots of slush, making more damage inevitable and problems of public safety arose. Moreover, another weekend, when exercises with vehicles on roads was forbidden, was approaching. It was a frustrating and sad conclusion but much had been learned. Warrior proved itself, albeit in a more limited fashion than was wanted.' Most of the Battalion took part in at least one battle, although No 3 Company having dug five defensive positions, participated in only one attack. The Divisional Commander, Maj Gen MJ Wilkes, declared exercise Iron Hammer a success. (He later commanded the Field Army in UK.) Some Grenadiers felt the name of the last big exercise in BAOR was a poor choice. It was referred to as 'Rubber Mallet' by some thereafter. But large divisional exercises in BAOR seldom lived up to expectation since keeping many thousands of soldiers engaged in battle for several weeks was impossible when environmental and financial considerations were paramount.

In 1989 the 'Warrior Sword' was awarded to the Champion Warrior company of BAOR. It was won by No 2 Company.

The highlight of the training season was a demanding all-arms live firing exercise held in Suffield on the Canadian Prairie in June and July 1989. It was a unique opportunity for all elements of the battlegroup to participate in realistic tactical exercises, firing live ammunition over an area the size of Dorset. Such ranges simply do not exist in Germany, let alone the UK.

The Grenadier Battlegroup consisted of The Queen's Company (Maj JP Hargreaves), No 2 Company (Maj JS Lloyd), A Squadron 14/20 Hussars, D Squadron 17/21 Lancers, N Battery 2 Field Regiment Royal Artillery, an engineer troop from 25 Engineer Regiment and numerous other minor attachments. No 3 Company (Maj AD Hutchinson) had undertaken an earlier exercise as part of the 17/21 Lancers battlegroup.

Before leaving Germany the Grenadiers carried out a two-week work-up exercise at Soltau.

Capt JA Sandison QGM, the Unit Emplaning Officer, oversaw the series of flights by RAF Tristar from Münster to Calgary. Regrettably, the two rifle Companies had to leave their Warriors in Germany; Suffield was equipped with the old, slow unreliable AFV 432.

The first seventeen days were spent on the training area. The first phase of five days involved special-to-arm training. Plenty of ammunition was available as the rifle Companies developed their infantry skills from section up to Company level. This phase culminated in a spectacular Battalion night attack in which tracer, mortar illuminating and MILAN anti-tank rounds lit up the night sky.

The second phase saw the start of the all-arms cooperation; mechanized infantry and armour trained together. In the final phase all elements of the battlegroup came together for two all-arms exercises with live artillery support on call. Obstacles were crossed with the support of the Royal Engineer troop. This final exercise was run by the British Army Training Unit Suffield staff. Sadly it was at this juncture that a tragic accident befell three members of the MILAN platoon. While digging a trench they hit a high explosive shell. The ensuing explosion resulted in the loss of both legs below the knee of L/Cpl J Ray, Gdsm A Hicks and Gdsm S Povey.

The exercise covered all phases of war and provided a comprehensive test of everything from command and control to the logistic resupply of over 100 vehicles and 1,000 men.

It can be seen from exercise Iron Hammer in Germany and the training at Suffield how important it was for regular officers to serve in Germany. The experience gained was invaluable. Officers serving too long on Public Duties 'missed out' and their careers suffered accordingly.

With the exercise at Suffield over, it was now time for more adventure training. 2Lt ARG MacLellan led a trekking expedition to the Banff and Jasper National Park. Lt AA Pollock and Lt AS Mackie enabled more Grenadiers to do everything from ski mountaineering to horse riding.

The most remarkable exercise took place in the Golden Ears Provincial Park, British Columbia and was organized by Capt RMT Reames, the Battalion Signals Officer. Twenty-three years earlier a prison officer, Mr L Evans, was tasked to find a suitable logging trail for prison inmates up Gold Creek Mountain. He and his eldest son were killed in an avalanche; their bodies were never found. The mountain was renamed Evans Mountain in their memory. Two younger sons, G and D Evans, joined the Grenadiers and by 1989 had risen to the ranks of C/Sgt and Sgt respectively. They helped Capt Reames organize the hugely successful expedition which included twelve members of the Signals Platoon.

Meanwhile 2Lt RT Maundrell with a team of twelve participated in the gruelling four-day Nijmegen Marches – 25 miles a day starting sometimes as early as 3 am.

The Battalion returned from summer leave in 1990 expecting a quiet period prior to the handover of Oxford Barracks and the Warriors to 1st Battalion Coldstream Guards. It was not to be. On 2 August Saddam Hussein invaded Kuwait. In response to American and Saudi requests for help, the Ministry of Defence launched Operation Granby. The Battalion was immediately ordered to produce 116 men and most of the Warrior fleet to deploy to the Gulf, with seven days' notice to move. Five weeks later they were in Saudi Arabia. Most of the Battalion was to follow. The liberation of Kuwait and defeat of Saddam Hussein's Army is described in Chapter 28.

In late March 1991 the Battalion was reunited in Münster. Looking back on the Regiment's last tour in BAOR it may be of interest to summarize some views on what had changed.

Maj EF Hobbs had served four tours in BAOR within twelve years. Interviewed in 1991 he said, 'There were dramatic changes. Starting with the equipment, Warrior revolutionized warfare as it demanded a new speed of thought and speed of action: secure orders must be quick. Soldiering has become so much more professional. There was a marked change in German civilian attitudes to us. We felt we were no longer wanted in Germany. We could no longer deploy in some streets in the rush hour and had to be in packets of only four vehicles. The Battalion was no longer "crashed out" as if a nuclear strike was imminent. I haven't seen the General Deployment Plan since 1983, perhaps because no firm plan existed. It was instead a question of being committed when the enemy threat was known. I recently travelled extensively in East Germany and beyond. They had kept some gardens beautifully but buildings' stonework was falling to pieces. The roads were in terrible order. On driving back from Czechoslovakia, I saw less than forty cars in over 100 miles.'

Mrs RC Radford, married to the RQMS, also recalls the changes. 'When I first went to Germany in 1972 I was very much "wife of Sgt Radford". I received no welcome and my married quarter needed a good cleaning, whereas when I joined this time the Commanding Officer's wife brought round some wine and flowers. Wives have now got 'phones, TV, microwave ovens, washing and drying machines, some of their own furniture and cars. Twenty years ago we had none of these things and very few local jobs. Few wives could drive and so we couldn't go far. Due to the IRA threat we are more security-conscious now. We are taught to close the curtains before turning the lights on. The older children search the cars for bombs and lock the doors. The funniest thing that happened to us was when Prince Philip visited us at a Grenadier Garden Party at Windsor almost ten years ago. My nine-year-old daughter asked him where he came from. "Windsor", he replied. "Do you know the Bricklayers' Arms Pub?" she asked him. "I don't know that one," he replied!'

One Company Commander's wife also commented on the differences: 'Comparing Münster in 1979 with ten years later, the type of Grenadier wives had changed significantly. By the late 1980s they had become more ambitious, more intelligent, more affluent, more everything. Why? Perhaps because they were better educated, had travelled more; they were not as apprehensive of being abroad. There was the same separation and the standard of living was always quite good due to the local overseas allowances, cheap petrol with coupons, and cheap drink – all better than England. But the changes in the wives was noticeable.'

Mrs Joanna Houstoun remembers that 'it was this change that instigated the Grenadier Wives Association set up in 1988. Previously, the Wives Clubs had been organized and run by the Officers' and Sergeants' wives, but I recognized there were many Guardsmens' and Corporals' wives, not only capable of organizing their own activities, but wanting to do so. The GWAC, as it became known, was an instant success. Voted democratically on to a committee, wives across the rank structure organized their quality of life in Germany – be it clubs, outings, creches, sports events, community centre activities etc. The most successful side was the welfare, as has been previously mentioned. No wife arrived in the Battalion without a warm welcome and hospital visiting was shared by everyone. Quite unforeseen, the GWAC really came into its own when the husbands were suddenly sent off to the Gulf. The infrastructure for support was already there.'

RSM CC Savage also noted the changes: 'The young Guardsman six years ago on coming to Münster had extremely limited interests when off duty – girls, drink and dancing. Gradually the explosive situation "down town" receded for a number of reasons. Because of the well known reputation for discipline which the Regiment has with the Royal Military Police, jurisdiction for minor incidents was passed directly to our Commanding Officer as they knew culprits would be dealt with. The German Civil Police soon adopted a similar policy. A large proportion of the Guardsmen settled down with regular German girlfriends and/or a circle of friends within the local community. Our young men became more familiar with the German language, to impress the locals perhaps: misunderstandings were therefore avoided. Also, a fair amount of men wanted to do something constructive with their spare time. And so the hierarchy of the Battalion set up lots of clubs and hobbies – with financial assistance where necessary. They included sailing and windsurfing, motor cycling, all ball sports, parachuting, microlighting, cycling, equestrian, computers, photography, stamp collecting, skiing, motor mechanics and much else. Some Germans joined our clubs which also helped sometimes strained relations. The Battalion's tour in Münster has therefore been enjoyable, productive and interesting: integration with the local community, although not total, has been progressive and rewarding.'

In October 1991 the Grenadiers handed over to the Coldstream and returned to London.

THE COLLAPSE OF COMMUNISM

The sudden collapse of communism in Russia, East Germany and elsewhere eclipsed even the events in the Gulf. Erich Honecker, the (East) German Democratic Republic Head of State had always maintained that the Berlin Wall 'will still be standing in 50 or a 100 years time'. The British Commandant in Berlin told a Grenadier wife that the wall would certainly not come down in his lifetime. Shortly afterwards, on 7 October 1989, President Gorbachev announced in Berlin that 'those who do not learn the lessons of history will be left behind in life'.

On 18 October Honecker was toppled from power; within three weeks the floodgates opened. The German Democratic Republic suddenly permitted East Germans to travel to the West. The Berlin Wall was rapidly demolished. In 1990 the communist system collapsed in the GDR, paving the way for German reunification.

With a suddenness no one had forecast the prospect of Soviet Forces flooding into Western Europe, with the inevitability of nuclear defensive operations and retaliatory action by the aggressors, evaporated – almost overnight. War in Europe which for so long had been central to all our military thinking and preparedness ceased to be a threat to our very survival. But no one rang church bells, no Lord Mayor thought of celebratory march pasts, no services of thanksgiving were ordered in the great cathedrals of the land. In just a few short weeks the greatest and most significant event in all our turbulent recent history faded into insignificance. The British Army of the Rhine, together with its allies, had provided for more than a quarter of a century a cast-iron deterrent and the armoured infantry formations, so often incorporating battalions of the Regiment, had provided the cutting edge for the protection of West Germany and the free world. The professionalism and dedication of those so closely involved together with their families accompanying them in

Europe was a fundamental that should not be forgotten, nor should the casualties suffered in the frequent and realistic training exercises. Many sacrificed their lives or limbs in the service of NATO and their country.

In March the first free elections were held in Eastern Germany for nearly half a century. Berlin began to knit together as one city again. At midnight on 2 October the German flag was raised in the front of the Reichstag to the cheers of 400,000 people. Germany was reunited. The post-war era was truly at an end. It was one of the great successes of British post-war foreign and defence policy.

The Berliners, whose courage and character had made it all so worthwhile, will be allowed the last word. They took full-page advertisements in the Berlin newspapers which read: 'To the Western Allies Danke! Thankyou! Merci! For 45 years you have enabled us West Berliners to live in peace and freedom. You stood by us in difficult times. Your aircraft kept us supplied during the blockade and thereby strengthened our resolution and courage. This gave us the possibility to raise our city which lay in rubble and ashes to a new state of well-being. From being the occupiers of Berlin, you became our protecting powers and friends. For this we thank you. We will never forget your help. In a united Germany you will always have a place in our hearts.'

CHAPTER 27

THE FALKLANDS FORTRESS

1990 – 1992

The invasion of the Falkland Islands by the Argentinian Armed Forces on 2 April 1982, followed by the military occupation of South Georgia, was an act of unprovoked aggression – a clear violation of international law and the fundamental principles of settlement of disputes by peaceful means.

The liberation of the Falklands, of which more shortly, was achieved more rapidly than even the optimists expected. Thereafter forces were required to deter a fresh Argentinian invasion. In November 1990 The Inkerman Company Group of the 2nd Battalion Grenadier Guards provided that deterrent, for four months during the Falklands summer. The Company, commanded by Maj ET Bolitho, was 230 strong. Capt RMT Reames was the Second-in-Command.

While relations between Argentina and the UK were improving, the threat still existed. The Falklands are only 280 miles from Argentina as opposed to 8000 miles from the UK. The Islanders who had been forced to undergo the trauma of occupation still needed reassuring. The emphasis of the tour was therefore very much on operations. A camp had been built at Mount Pleasant on East Falklands, 35 miles from Stanley to which it was connected by a treacherous gravel road. The camp had an airport, accommodation block and over 3½ miles of corridors. The military garrison numbered 2000 and so almost equalled that of the civilian population. The RAF were at constant readiness with bombs on aircraft in hardened shelters. Rapier anti-aircraft systems were permanently manned. The Royal Navy's presence included a guard ship and the threat of submarines.

Maj Bolitho had five platoons under command, two from The Inkerman Company (2Lt RL Fanshawe and 2Lt SW Gammell), one from No 1 Company (2Lt JA Heroys), one from No 2 Company (Lt JW Walker then

325

NC Argles) and Support Platoon. This was commanded by Lt RJ Long and contained both 81mm Mortars and MILAN anti-tank weapons. The platoons carried out a series of tasks, changing weekly. These included guards and patrols around the islands, getting to know them and the local population, use of the Onion Range, a huge field-firing range the size of Salisbury Plain, and adventure training. The fifth platoon had a more relaxed time, carrying out adventure training, liaison visits to other elements of the armed forces, or visits to the capital, Stanley, or other parts of the islands. During one such period, L/Cpl Barber and Gdsm Pitchford successfully stalked a bull that had gone berserk on the aptly named 'Cow Island', shooting it with one round from a sniper rifle and finishing it off with a belt of ammunition from the GPMG, just to be sure. On another occasion, a section under L/Sgt Till helped build a tourist lodge at San Carlos.

The Company's mission was to provide close protection of the airport and to hold it while reinforcements were flown in.

'The tour was excellent,' recalls Maj Bolitho. 'The prolonged period of training away from England and Public Duties was particularly good for developing leadership at all levels, and we had a great opportunity to see how the other Arms and Services operated and to work with them. Relations with the locals were also very good, and they were almost always extremely hospitable to patrols visiting the outlying settlements; in return we would try and give them things that were hard for them to get hold of such as newspapers, bacon and fresh vegetables. Meat, in the form of mutton, was very plentiful but, apparently due to EEC regulations, all that we ate was imported from New Zealand. The Falklanders were still deeply distrustful of the Argentinians and when two of them tried to land on Boxing Day, they were refused permission to disembark from their cruise ship.'

The Grenadiers borrowed horses from a local farmer and covered 120 miles along the northern coastline of East Falkland in four days. Led by Lt Fanshawe, the 'Inkerman Light Horse' distributed its backpacks in saddlebags, panniers and on a packhorse, and included a member of the local Falkland Islands Defence Force. Lt Fanshawe had earlier won the Military Race at the annual Stanley Boxing Day meeting. The race started several hours late as the starter was too drunk to blow his whistle. As well as horse races, a variety of others took place including a three-legged race and sack race. Gdsm S Moran became the mile champion, beating a local who had run in the Commonwealth Games; he surged past Moran, flung his hands up in triumph and stopped, only then to realize that he had completed just half the laps. The training culminated in two Company exercises. They started with amphibious landings in San Carlos Water from HMS *Cumberland*, then the Falkland Islands Guard Ship. On the second occasion the Guardsmen showed no signs of the sea-faring traditions of the Regiment: the gale was truly dreadful. The platoons then marched and shot their way across the Falklands, moved, harried and resupplied by the RAF aircraft, using up great quantities of ammunition.

The 2nd Battalion was planning a Battalion exercise in the Falklands. Preparations were hectic, interspersed with Royal Guards. Sadly all was to change. As the advance party touched down on Ascension Island to refuel a flurry of signals cancelled the exercise. It was just as well that the Commanding Officer, Lt Col GF Lesinski, had eight further hours' flying time in which to cool down before meeting Commander British Forces, Falklands, who was most apologetic. The cancellation was related to the Gulf War (Chapter 28). But the visit did allow more members of the Battalion to see the Islands. The Commanding Officer and his three Company Commanders (Majors Bolitho, Lord Valentine Cecil and Wauchope), 'expert off-road drivers all,' managed to get thoroughly bogged down in three Land Rovers in the peat in a remote area. They were finally unstuck by an RE tracked vehicle, summoned by L/Sgt Cofre, a Grenadier brought up in the Falklands, who knew from where he could get assistance.

THE 1982 FALKLANDS WAR

Grenadiers saw much evidence of the 1982 fighting in the Falklands – crashed aircraft, old Argentinian cookers in rusty trailers, odd boots, trenches, spent ammunition and the remnants of massive Argentinian anti-tank weapons. Battlefield tours were run and commemorative services were held at the cemeteries.

Before the decision had been taken in 1982 to send the Scots and Welsh Guards to the Falklands, Lt Col HML Smith, then serving in HQ London District, had recommended to the Major General Commanding the Household Division (Maj Gen HDA Langley) that both Grenadier Battalions should go. Both had recently returned from Germany, were highly trained and commanded by two officers of great ability, both of whom became Generals (Lt Cols J Baskervyle-Glegg MBE and AA Denison-Smith MBE). However, the Scots and Welsh Guards were chosen.

The Falklands War proved to be a particularly interesting time for those Grenadiers serving in the Ministry of Defence. One of them saw the Foreign and Commonwealth Office signals being despatched late at night to British

Ambassadors and High Commissioners across the globe. The replies were gathered together before dawn, in time for the morning briefings when senior officers sat in their offices watching the in-house TV MOD broadcasts, updating them on the, usually worrying, situation.

As the Fleet sped towards the Falklands, a Grenadier was one of those responsible for a twice-weekly summary of the pros and cons of half a dozen alternative options, other than fighting across the East Falklands. They included, for example, the naval and air blockade of the Argentinian coastline instead. But ultimately reconquest of the Falklands was decided upon, despite the inevitable bloodshed.

A number of Grenadiers up to the rank of Major fought with the SAS in the war. Intelligence was obtained from SAS and Special Boat Section patrols which were landed on East and West Falklands from the Naval Task Force three weeks before the main landings. Working close to the enemy and living rough in conditions of extreme discomfort and danger, they conducted reconnaissance of the beaches and enemy defences. By mid-May 1982 sufficient intelligence had been collected and assessed to enable a decision to be reached on where the initial landings should be made. Due to the SAS involvement in highly classified operations in many parts of the world, it is the policy not to reveal the names of those who serve in the SAS, nor to provide more details.

Everyone had been full of admiration in particular for the 2nd Battalions of The Parachute Regiment and Scots Guards who had captured Goose Green and Tumbledown Mountain respectively. There had also been considerable sympathy for the Welsh Guards who were hit by an air strike when about to disembark at Fitzroy.

The Inkerman Company Group, nine years on, had plenty of opportunity to see the abundant wildlife on the islands, despite the operations and training. Huge elephant seals lay on the beaches near King Penguins and Rockhoppers.

Relations with HMS *Cumberland* had been excellent throughout the tour through the labours of Capt Reames and CSM Hague. Seven NCOs and Guardsmen had travelled down to the Falklands on board, via Gibraltar and St Helena. These had included L/Cpl Collinge from the 1st Battalion's Corps of Drums, who on arrival was immediately despatched back to the Battalion and the Gulf War.

The Inkerman Company returned to Caterham in early March 1991 after an extremely successful four-month tour. Fourteen Grenadiers travelled back on board HMS *Cumberland*. They were looked after magnificently; tales of their two months' adventures in Valparaiso, Panama, the Caribbean, Florida and Bermuda were mouthwatering for those finding Public Duties in London.

328

No sooner had the Company returned to Caterham than Lt Col Lesinski announced that deployment changes resulting from the Gulf War meant that the Battalion would have to send a company group back to the Falklands a few months later. Thus, at the end of October 1991, Maj JP Hargreaves and No 2 Company deployed there, with many Guardsmen going for a second time, having enjoyed the first so much, or to avoid another English winter.

4 Platoon (2Lt JA Waldemar-Brown) initially guarded the airport; 5 Platoon (Lt JM Lindsay) was based on a hutted camp for adventure training; 6 Platoon (2Lt AL Grinling) went live-firing at Onion Range, while 7 Platoon (2Lt BGT Stephens) acted as the Battalion's quick reaction force. As previously, the platoons switched commitments weekly. Support Platoon was commanded initially by Captains RJ Long and then RL Fanshawe. The Company Second-in-Command was Capt MGP Lamb.

The adventure training consisted of rock climbing, hill walking, canoeing, wind-surfing and water skiing. C/Sgt TA Sentance found the fishing in streams for brown and sea trout 'unbelievable'. They were cooked on makeshift fires when the weather was good enough. So many fish were caught that many were returned to the streams.

Patrolling the islands was most enjoyable. Patrols lasted four days at a time, being dropped off and collected by helicopter or boat. Water was obtained from isolated settlements. The soldiers carried sleeping bags and slept beneath small tents. It was possible to visit the more distant islands in a light aircraft. The locals were nicknamed 'Bennies' after a TV serial. They, in turn, called the soldiers 'When I's' or 'Have-Beens' because the soldier so often said, 'When I was in Northern Ireland . . .', 'when I was in Belize . . .'; alternatively, 'I have been to Germany. . . . I have been to . . .' CSM A Prentice went on operations on HMS *Jupiter*. Another patrol, which was being deployed to an island in a fisheries protection vessel, was diverted for four days in high seas and a force 10 gale looking at Argentinian ships which were fishing illegally.

No 2 Company made a great name for itself, as had The Inkerman Company. They came first and second in both the march and shoot and the falling plate shooting competitions. However, this time the Company Commander was only able to manage third place in the military races in the Stanley Boxing Day Race meeting.

The Company live-firing exercises were again held – each over three days. Both involved platoon and company attacks with direct support from the Mortar platoon and fighter ground attack aircraft together with resupply by air drop.

* * *

On both tours the Grenadiers maintained British Sovereignty in South Georgia – some three and a half days steaming (or 800 miles) south-east of the Falklands. It is the most isolated British garrison in the world. The climate is inhospitable with an intense wind. The Island is largely covered in ice, the remainder being high mountains of very brittle rock.

On the first tour the garrison, numbering forty, was commanded by Maj PZM Krasinski with Lt W Coulson commanding the recce platoon. On the second tour the commanders were Maj TW Jalland and Lt JW Walker. Eight Royal Engineers and four soldiers in the Royal Signals were among the garrison.

The accommodation consisted of a long block which looked as if it had been made out of MFO boxes. Two resident Royal Marine instructors ran a course on mountain and arctic warfare. The first week was spent learning how to survive in the Arctic environment, together with short patrols carrying extremely heavy weights. During the next six weeks sections alternated between military training and the arctic warfare course.

Every two to three weeks a Hercules aircraft dropped mail into the sea where it floated for 15 minutes enabling it to be collected. If the RAF misjudged the wind, the mail landed on the inaccessible shore which made it difficult to collect.

Every six to eight weeks cruise liners visited South Georgia with American and German tourists. They came ashore for up to seven hours to see the frontier spirit atmosphere. Some Grenadiers, in combat dress, were invited back to the liners to enjoy the caviar, champagne and telephones which enabled them to ring their homes. The Americans seemed obsessed with the British soldiers' accents. One tourist from Manchester, a former Commando Lance Corporal, told everyone that he had made a fortune making bomb-proof dustbins for royal palaces.

Time passed quickly. On the second tour a hurricane destroyed all communications for a week. Communications were difficult at the best of times and two soldiers were permanently on duty on the radio.

Reindeer were occasionally shot with sniper rifles to provide fresh venison. The ration of beer was two tins a night. Roderick Newall, a former officer in the Royal Green Jackets, lived with the Grenadiers for almost three months, apparently surveying the coastline for the British Government to improve the maps. He had earlier murdered his parents for which he was convicted in 1994.

During the weeks leading up to Christmas 1991 the Grenadiers learnt many new skills including top-rope and abseiling, mountain rescue, snow and ice climbing and crevasse rescue. After a very special Christmas there was a three-day garrison exercise before the patrolling and training resumed.

It was a beautiful morning when they left in March 1992. Gradually the mist came down as they sailed away, and South Georgia had gone. The Grenadiers were left with many memories of things achieved: memories that will last a long, long time.

CHAPTER 28

THE GULF WAR

1990 – 1991

On 2 August 1990 Saddam Hussein of Iraq invaded Kuwait. The immediate source of conflict was the Rumaila oilfield which straddles the Iraq/Kuwait border. Kuwait City quickly fell to the invaders; this was the first time in modern history that an Arab nation had invaded and totally controlled another Arab state.

Saddam Hussein, with the fourth largest Army in the world, had further ambitions: he massed 100,000 troops on the border with Saudi Arabia. The United Nations Security Council immediately condemned the invasion and demanded Iraq's withdrawal. The oil in the region is essential to the industrialized world. Vital interests were at stake. They were worth fighting for.

Within a week the United Kingdom announced that it would follow America's example in sending forces to the Gulf. The initial objective was to deter any further Iraqi aggression. The countdown to war had started. The objective was expanded to securing the complete Iraqi withdrawal from Kuwait.

*   *   *

The 1st Battalion Grenadier Guards in Münster was preparing to leave Germany for Wellington Barracks. The pre-advance party had already returned to London; the families were packing their crates; family pets were in quarantine and wives' cookers were being cleaned ready for handover.

On 14 September the Battalion, commanded by Lt Col EH Houstoun, who was about to hand over to RG Cartwright, was tasked to provide eighty-five men to reinforce the 1st Battalion The Staffordshire Regiment and bring them up to wartime establishment. Also required were twenty-three as Battlefield Casualty Replacements. As the pioneers of Warrior, the

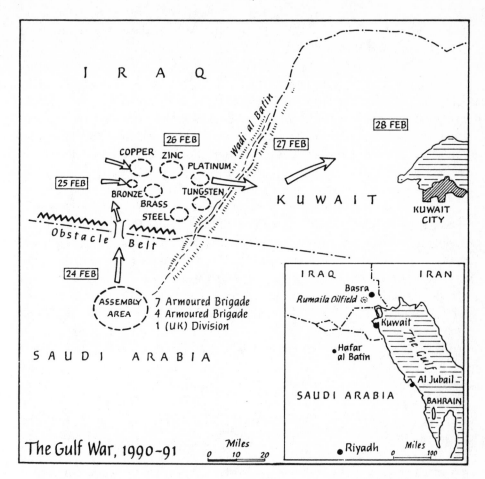

The Gulf War, 1990-91

new armoured personnel carrier, it was desperately disappointing for the Battalion that it was not chosen as one of the battlegroups to go to the Gulf. Lt Col Cartwright does 'not believe that preparations for our return to the UK had anything to do with the choice of infantry battalions for the Gulf. It was because 7 Armoured Brigade was chosen and not our 4 Armoured Brigade. The Battalion then lost so many men and so much equipment to 7 Brigade that, when 4 Brigade was needed, the Battalion was beyond reconstitution.'

As the equipment and vehicles, ready for handover, were of an extremely high standard, it came as no surprise when most of the Warriors were taken by 7 Armoured Brigade for the Gulf.

The battlegroups destined for the Gulf were brought up to war establishment without calling out the individual reservists, the necessary increments being found from other Regular units in BAOR. Hence the breakup of some units to achieve this.

The difficulties of evicting the Iraqis from Kuwait appeared formidable. One Brigadier who had commanded the British Army Training Unit in Suffield, Canada, warned that: 'In BAOR we are accustomed to a defensive posture; therefore the Engineers will have a major problem on crossing the Iraqi obstacle belt in the attack, particularly as they are not fully NBC equipped. The Iraqi artillery has a considerable range, and in the desert our fuel consumption will be double that in BAOR. The enemy will be very, very well dug in.'

Some ninety Iraqi brigades were established in and around Kuwait, opposed by only twenty-five Coalition (allied) brigades.

The most favoured option for the Coalition attack was a wide left hook out into the desert to the west and back into Kuwait to the north of the city. But there were insufficient ground forces to carry this ambitious manoeuvre through. The American Government therefore agreed to huge reinforcements. The British decided that their contribution should also be increased – to divisional strength with the inclusion of 4 Brigade and an artillery brigade – all to be operational in Saudi Arabia within ten weeks.

The Grenadier Battalion had already lost over 100 men and all its Warriors and so there could still be no question of sending a Grenadier battlegroup. Two armoured infantry companies, each of 148 men, were therefore formed, largely by disbanding No 3 and Support Companies. The Queen's Company (Maj GVA Baker) was earmarked for 1st Battalion The Royal Scots and No 2 Company (Maj AC Ford) for 3rd Battalion The Royal Regiment of Fusiliers. More Grenadiers were required for HQ 4 Brigade's Guard Platoon. Lt FS Acton commanded it initially but had to be invalided home with eye problems. He was replaced by Lt MG Newton. Grenadiers also reinforced 1st Battalion Scots Guards. The Senior Major, CRJ Wiggin, and his successor, REH Aubrey-Fletcher, were despatched to HQ British Forces Middle East in Riyadh to serve on the Land Operations Staff, leaving a dwindling rear party. By early January 1991 Lt Col Cartwright, Maj EF Hobbs and RSM CC Savage were also en route to the Gulf, as part of the Prisoner of War Monitoring Cell. They were followed by Capt S Tuck BEM and C/Sgt GJ Couzens who, at seven days notice, were trained, documented, jabbed and in Saudi Arabia to organize a Scandinavian Hospital! Grenadiers also served with the SAS in the Gulf. Lt CTD Hewitt was to join the UN observer mission in Kuwait.

By early February some 500 Grenadiers and attached had been divided between fourteen different units in the Gulf. Gen Sir Peter Inge, Commander-in-Chief, BAOR, later told a Grenadier that, with the benefit

46. Training with 1st Battalion, the Staffordshire Regiment, prior to moving towards the Kuwait border and the start of the Gulf War ground offensive in 1991.

47. Company Headquarters of The Queen's Company in The Gulf in 1991.
Standing in front: Lieutenant M Manning, RHG/D (Gunnery Officer), CSM D Barrett, Major G V A Baker, CSM K Smith, CQMS V Overton, Captain G K Bibby.

48. The planting of the Hornbeam avenue at Windsor in 1992 to celebrate The Queen's fifty years as a Grenadier. The Queen and the Colonel speak with (left to right) Major General B C Gordon Lennox, Brigadier M S Bayley, Captain B D Double, Major E C Weaver and Mr B Flemming.

49. A Lynx helicopter at Crossmaglen base after being struck by an IRA mortar bomb in March 1994. The helicopter crew successfully crash landed. As the Grenadiers fought the fire, ammunition in the burning helicopter exploded.

of hindsight, it would have been better if formed sub-units of one cap badge had been kept together, a view with which all Grenadiers would agree. 'The way in which the Battalion was broken up was outrageous,' wrote Maj Baker afterwards. 'It demonstrated (to us at snail's eye level!) a disturbing lack of clarity of thought up the chain of command. The firm impression was of panic-stricken crisis management.'

Preliminary training for both The Queen's and No 2 Companies in Germany had a great sense of urgency. Time was short. Warriors were obtained from other units throughout BAOR. No 2 Company gathered theirs, in various states of order, from the remnants of the 1st Battalion and from 1st Battalion Scots Guards at Hohne. With considerable grace the Scots Guards Commanding Officer donated a 30mm cannon taken from his own vehicle, to make good No 2 Company's deficiencies.

The Warrior crews were hurriedly despatched to Bergen Hohne ranges, and the dismounted sections to Sennelager for their respective firing packages. Both Company Headquarters spent two days at the Battle Group Trainer as part of the 14/20 Hussars battlegroup with whom they were to serve on exercises and in the Gulf, despite being attached to the Royal Scots and Fusiliers respectively. On return to Münster, three weeks were spent on individual skills such as fitness, first aid and nuclear, biological and chemical (NBC) warfare. Everyone received five days' leave to allow them to say goodbye to their families in Germany or England. 'The support for us in England was very strong and we felt very proud on returning to Germany,' noted 2Lt CIT White-Thomson.

Advance parties left for Saudi Arabia on 20 December. They were based in tented camps at Al Jubail.

On Christmas Day Lt Gen Sir Peter de la Billière, commanding the British forces in the Gulf, visited the Staffords where he found 'a platoon of the Grenadiers which had been blistered on to the Staffords to make up manpower shortages,' he wrote in his memoirs *Storm Command*. 'The Grenadiers had tremendous pride in their own Regiment, of course, but they had fitted themselves into a strange unit with typical application. One of their NCOs told me that, far from settling for a second-class job, they were doing things the Brigade of Guards' way, which, naturally, was better than the Staffords' way and in fact they were showing the Staffords how to carry on!'

By mid-January The Queen's and No 2 Companies were complete with their Warriors in the Gulf. Several weeks of hard training in the desert followed, starting at Company and graduating to battlegroup, brigade and divisional level. The Warriors were given add-on armour to defeat shoulder-launched anti-tank weapons.

Tactically the training was designed to reduce every manoeuvre to a set drill so each man would instinctively know his part. For example, The Queen's Company therefore usually had 1 Platoon (2Lt CIT White-Thomson) forward left, 2 Platoon (Capt MCJ Hutchings) in reserve and 3 Platoon (Lt CGF James) forward right. Hard though the training was, it was prejudiced by lack of track mileage, training ammunition, range space and time. The Companies carried out less manoeuvre training and less 30 mm Rarden and Chain Gun live firing than they would have wished.

While such shortages were thoroughly unsatisfactory, they were understandable as a result of the need for economy. The British Army had had no plans to send an armoured division from BAOR overseas; after the Falklands the emphasis remained on lighter, airmobile and amphibious forces for areas beyond Europe. The despatch of an armoured division to the Gulf with its special conditions and terrain as well as the perceived type of warfare, placed unexpected demands on both manpower levels and equipment. BAOR was stripped of many vehicle parts to provide adequate reserves and as a result its combat effectiveness was severely reduced.

Navigation was a concern as, unlike Germany, there were very few landmarks and it was almost impossible to see anything when closed down in the turret, nor could a compass then be used. Fortunately key vehicles received satellite navigation systems which gave their location and enabled the commanders to go from one point to another. However this equipment broke easily and was ineffective in the early morning and early evening as the signal from the satellites was too weak.

By now air supremacy was complete: RAF morale had been shaken by their initial losses. It was known to the most senior Coalition planners that the Iraqi re-supply ability had collapsed. Moreover the Iraqi world-wide intelligence-gathering capability had been snuffed out like a candle long before the ground war was about to start. Their communication intercept Headquarters, which had been considerable, was destroyed by bombing.

But not everything was going to plan. Lt Gen de la Billière felt that unless British Ministers increased the flexibility of the rules of engagement, the RAF would have to withdraw their aircraft from the American and Saudi plans. The rules of engagement for the Navy were equally difficult. He could not get the decisions he wanted from the Ministers. Rules of engagement in the Falklands War had been much less of a problem since there was no ally to consult.

Everyone was horrified when the BBC announced that an RAF Wing Commander in London had had a lap-top computer stolen from his car. It contained an outline version of the American war plan. The story was blown by the *Irish Times*.

As in the past Saddam Hussein had used nerve gas, mustard gas and blood-poisoning agents in battle and was believed to be prepared to use bacteriological warfare agents; preventative measures were taken. Inoculations were given against various BW threats. They, as with any vaccine, gave some personnel side effects such as flu-type symptoms. In addition Nerve Agent Protective tablets were taken as a precaution. There were some concerns among the Guardsmen about the unfamiliar injections.

In January 1996 the Ministry of Defence ordered a fresh study into claims by 700 former Servicemen and women that they are suffering from 'Gulf War Syndrome'. Evidence of abnormalities among babies born to Gulf veterans will also be examined. The 700 have complained of a range of illnesses including chronic fatigue, kidney failure and dizziness.

Both Companies participated in a two-day Brigade exercise from 23–24 January before moving to the concentration area. The vehicles were carried forward by low-loaders on the 25th while the Companies slept on the road-side in torrential rain and a bitter wind. The next morning they moved forward by truck, Hercules aircraft and Chinook helicopter. Maj Ford was with thirty Guardsmen on one of the Chinook flights. He had received no map, grid reference or orders, so was rather surprised to be dropped by the RAF in featureless desert clearly nowhere near his anticipated destination. Using a compass he led his men seven kms southwards, away from the border, in the blazing heat hoping to meet allied troops. The first encounter was with an Egyptian convoy which was entirely lost. The second was with an American group, equally lost, and finally he met the Fusilier Operations Officer who, although lost, had a map and radio!

By 27 January everyone was in the concentration area, reunited with their vehicles. They dug fire trenches and sleeping areas in case they were attacked by Iraqi air force or artillery. A ten day period followed for final preparation for battle, and for the last rehearsals which proved to be fairly chaotic. The problems of trying to move two brigades and their logistic elements quickly through a major obstacle, and then through an American Division possibly under chemical attack and at night, can be imagined.

Despite the dangers, the media was everywhere. One unexpected figure who appeared close to the battlefields was HVL Smith who wore the rank of a Grenadier captain although he had left the Regiment and Army long before. He had arrived to video the campaign. He was discouraged from going forward, for his own safety, but did so.

The Queen's Company had a morale-boosting visit from Lt Col Cartwright and RSM Savage, both on a short break from their Prisoner of War Monitoring Team. One of the lessons from the Falklands War was that there should be such teams. Lt Col Cartwright was able to dispel many rumours and give reassuring news of the families being looked after in

Münster by Maj NJR Davies, Lt RAJ Phasey (the Families Officer) and Caroline Cartwright.

Maj Davies's rear party included 192 wives spread over 18 kms, with a further twenty-five in England many of whom felt so cut off that some returned to Germany. Many fatal casualties were expected. Maj Davies therefore had forty assisting visiting officers standing by, usually with wives ready to accompany them. Branches of the Grenadier Guards Association in England had designated members on standby to make follow-up visits to casualties' relations.

The rear party and families in Münster faced a few demonstrations by Germans against Britain's involvement in the Gulf War (although Germany needs oil as much as others).

Mrs Harper recalls that 'It was unpleasant to find that Arab scarves had been left by some Germans for us in some 'phone boxes. "Blood and guts" had been painted on a few others.' Some German houses hung white sheets over balconies near the Grenadiers' barracks. White doves, symbolizing peace, had been painted on buildings.

Other Germans, particularly those of an older generation, were more sympathetic; typical comments to Grenadier wives in buses were: 'I'm very sorry we can't do more,' and 'How sad that it takes a war to bring us together.' The German armed forces, although fully committed to NATO, were then constitutionally bound not to serve outside Europe.

The wives did much to look after each other. One wife remembers the Grenadiers' policy that 'every person was a Gulf wife, regardless of where the husband was, or what he was doing. The Royal Corps of Transport had wives whose husbands were in Ireland and so they weren't allowed by the RCT to go to the Gulf wives' luncheons. We said we are all in the same boat regardless.'

'Prince Philip, the Colonel, came out to Münster and sat amidst mainly Grenadier wives,' recalls another. 'He was very good, sympathetic and constructive – not "gung-ho". He laughed and joked and cheered everyone up. He'd come earlier than anyone expected – a brilliant morale booster. He was amazed that so many Grenadiers were serving with different units. He was marvellous at talking to people, including children. Sally Rose and Bernard Gordon Lennox (the Lieutenant Colonel) arrived later when the fighting was at its climax – another great morale booster. She visited some married quarters which the wives really appreciated.'

*　　*　　*

On 14 February, under cover of another Divisional rehearsal, both Companies moved north to the assembly area north-east of Hafar Al Batin. As part of a deception plan, the exercise radio traffic was taped, edited and then replayed to the Iraqis to simulate the Division moving back elsewhere.

By February many of the Grenadiers, earmarked as Individual Battle Replacements, had been in the Gulf for over three months. Conditions in the camp at Al Jubail were initially thoroughly unsatisfactory. There was no equipment with which to train and nobody knew what was expected of them. The Grenadiers took a grip where possible, later doing their best to join formed sub-units such as the two Grenadier Companies. Some were successful and saw action rather than remain in a pool waiting to be called forward to replace the dead or wounded if necessary.

New equipment made its appearance. A rifle-launched bunker-busting grenade with a range of 100m, called CLAW (close assault weapon), reached the battlegroups, as did an explosive cable capable of clearing a wide lane through an anti-personnel minefield. It was called RAMS (rapid anti-mine system). CLAW broke many noses when fired as it had a massive kick.

News of a Soviet peace initiative caused hopes that a diplomatic solution would be found at the eleventh hour. But most knew that, unless the Iraqis withdrew unconditionally to abide by UN resolutions, then the ground offensive would be launched. Everyone wanted to get on with what had to be done.

At last, at midday on 23 February, The Queen's Company regrouped to 14/20 Hussars prior to invading Iraq. No 2 Company remained with the Fusiliers. Maj Ford was able to prepare a comprehensive set of battle orders and gave them formally to the Company O Group in the assembly area on the 23rd. The 1 (UK) Armoured Division plan was to protect the right flank of 7 (US) Corps as it thrust northwards to destroy the Iraqi Republican Guard Force. 7 Brigade was to be on the left and 4 on the right to attack through 1 (US) Infantry Division (who had to secure the breach) in order to defeat the Iraqi tactical reserve. The Fusiliers were to be initially rear left with 14/20 Hussars forward centre and The Royal Scots rear right. The mission given by the Fusilier Commanding Officer was to 'destroy all guns and armoured vehicles within boundaries.' No 2 Company's objective was 'to destroy four batteries of guns centred on Grid Square 3253', following a passage of lines, a move to a forming-up point and flank protection tasks for A and C Companies of the Fusiliers. Detailed tasks were given to the platoon commanders.

The general concept of operations had been known down to platoon commander level for about ten days but Maj Ford recalls a sense of great anticipation and some tension among his audience as the reality of what was required sank in. The Company had rehearsed passing through obstacle belts; it had seen the detailed layout of Iraqi positions from the satellite imagery provided, and all were aware of where potential Iraqi counter-attacks might materialize from. As a body of soldiers they were supremely confident, but there was a perception that they had reached the edge of an abyss and were unable to see what lay ahead.

In the broadest terms, working from right to left (east to west), the plan was that the seaborne US Marines should make a feint from the Gulf. The Coalition Arab units would then attack Kuwait from the south. Next to them, to the west, the US Marine Corps would engage the Iraqi forces due north of them in a major diversionary attack. To their west the Egyptians and Syrians would push to the north and then swing right-handed, to come into Kuwait City from the west.

The main allied assault involved the Americans and the British Division, part of the 7 (US) Corps, starting far out in the west. After driving north they also would swing right-handed, sweeping in behind Iraqi defences, to attack the élite divisions of the Iraqi Republican Guard and cut the front-line troops off from their communications and supplies.

The ground offensive started at 1 am on Sunday 24 February, deliberately in the way the Iraqi forces appeared to be expecting. Naval gunfire and amphibious feints from the Gulf, with frontal attacks through the defensive fortifications in southern Kuwait, were aimed at distracting Iraqi attention from the main thrust into south-eastern Iraq.

Grenadiers braced themselves for action. What would be waiting for them – minefields, barbed wire rigged with booby traps, then more mine-fields? Then tanks? Strong points consisting of dug-in tanks and trenches from which infantry would shoot? Tank traps, barriers of sand, fire-trenches consisting of ditches flooded with oil and designed to confront the attacker with a wall of flame?

On the 24th No 2 Company held a Company Church Parade in the centre of their leaguer position. The Padre attached to the Fusiliers, Adrian Pollard, said appropriate words to a congregation that seemed to hang on his every word. In the background armour was moving forward towards the breach, about 40 kms to their north. Maj Ford took the opportunity to address the Company group (that consisted not only of Grenadiers but Coldstream, Welsh Guards, Fusiliers, Queens Own Highlanders, Devon and Dorsets and REME Craftsmen), and concluded that they should have confidence in the Allied plan, that all ranks should think two levels up in order to achieve Brigade objectives, and above all else that they should enjoy themselves. The Guardsmen seemed to relish the imminent prospect of violence which caused bemusement to the Padre who had just witnessed his apparently reflective flock transform themselves into something altogether less holy! No 2 Company's Second-in-Command was Capt JCM Walters, Royal Australian Infantry, who had been serving with the Battalion beforehand as the Operations Officer. He had volunteered for war service with the Regiment at a time when he could have returned to Australia. All ranks of No 2 Company were most grateful that he did so.

Throughout the 24th 1 (US) Mechanized Infantry Division conducted breach operations against light enemy defensive positions. 1 (UK) Armoured Division remained at two hours' notice to move once sixteen lanes had been cleared through enemy minefields.

By now the enemy had been paralysed by long-range artillery, air attacks and electronic suppression of Iraqi command and control.

That afternoon No 2 Company moved to the start point and travelled the 40 kms across the desert to staging area one where they joined the 3rd Battalion Fusilier Battle Group at about 4 pm. They passed through US taxi ranks of armoured vehicles and 155 mm gun lines being 'largely serviced by black men working flat out, singing, despite the intense heat, looking like a chain gang.'

No 2 Company heard there that operations to their front were going well for the Americans and that the Iraqis were apparently content to surrender after an initial firefight. The staging areas were akin to taxi-ranks of armoured formations that could be called forward as necessary to make up appropriately balanced forces. Earlier they had been made aware of the Divisional Commander's plan for this phase: if 1 (US) Infantry Division was successful in their breach operations, then 7 Brigade (which was largely armour) would lead 1 (UK) Armoured Division. But if the US assault had faltered and 1 (US) Infantry Division tactical reserve was deployed, then 4 Brigade (which was largely infantry) would lead. Similarly, it was planned that 4 Brigade would lead 1 (UK) Armoured Division should US troops fail to breach the obstacle belt.

Maj Gen RA Smith commanded the Division. His plan was to use his two Armoured Brigades like a boxer's fists, punching first with one, then with the other. His artillery, heavily reinforced by American guns, acted in direct support of the Brigades. His objectives, designated by metals – COPPER, BRASS, ZINC, PLATINUM, STEEL and TUNGSTEN – were not features in the desert, which was almost featureless there, but rings drawn on the map round concentrations of enemy. His aim was never to capture ground, but to knock out enemy armour and guns, and press ahead. The reader will wish to follow the map on page 333 closely since it shows the objectives.

At 2 am on 25 February 16/5 Lancers, a medium recce regiment, crossed into Iraq, followed at 1145 am by 7 Brigade with the Queen's Royal Irish Hussars and 1st Battalion The Staffordshire Regiment. Let us follow the experiences of three Grenadiers who were amongst those under the Staffords' command.

341

Capt DJC Russell-Parsons was the Second-in-Command of B Company of the Staffords: 'The Company had been able to train hard from the outset, without the limitations of low track mileage and ammunition, from late September to Christmas. The operation order was disseminated early so the plan was well known. Levels of confidence were high with the more vociferous Guardsmen eager to prove that "going in was the quickest way home". The (Stafford) Grenadiers had been in the desert five months now. Many commanders had taken an opportunity to talk to the Company, including General Sir Peter de la Billière who inspired us all late on Christmas night, and Brigadier PAJ Cordingley, the Brigade Commander, who quoted Kipling at length.

'As well as the training, discipline was good. Radios were switched on to minimum power and line was dug in whenever possible to cut down on transmissions. By night there was no light or noise and my knowledge of astronomy grew with the aid of Patrick Moore's reference guide as I did my rounds of the sentries. My own Warrior was a hotchpotch of capbadges. It was driven by a Welsh Fusilier, gunned by an ex-RSM of the Cheshire Regiment and had Green Jacket and Stafford signallers in the back. Over five months of work together had moulded a tight team keenly aware of its responsibilites, if not a little bored of the steak and kidney 10-man ration pack.

'The Brigade moved together on the 23rd to its start point, and the Company continued on the 25th to its forming up point beyond the initial Iraqi positions cleared by the Big Red One. Our lane through the minefield was well-marked by white letters on red boards but the obstacle was much thinner than anticipated. We had our NBC suits and boots on throughout, ready to put on respirators at this stage. As the commanders peered out of the Warrior turrets we saw American sappers clearing mines and our first enemy – a group of seven Iraqis in a POW cage, with their officer well segregated behind barbed wire.

'As we passed the last Americans I focused on our part of the battle. The Brigade was in arrowhead, with its 120 tanks spread across a 10-mile frontage and the Queen's Royal Irish Hussars battlegroup was a very formidable arrowpoint. I felt we were unstoppable as we progressed at about 25 kms per hour and the US A10s continued to fly very low, very fast past us. The weather was deteriorating through the night and our first objective COPPER proved to be abandoned. I had a great amount of information coming through on both radio nets as we were reassigned to ZINC. At 1 am on the 26th there was an awesome thunderous roar as our Multi-Launch Rocket System and artillery opened up again. The ground trembled and the air cracked before we heard massive explosions on the objective and saw great arcs of white light. A Company were first to clear a fortified position on our left.

'B Company's first proper action came as we had continued through the night and were tasked onto PLATINUM. All our aggression focused on a static defended position and a solitary truck trying to get through us. It was destroyed by a Milan missile. Sgt Harding with his Grenadier section dismounted to pick up the pieces. An unknown number of enemy were killed as the Company cleared the position with grenades. Numerous POWS were taken. One Warrior later had a miraculous escape when it survived a direct hit from a Challenger main armament round at less than 500m. I saw evidence of an Iraqi officer who had been shot by his own men – a single entry and exit point, lying face down in the sand, very dead. We did not linger. Burning enemy tanks were blowing up beside us with huge force and bits of turret were flying about.

'Later that night we watched the C Company action as we stood ready to provide support. I did not know at this stage that we were to be the first Grenadiers into Kuwait.'

L/Sgt DA Ibbotson was the senior medical NCO in A Company: 'One Stafford had shot his best friend accidentally in the chest on the start line. I put him in the back of the ambulance. Apart from tracer, I couldn't see anything, it was so dark. After an hour a helicopter collected him but he died soon afterwards. Later, a Corporal threw a grenade too close to a bunker and got a shrapnel wound in the stomach. I put him in the Warrior ambulance, the back door of which had fallen off at the start line. It was never replaced. The 15-stone corporal was semi-conscious but I got him to the Regimental Aid Post and he survived. Another soldier sustained minor injuries when an Iraqi round hit the empty magazine he was carrying. I picked out the bits from his chest with tweezers. I treated lots of Iraqis, including one who had to have his gangrenous leg amputated.'

Gdsm DJ Chant was a Warrior gunner with C Company: 'My vehicle commander, a Corporal in The Duke of Edinburgh's Royal Regiment, fired over the heads of the surrendering Iraqis. We were told to debus and get them. The enemy position was a maze of bunkers, buildings and vehicles. Suddenly the Iraqis fired an RPG-7 rocket which passed through Private Moult before hitting my Warrior. I saw my trousers were alight and that the vehicle commander was on fire and unconscious. I knocked him hard to bring him round. The whole Warrior was in flames until I got a fire extinguisher to put them out. I felt desperately that I must find my steel helmet. The face of the dead Private Moult was perfect but he had lost the side of his body, and both his arms had blown off. The lads on each side of him were in shock.'

* * *

Let us now turn to 4 Brigade fighting to their south-west. Their objectives were BRONZE, BRASS, STEEL, and then TUNGSTEN. The Brigade consisted of 14/20 Hussars, 3rd Battalion The Royal Regiment of Fusiliers and 1st Battalion The Royal Scots.

In staging area 1 the Company was able to prepare themselves finally for war; the Guardsmen cooked and slept, the vehicles were topped up with fuel and, incredibly, C/Sgt PM Ladd MBE arrived with a mail delivery. Maj Ford received two letters from former Grenadiers – Lt Col A Heroys and Lt Col PR Holcroft. Both sent welcome notes of encouragement. C/Sgt Ladd, by the same mail, had been sent practical help; he opened a parcel that contained a nappy with a label on it suggesting that he wore it as he crossed the start line!

On Monday 25th No 2 Company moved towards their designated lane. In column they passed the Fusilier Commanding Officer, who placed a poncho against his Warrior with the words 'Good Luck and God Bless You All' highlighted upon it. As the Company passed him the vehicle commanders gave him an immaculate salute.

No 2 Company entered the breach about five kms on at about 3 pm. US Engineers were widening the lanes and busy improving the routes. They had placed a large sign at the entrance stating, 'Welcome to Iraq courtesy of the Big Red One'. Closed down in the Warriors, the vehicle crews began to see the results of modern war through their sights and vision blocks. Enemy vehicles, guns and tanks were burning. Evidence of destroyed Iraqi defensive positions were all around, yet against this background they witnessed US Servicemen standing on the side of the lane attempting to trade their personal equipment for British items. The British respirator was much in demand.

Much depended on accurate information on the enemy. Strategically the intelligence was thought to be very good. Information about Iraqi troops' positions was relayed back from satellites and surveillance aircraft which helped compensate for a lack of agents and spies in Baghdad. General Norman Schwarzkopf, the brilliant Commander-in-Chief of the American forces, later told a Congressional Hearing that Intelligence provided to him was 'so seriously flawed as to be almost useless. The US Army and Navy could not share Intelligence because their computers were incompatible.' It should not be forgotten that Western Intelligence had failed to predict the Iraqi invasion. As a result British Servicemen serving in Kuwait before the war had been taken hostage. However, Maj Baker was impressed with the Intelligence: 'I was issued with a 1 : 50,000 map which showed BRASS with individual enemy vehicles marked on it. This turned out to be almost entirely accurate.'

The Grenadier Companies received no further intelligence on the enemy after leaving the forming-up point. The Queen's Company moved off in column from the FUP at 7.30 pm on Monday 25 February.

Visibility was appalling due to the heavy rain. Night sights were blinded by the primitive Identification Friend or Foe lights on the rear of each vehicle, and the satellite navigation equipment was unable to pick up the signals of sufficient satellites to give a 'fix' – not an auspicious start. As the Companies advanced north, logistic vehicles of 7 Brigade in front approached from all angles, hell-bent on getting to their destinations. Maj Baker's satellite navigation system blew a fuse which meant that the batteries had to be renewed every thirty minutes until CSM Barrett came up with a spare fuse. The Spartan of 2Lt CS Allsop, the Company Liaison Officer, broke down at this point, for the umpteenth time during the campaign, and had to be towed thereafter by the Light Aid Detachment.

At 11pm B Squadron 14/20 Hussars reported heat sources seen through the thermal imaging sights. They opened fire. Shortly after midnight The Queen's Company was ordered to clear the position. Moving up behind the tanks, the Iraqis were seen to be in no mood for a fight. As a result Maj Baker gave orders that bunkers were not to be cleared for fear of booby traps and to avoid causing unnecessary casualties. Under Warrior's headlights the Iraqis were made to strip half-naked, in case they were carrying explosives strapped to the chest, before their surrender was accepted. The eighty prisoners were then marched off to the rear. A similar action took place at 4.15am. Some prisoners had to be carried on Warriors to the rear until Lt Newton, still commanding the Brigade Defence Platoon, was able to collect them.

BRONZE AND BRASS

The next main 4 Brigade objective was BRASS. The plan was that the Royal Scots would destroy the infantry battlegroup in the west of the area after which 14/20 Hussars with The Queen's Company would attack two tank battalions to the south. Subsequently the Fusilier Battalion, with No 2 Company under command, was to sweep through the artillery positions to the east.

The Queen's Company formed up behind the armour once more while artillery fired upon the Iraqis. The 14/20 Hussars destroyed most of the Iraqi T-55 tanks after which the Grenadiers advanced firing machine guns at the enemy bunkers. Platoons did not dismount because there was no dug in Iraqi infantry; many of the Iraqi vehicles were found to be unmanned. By now the Company was very tired, having been on the go for over twenty-four hours. The Guardsmen had largely remained cramped in the back of Warriors with little respite. But morale was high. Everyone had confidence in their training and each other. Communications were working well. 2Lt White-Thomson found that 'the trick for five minutes sleep was to light a cigarette, rest your head against the sights, and when it burnt you, you woke up – five minutes kip and no more.'

A résumé of No 2 Company's major battles to date is now required. Their main objectives to far had been BRASS and STEEL while under the command of the Fusiliers.

Leaving their lane, No 2 Company took position in the Fusilier forming up point which had been secured by 366 (US) Infantry Battalion. It was pitch black and beginning to rain. There was intense artillery noise as friendly batteries of guns and Multi-Launch Rocket Systems engaged targets to the north. Red recognition lights displayed by allied vehicles seemed to be everywhere. Radio silence was lifted and new orders were received for operations in BRASS and BRONZE. During the next twelve hours they moved quickly with the Fusilier Battlegroup and seized their objectives within BRASS and BRONZE; on each occasion formal assaults were mounted with armour and artillery support but very little enemy opposition was encountered.

The ground was a flat and featureless gravel plain, with little except wrecked Iraqi vehicles dotted about to act as landmarks. All the British vehicles were displaying their standard fluorescent recognition panels on their upper surfaces as well as inverted black Vs on the sides of their bodies and turrets.

STEEL

At about 12.45 pm on the 26th, preceded by an artillery plan that made use of six batteries of guns in direct support (during which the LAD detachment repaired a powerpack), No 2 Company and D Squadron 14/20 Hussars assaulted STEEL. The noise was intense as the Company advanced with two platoons forward (Lt RAC Duncan and 2Lt CAG Hatherley), in line abreast, out of the forward assembly point; visibility was poor but targets were being engaged with great accuracy. The radio nets were surprisingly quiet until the Gunnery Captain (Capt D Ridley), who was sitting next to his Company Commander and thus situated forward centre, suddenly noticed three Warriors from a flanking company reversing towards them, away from their own artillery rounds, at high speed into the path of an attacking friendly company. On the Battlegroup net clarification was sought but none was forthcoming. At this point a number of Iraqi T62 tanks were claimed to be ahead which caused some consternation (in fact they were Challengers). The incident served to illustrate how confusion is easily caused in the stress of battle.

Maj Ford recalls that 'the whole area of Objective STEEL consisted of dug-outs, trench systems and artillery pieces and an enemy that could not surrender fast enough, although the latter point was not immediately obvious to the crews of closed-down armoured vehicles travelling at pace. Much equipment was destroyed (especially by Lt BM Weiner's platoon which was responsible for the detailed destruction of a battery of D30 guns and associated equipment) and much damage was inflicted during this attack. We collected about 150 prisoners, many of whom were severely wounded. The Guardsmen displayed a typically generous attitude and distributed carefully conserved chocolate to the captured Iraqis. During this chaotic phase we were briefed that two Warriors had been destroyed about three kms to our rear. Shortly afterwards, whilst re-grouping, we passed the burning vehicles and noted the British casualties, victims of a US air attack, being evacuated.'

Grenadiers were not unduly shaken by the Fusiliers' casualties for it seemed that the war had just begun. Many more fatal casualties were still expected. No 2 Company had got on very well with the Fusiliers who later wrote in their Regimental Magazine: 'The relationship had been very successful and No 2 Company had become closely integrated into the Battalion. Their contribution to the success of the battlegroup on operations had been significant and they had been outstanding ambassadors for their Regiment.' The Grenadiers in turn respected the Fusiliers and later gave them a silver statuette to be presented annually by their Commanding Officer to the best NCO in the Fusilier Battalion.

TUNGSTEN

That night (the 26th) No 2 Company was regrouped to 14/20 Hussars battlegroup which entailed a full replenishment, followed by a 15 kms night march southwards, barrels reversed and red lights on, to join them in an assembly area. There they met up with the Queen's Company. Lt Col MJH Vickery, commanding 14/20 Hussars, found himself commanding no armour (all his tanks having been detached) but in their place two Companies of Grenadiers! Through the night of 26/27 February both Companies were held in reserve as an attack was mounted on TUNGSTEN. It proved to be a long night as both Companies, led by No 2 Company, moved through difficult terrain in order to maintain a tactical bound behind their forward troops.

At dawn on 27 February, still in reserve, The Queen's and No 2

Companies crossed the pipeline that marked the forward edge of TUNGSTEN. Resistance had been minimal and the Brigade Reserve, the Grenadier Companies, were not called for or needed. An extremely large oil installation was in flames. Many Iraqis waved white flags. Particularly impressive was the use of the 8-inch guns and the Multi-Launch Rocket System which caused a really spectacular explosion in the enemy position as an ammunition dump blew up in a massive orange fireball. More prisoners were taken.

Capt GK Bibby had earlier come across an Iraqi Captain who had trained at Sandhurst: 'He was terribly nice and very relieved that we had appeared. He said he would have deserted months before if he had had a white flag. His senior officers had gone off to an orders group and never returned: that's why he knew the Coalition forces must be close by.'

CSM Eldershaw adds, 'We were so hyped up about the Iraqi army, it came as a surprise to me that they had only one respirator between three men. I came across one POW – an architect – who had been in Baghdad only two days before. I had earlier been told to collect ammunition so expected a landrover. Instead three 8 tonners full of ammo turned up for us, together with a fork lift truck to unload it. We found space for it all somewhere.'

After the capture of TUNGSTEN by 4 Brigade the Grenadier Companies carried out essential maintenance on the vehicles, fed and rested.

Shortly after dawn on the 27th No 2 Company re-grouped to the Fusiliers, taking in their stride some 1500 prisoners. The battlegroup moved almost immediately in preparation for a counter-penetration task astride the Wadi Al Batin to prevent any movement south. The task was cancelled as they had just started to prepare the position. They had their first night's sleep since the 24th.

By now the Iraqis were fleeing towards the north-east. At 6 am on 28 February 7 Brigade headed east at high speed to cut the main road to Basra, north of Kuwait City, to prevent the Iraqis escaping. The rest of the Division followed in column over difficult country still dotted with Iraqi positions and sown with mines.

The battlegroups containing the Grenadier Companies covered some 60 kms into Kuwait in about two hours. The advance across the Wadi Al Batin was fast and exhilarating. In all, the Division advanced over 300 kms, destroyed most of three Iraqi armoured divisions and took more than 7000 prisoners.

At 10 am all forward movement was stopped. A ceasefire had been announced. Some soldiers, inevitably, had a feeling of anti-climax. 'The ground war proved to be short and sharp thanks to allied air power and the unexpected direction of the attack, largely against a non-playing enemy,' wrote Maj Baker later. 'In the course of the odd very minor skirmish most of The Queen's Company managed to fire a shot in anger, although none of us experienced any bullets whistling around our ears. The war altered peoples' perspective of what is important in life. We all grew up a lot.'

The Companies moved into leaguer and began to relax. Lt Col Vickery congratulated the Guardsmen on their fine performance. The Grenadiers, in turn, were full of praise for the 14/20 Hussars.

The war was over. The United Nations mandate had authorized the Coalition forces to use all necessary means to eject the Iraqis from Kuwait. This had been accomplished. They did not have the mandate to invade Iraq. If the Coalition had tried to do that, the Arabs who had fought with them would have opposed it.

The final weeks in the Gulf were spent on sport, training and preparing the equipment for the return to BAOR. Everyone longed to return to their friends and families in Europe. Some fired the Soviet AK47 rifle which, unlike the British SA80, worked perfectly, despite the sand and dust. Some Grenadiers also felt that their hand grenades were insufficiently powerful; the new light support weapon was not as good as the old general purpose machine gun; and the Warrior's Image Intensification sight, with a range of 500m, was mismatched with Challenger's Thermal Imaging sight with a range of 2–3 kms. Warrior, on the other hand, had performed well.

Those Grenadiers who went into Kuwait City were amazed at the devastation – the burnt-out British Airways aircraft at the International Airport; T55 tanks hanging over partly destroyed bridges and burnt-out shells of buildings. The sun was seldom visible due to burning oil. Grenadiers heard a subdued roar which filled the air. It was the noise of oil wells burning all round the horizon.

There was a continuous flow of visitors including a further visit from Prime Minister, John Major, who spoke very well.

The return home began in mid-March. Despite arriving in Münster in many different groups, a wonderful reception was arranged for everyone. Wives and children were waiting in the barracks whatever the time. They had gone through great stress due to the media forecasting heavy casualties and the impact of seeing news reports of the fighting on TV.

The Queen's Company really tested the system by arriving at 3.30 am on 4 April but even this did not dampen the excitement or dry the tears. It was a wonderful home-coming. For some there was even a German band playing 'The Grenadiers Return'.

Representatives of all ranks gave their reminiscences of the campaign on tape. They will be of considerable interest to historians and others in the years to come. The tapes are in Regimental Headquarters. Gradually all the Grenadiers met up again. They included Capt RT Maundrell after an attachment with the Scots Guards, and men such as Sgt DA Trayner who had done outstandingly well in the Force Maintenance Area at Al Jubail. As they all left the Gulf, Capt RG Adams was sent there to participate in cleaning up operations.

Grenadiers were well represented in the Victory Parades in America and London. Even so, there was a definite feeling in some areas that the Battalion had not got the recognition it deserved.

Forty-seven British servicemen had lost their lives in the Gulf, many of them in accidents. 'I believe strongly that Grenadiers did not suffer significant casualties during or after the conflict on account of our strong code of discipline,' wrote Maj Ford. 'During the war we did not dismount our Guardsmen to do something that turret crews could do from behind the protection of armour. Similarly, after the conflict we did not allow our Guardsmen to collect equipment and run the risk of encountering booby-trap devices. There were very few negligent discharges with personal weapons. Professional military practice, particularly by our NCOs, helped ensure that everyone who deployed to the Gulf returned safely to Germany.'

The war showed that different forces can fight under a Coalition flag, and, if really necessary, with different capbadges within a unit. The United Nations proved that they can respond effectively, on occasions, in the face of unprovoked aggression.

Although no Grenadier was killed and the Regiment's Colours now proudly carry the Battle Honours 'The Gulf' and 'Wadi al Batin', it is worth remembering that, in many ways, it was a deeply unsatisfactory time for the Regiment. The Battalion, acknowledged to be the most experienced Warrior battalion in BAOR, was broken up and split between a host of other units. Going to war is difficult enough, but the cohesion produced by regimental pride and spirit makes it easier. The strain on those deployed to other regiments was considerable, as it was for those Grenadiers and families who saw them leave Münster for a double unknown. While Grenadiers were not the only ones to suffer in this way, all would agree that it was not the right way to go to war and its necessity was questionable.

CHAPTER 29

REGIMENTAL HEADQUARTERS AND THE 2ND BATTALION

Chapter 10 covered Public Duties and the London scene up to 1961. This chapter summarizes in turn the funeral of Sir Winston Churchill, athletics, the Guards Battalions' war role in London District, the changing responsibilities of Regimental Headquarters and the commitments of a Public Duties Battalion in London up to the mid 1990s. The emphasis is on the 2nd Battalion throughout.

SIR WINSTON CHURCHILL'S FUNERAL

On Sunday 24 January 1965 Sir Winston Churchill died in London. He had served in the 2nd Battalion near Neuve Chapelle in November 1915. The Queen directed that he was to be accorded a Lying-in-State, followed by a State Funeral – the first commoner to receive such an honour since the Duke of Wellington. On the 26th Sir Winston's body was taken to Westminster Hall where the Bearer party formed by Grenadiers of the 2nd Battalion took the coffin into the vast gloom of the historic hall. The great pageant of State, upon which were fixed the interest and emotions not only of his own countrymen but of people throughout the world, had begun. During the next three days, despite the bitter cold weather, over 320,000 people passed the catafalque.

Those on duty at the Lying-in-State stood like statues, heads bowed, hands clasped on the hilts of their swords. The Grenadiers were Majors GW Tufnell, DJC Davenport, Captains the Hon JDAJ Monson, DH B-H-Blundell, CR Acland and Lt NA Thorne.

On 30 January the Bearer Party, commanded by Lt AC McC Mather and led by CSM W Williams, carried the coffin from the Great Hall. It was draped with the Union Jack and surmounted by the insignia of the Order

351

of the Garter. The coffin was placed on the Naval Gun Carriage and taken to St Paul's under the watchful eyes of the Major General (John Nelson) the Chief Marshal (Col FJ Jefferson) and the procession Regimental Sergeant Major (RSM D Randell). Maj the Hon DH Brassey commanded the Grenadier detachment. On arrival at St Paul's the Bearer Party carried the coffin on their shoulders up the long flight of steps and through the Great West doors of the Cathedral. The service over, again every eye and every camera was on the Grenadier Bearer Party. Step by step, slow but so very sure, the coffin was returned to its gun carriage, watched by The Queen. The procession continued to the Tower, where the Bearer Party carried the coffin on to the launch *Havengore* before Sir Winston's body was taken on the last journey home to Bladon churchyard near Blenheim.

'It was a sad day but very far from a gloomy one.' noted Col IA Ferguson, Scots Guards. 'It was also, inevitably, a day of many memories and of these there can be no prouder one than that of the Grenadier Guards Bearer Party. Theirs was a duty to the Nation, and in carrying it out with such dignity and with such bearing they not only earned tributes from every part of the world but brought the warm glow of pride to every member, past and present, of the Household Brigade.'

That night Lady Churchill and her daughter, Mary Soames, watched a replay of the funeral on television. 'My mother got up to go to bed,' wrote Lady Soames later. 'As she reached the door she paused, and, turning round, said, "You know, Mary, it wasn't a funeral – it was a Triumph."' It was the privilege of the Grenadiers to have helped to make this so.

*   *   *

Three years later, at the very beginning of the troubles in Northern Ireland, the 2nd Battalion found itself at short notice in Londonderry (Chapter 14). Tours followed in England, British Honduras (Chapter 16), Northern Ireland (Chapter 17), Hong Kong (Chapter 18) and Belize (Chapter 25).

ATHLETICS

Ever since the end of the Second World War Grenadiers have played a prominent part in sport. However, one prize constantly eluded them – that of winning the most prestigious Army Athletics Cup. Despite operational tours overseas, the 2nd Battalion had been the runners up in 1950, 1951, 1952, 1953, 1960, 1961, 1968, 1974 and 1983. For nearly thirty-five years the Battalion, together with the Irish Guards, had been in the forefront of Army Athletics. The 2nd Battalion, now commanded by Lt Col JVEF O'Connell, was determined that 1984 would be an outright victory for the

Regiment. Other units had frequently seen Grenadiers from both Battalions and the Guards Depot competing against each other. Training started in earnest in January in Germany. On returning to England the team joined a local sports club, the Belgrave Harriers. In the London District athletics match, the Grenadiers were only six points short of making a clean sweep of the entire competition. But at the start of the last race in the Army finals at Portsmouth, the team was only three points in the lead, and had to win to clinch the prize.

The relay team rose to the challenge and stormed to victory to the delight of numerous former Grenadiers who were cheering them on. Success led to success – a vast boost to morale. The Battalion was determined to prove equally successful in 1985, but a five-month tour in Belize intervened. Athletic facilities were non-existent there and the Battalion was split among remote jungle camps.

On return to England the athletics team competed in a friendly match against RMA Sandhurst which did much for the Regiment's reputation among the officer cadets. The team also won the Zone A finals. Their principal opponents at the Army finals were the formidable BAOR champions – 50 Missile Regiment. Luck was against the Battalion: in the 200 metres L/Sgt Steaman pulled his hamstring and crashed to the ground in agony. At the end of the day the team had to be content with second place to the Gunners.

Six months later they started a tour of almost two years in Ireland and were unable to be serious contenders again for the Army Cup. In the years ahead the same may be said for all infantry battalions: the very large Corps – the Royal Artillery, Royal Signals and others are apt to put their outstanding athletes into one unit with which it is impossible to compete on equal terms.

*   *   *

After the long tour in Northern Ireland (Chapter 23) the 2nd Battalion returned to Caterham Barracks in 1987. The usual busy cycle for a Public Duty battalion consisted of Spring Drills, shooting in the London District Rifle Association meeting at Pirbright, the Major General's inspection, Public Duties, including the Queen's Birthday Parade, and training. In 1988 the Battalion, commanded by Lt Col AMH Joscelyne, formerly Scots Guards, moved to Salisbury Plain for a week to concentrate on Nuclear, Biological, Chemical drills; fitness; map reading and first aid tests. The second phase was at Lydd for a skill-at-arms camp. All Companies underwent the first two phases of the Annual Personal Weapons Test on the SA 80 rifle (with very good results), as well as the general purpose machine gun and the light support weapon.

Two unusual commitments followed in the autumn; two Companies guarded the perimeter of a civil prison in Blackdown, before the whole Battalion deployed to Salisbury Plain to guard Cruise missiles from protestors, providing an outer cordon for a United States Air Force exercise. The Recce and Signals platoons, commanded by Captains TW Reeve and JPW Gatehouse respectively, participated in the Cambrian marches. The whole Battalion was then deployed for ten days on a Home Defence exercise in the north-east.

HOME DEFENCE

The usual war role of Guards Battalions, when stationed in London, was guarding key points – vital areas, buildings or installations the loss of which would seriously prejudice the war effort. In the 1970s and 1980s Home Defence was taken fairly seriously and so deserves some explanation.

The basic purpose of Home Defence is to achieve a state of national preparedness for the unusual situations that are likely to arise in war and in periods of increasing international tension preceding a war. By the 1970s there were ten Home Defence regions which matched to some extent local government organizations. Plans were drawn up to ensure the survival of as many people and resources as possible following enemy nuclear strikes on the United Kingdom. It was recognized that no plan would provide adequate protection for the population. It was too expensive to build fallout shelters for them. But it was hoped that casualties could be reduced by giving the people as much warning as possible so that they could seek the best available cover. There was, therefore, a United Kingdom warning system.

Nor was there an adequate plan for mass casualty care. Scientific advisers said that three to four million people were likely to be killed by the direct effects of nuclear attacks and another five to six million seriously injured.

In the post-attack period Home Defence plans tried instead to address the problems of care of the homeless, provision of emergency food supplies, water, power, measures for public health and, perhaps most important of all, the maintenance of law and order.

The official policy was to 'stay put', as was the case in the Second World War had the Germans invaded the UK. But should nuclear attack threaten, uncontrollable exodus from cities was to be expected. Millions of people would have to be looked after. Local authorities were expected to make plans for the conservation and distribution of food. Some water supplies would be contaminated by radioactivity and dependent on electric power to operate pumps. There were said to be plans to reduce the effects of an attack on electricity, gas and oil installations. Survival and recovery depended on maintaining public health and law and order. But disease would spread with frightening rapidity; the police and the armed forces, despite operating jointly, would be insufficient to cope. Maintaining the morale of the civilian population after nuclear strikes would be yet another almost insurmountable problem.

The BBC TV made a film called *The War Game*. It gave such a horrifying picture of events following a nuclear strike that, although shown to official audiences, it was never seen by the public.

Some local authorities attached little importance to Home Defence, partly because the general public was apathetic; the likelihood of nuclear war was too remote. Some Left Wing councils announced that they were 'nuclear free zones' and shunned anything to do with Home Defence.

The Army's joint planning with Civil authorities gradually decreased. Home Defence exercises for Guards Battalions did not concentrate for long on guarding key points because such exercises would be boring, few lessons would be learned and the areas or buildings to be guarded in war were secret. HQ London District Home Defence exercises started with the Headquarters moving outside London to its regional concrete bunker which was damp, unhygienic and rather claustrophobic. It served its purpose for exercises but was built prior to the Second World War and was no doubt known to the Russians.

Largely paper exercises were played out, communicating by telephone, wireless and telex. Regimental Headquarters, elsewhere, were put through their paces. The exercises invariably ended with reports of widespread sabotage, terrorism and the enemy's use of nuclear weapons. Maj MH Wise, serving in HQ London District in the late 1970s and early 1980s, was responsible for Home Defence and Mobilization.

Following a typical exercise, the whole of the 2nd Battalion deployed for ten days in October 1988 to North East District. Battalion Headquarters was based in Wathgill Camp, Catterick, while the Companies, commanded by Majors REH Aubrey-Fletcher, Lord Valentine Cecil, DL Budge and AH Drummond, were spread throughout the District. They set up defensive positions around obscure key points, before handing them over to the now disbanded Home Defence Force.

An assortment of civil emergencies and those playing the part of 'Soviet Special Forces' kept everyone occupied. The Corps of Drums (D/Maj RA Hulse) were in Ellington Barracks beating off increasing numbers of 'enemy' drawn from 1st Battalion the Prince of Wales's Own Regiment of Yorkshire.

Regimental Headquarters was also deployed in the London area throughout the exercise.

REGIMENTAL HEADQUARTERS

Until the mid-1980s the Lieutenant Colonel Commanding the Regiment, based on RHQ in Wellington Barracks, had considerable command responsibilities. In the event of a major ceremonial event such as a State Funeral or Coronation he was given an important command appointment. He usually chose to command the Queen's Birthday Parade (perhaps to the disappointment of the Battalion Commander) when the Colour of his Regiment was being trooped. In war he was responsible for the tactical area of operations of TAOR North which covered the City of London and Westminster and the boroughs of Camden, Islington, Haringey, Barnet and Enfield, probably the busiest and most densely populated area in London. He could also be responsible for this area in the event of strikes (Chapter 21).

The Lieutenant Colonel ran a test exercise for his battalion if it was stationed in London District, although RHQ had insufficient staff and no vehicles or communications of its own. Nor did he, or HQ London District, usually have the influence to obtain gunners, engineers or helicopters for an all-arms exercise. The Lieutenant Colonel was also responsible for running a practical promotion exam in the field, for organizing regimental events, for inspecting Combined Cadet Forces and Army Cadet Forces, and for the recruiting, manning, postings and promotions within the Regiment. He inspected the battalion if it was in London District and formally visited it if it was elsewhere. Some of this work fell upon the hard-pressed Regimental Adjutant; the Assistant Regimental Adjutant concentrated on being Equerry to Prince Philip and recruiting, but recruits began to be sent to units under MOD manning policies and so it was increasingly difficult to influence the recruiting of Guardsmen.

In the mid-1980s too, and as Army manpower was ever squeezed, there was an increasing feeling that Foot Guards Lieutenant Colonels did not justify the automatic rank of full Colonel. (The glossary explains this rank structure to those unfamiliar with the Household Division). Gradually the strength of Regimental Headquarters was decreased. First there was a serving Lieutenant Colonel (Lt Col A Heroys), and then it was decided not to have a full-time Lieutenant Colonel but, as with other Regiments, either a retired or serving senior officer as Regimental Lieutenant Colonel. The first of these was Maj Gen BC Gordon Lennox CB MBE who, as a retired officer, did the job for 6½ years. The second is Brig EJ Webb-Carter OBE, a senior serving officer.

When Col ATW Duncan LVO OBE was the last serving Lieutenant Colonel to be a full Colonel, Grenadier records were moved in 1986 from Regimental Headquarters to the Combined Foot Guards and Infantry Records and Manning Office in York.

Although promotion and appointment boards became the formal responsibility of this centralized Records Office, Regimental Headquarters continued to brief and often to attend the Foot Guards manning boards, and so to play a major role in the promotion and manning of the senior appointments within the Regiment. From this time onwards Regimental Headquarters must exercise much of the influence over some appointments by persuasion rather than by direct contact.

Since 1985 therefore the formal military responsibilities of RHQ have been reduced, and so, dramatically, has its staff. However, informally the Lieutenant Colonel and Regimental Adjutant still have to undertake most of their former Regimental duties, and these include a wide spectrum of activities. Moreover, their influence upon the standards and culture of the Regiment remains of crucial importance, as does that role in influencing the careers and appointments of all within the Regiment.

In 1986, in accordance with the latest policy of having all units in the UK under a Brigade Headquarters, 56 London Brigade was formed with its headquarters in Horse Guards. (The Brigade was named in memory and recognition of 56 Division of Second World War fame.) The Foot Guards Lieutenant Colonels therefore ceased to command their Regiments. The Brigade Headquarters, commanded first by Brig DH B-H-Blundell, was made responsible for administration and training for all London District units.

The Regimental Adjutant (Maj CJE Seymour followed by Lt Col TJ Tedder) became responsible for the day-to-day running of the Regiment, aided by a reduced Headquarters staff and the newly formed Regimental Council, which is chaired by the Colonel, The Duke of Edinburgh. RHQ was also fortunate in having a strong Council, these two senior and experienced Regimental Adjutants in turn, and also in two retired officers then in RHQ – Maj EC Weaver MBE, the Honorary Regimental Archivist and Capt BD Double, the Regimental Treasurer and General Secretary of the Grenadier Guards Association.

It will be seen, therefore, that the pre-1985 responsibilities of the Lieutenant Colonel Commanding the Regiment have largely passed to others.

One of their earlier responsibilities, that of commanding an area of

London for Home Defence, by 1995 seemed rather irrelevant anyway because, with the collapse of the Warsaw Pact, there was no credible threat of invasion any longer. Three years earlier the Home Office started selling the 'secret' bunkers including the regional government headquarters. The threat of nuclear bombardment had, mercifully, receded.

THE 2ND BATTALION 1989 – 1994

In early 1989 the 2nd Battalion was lucky enough to be sent to Kenya for six weeks' training. As the last vehicles of 1st Battalion The Devon and Dorset Regiment left Nanyuki Camp, the Grenadier advance party went to work: blue-red-blue signs went up everywhere. The positioning of all the tents was altered until the Quartermaster, Capt PT Dunkerley, was totally satisfied. RSM KR Fairchild, who had just taken over from RSM BMP Inglis, ensured that reveille started punctually at 6.30 am. The rifle companies, meanwhile, settled into a training circuit of revising basic skills, field firing, jungle warfare and, finally, company attacks. The other senior Warrant Officers serving in the Battalion at this time were RQMS D Bradley BEM, ORQMS R Le Louet BEM, D/Sgt DL Cox, AD/Sgt SD Marcham, WO2s DJ O'Keefe, JF Rowell, A Bradley, D Newton, L White, LW Elson and J Swain.

Grenadiers in the 2nd Battalion travelled far afield in 1990. No 1 Company and the Recce Platoon spent a valuable six weeks in Belize. In August the whole Battalion moved to Camp Wainwright in Alberta, Canada. The Companies occupied bivouac sites around the training area. A three-day battle run included participation by C Battery Royal Horse Artillery and the firing of mortars and MILAN anti-tank missiles. Throughout the exercise an adventure training camp was run in Jasper National Park by Maj TH Breitmeyer and CSM R Smith. The package included a 50-mile cycle ride, horse riding, canoeing, climbing, trekking, watersports and whitewater rafting. The climax of the stay in Canada was the Battalion exercise. The Corps of Drums and Support platoons commanded by Maj Sir Hervey Bruce and Capt PZM Krasinski acted as enemy. The exercise finished with an amphibious crossing of Border Lake and the enemy's destruction at Patricia Hill. The exercise broadened everyone's professionalism and was a timely reminder that soldiering can still be fun.

In October The Inkerman Company went on leave prior to their tour in the Falkland Islands (Chapter 27).

* * *

By mid-1991 it was evident that cuts in the Army's budget had undermined operational effectiveness and damaged the morale, motivation and welfare of the Army, civilian staff and the families. To save money ranges were closed at weekends. The number of recruits permitted to join was so reduced that ten battalions were more than 100 men understrength. Training directives could not be completed partly because there was insufficient blank or live ammunition, smoke grenades or thunderflashes. In 1990 and 1991 there was a 50% cut in the programme of overseas training exercises. Even more worrying to some, the average infantry soldier was spending 225 days each year away from his barracks and home. Across the Field Army as a whole, soldiers were spending twenty-four hours on guard every five days, partly due to the terrorist threat. Units were meant to have twenty-four months between tours in such places as Northern Ireland but the average tour interval had dropped to twenty-one months.

What did all this amount to as far as the 2nd Battalion at Caterham was concerned in 1992/1993? What were the training priorities and to what extent were they realistic and being met?

Headquarters 56 London Brigade was now commanded by Brig AG Ross, formerly Scots Guards. (The Headquarters was disbanded in 1993.) In his 1992/93 training directive he gave both Grenadier Battalions the following order of priorities: first, training for emergency operational tours; second, Public Duties; third, recruiting and retention; fourth, leave; fifth, overseas and adventurous training; sixth, individual and support weapons training including specialist training as medics, drivers and signallers; and, finally, collective training. Battalions were also encouraged to enter teams for the London District Rifle Association meeting which had been cancelled in 1991 and 1992, the Cambrian Patrol competition, the United Kingdom Land Forces Support Weapons Concentrations (for those trained as snipers, anti-tank, mortars and on the general purpose machine gun in the sustained fire role).

Lt Col GF Lesinski, now commanding the Battalion, considered each priority in turn. He was fairly confident that the Battalion would acquit itself well on a Northern Ireland tour only because of the very thorough pre-training package. In 1991 the Battalion had been on Operation Spearhead but the pre-training week at Hythe ranges beforehand had regrettably been cancelled. (The Spearhead battalion is the first to deploy in a crisis.)

Lt Col Lesinski regarded his first priority as Public Duties. By August 1992 the 2nd Battalion was doing ten duties per month which was a heavy commitment involving Guardsmen being on duty for thirteen nights a month. At one stage five consecutive Queen's Guards were done at Buckingham Palace and St James's Palace. The Battalion was eighty below strength, the equivalent of almost one company. This reduced the scope for training.

The Commanding Officer regarded his second priority to be leave: the men were only just getting their entitlement and only then because some weeks in the winter months, earmarked for Battalion training, were instead spent on leave.

His third priority was adventure training. Platoon commanders were encouraged to get their platoons well away from London District. His final priorities were individual and collective training. Much could be achieved on NBC, first aid, fitness and shooting, despite Public Duties, but there was quite insufficient time for collective training. Companies had mastered platoon and company attacks, but little tactics had been practised at battalion level.

Fortunately the calibre of the Grenadiers joining the Battalion from the Guards Depot was very satisfactory and the School of Infantry courses for platoon commanders were very thorough.

By April 1992 the 2nd Battalion was reunited at Caterham on completion of the commitments in the Falklands and South Georgia. Both tours were an unqualified success.

On 26 May The Queen presented New Colours to the 1st and 2nd Battalions on Horse Guards Parade. The 1st Battalion's Old Colours were carried by 2Lt JA McDermott and Lt MA Griffiths and the New Colours by 2Lt AJH Watkins and JJR Hunt-Davis. The 2nd Battalion's Old Colours were carried by Lt JW Walker and 2Lt AFR James and the New by 2Lt MP David and BGT Stephens. On the glorious day the capacity crowd saw both the Battalions together march past their Sovereign for the last time. Members of both Battalions, their families and friends and members of the Association joined for a reception held afterwards in Earls Court and were honoured by the presence of The Colonel.

In October Lt Col REH Aubrey-Fletcher took command of the 2nd Battalion. He is the fourth generation of his family to serve in the Regiment. (Lt Col Seymour is also a fourth generation Grenadier.) The following month No 1 Company group of 122 Grenadiers commanded by Maj RHG Mills deployed to Botswana for a four-week exercise.

Three days of bush tactics were followed by a two-day inter-platoon patrolling exercise with two platoons of the Presidential Guard Company of the Botswana Defence Force. This phase was followed by a very enjoyable nine-day range package arranged by the Training Officer, Capt DW Ling. Each platoon completed a three-day survival course which included trapping, solar skills, navigation and emergency shelters. By the third day Guardsmen were munching on lizards, snakes and scorpions, the last being considered 'not bad but a little crunchy'. Heat was the main enemy. Several Guardsmen succumbed temporarily in temperatures exceeding 48°C. Snakes were not too much of a problem until the medic, C/Sgt JJ Seymour, confronted a large puff adder as he was about to get into bed. NCOs from 9th Parachute Squadron Royal Engineers rigged up an ingenious power shower system that removed Kalahari dust in seconds, while another soldier from 216th Parachute Signals Squadron maintained a satellite communications link with England.

The training ended in a four-day exercise. Each of the rifle platoons, commanded by Lieutenants SRA Miles, JA Waldemar-Brown and 2Lt DNS Freeland, participated in a company defensive shoot, a camp attack, and, as a gruelling finale, a mountain clearance operation.

After that everyone was ready for relaxation, a visit to Victoria Falls, game parks, and, for the adventurous, an excellent day of whitewater rafting. Everyone looked back with fond memories of this hard but worthwhile Botswana exercise.

Earlier in this History a brief summary was given (Chapter 26) of the career of a Commanding Officer and Regimental Sergeant Major. In Botswana the CSM was A Thomson. Few Warrant Officers' careers followed a similar pattern, but his may be fairly typical of a high-grade Grenadier. CSM Thomson joined the Army in 1971 as a Junior Guardsman at Pirbright. After service in West Germany and Northern Ireland, he went on overseas exercises in Kenya, Sudan and Jamaica before playing his part in the firemen's strike. Early in 1978 he went to the mortar platoon, serving in the close observation platoon in Northern Ireland, doing four tours there in all. After Berlin, instructing at the Guards Depot and service in Hounslow and Germany, he was posted to RMA Sandhurst as a Colour Sergeant. He then saw service in the Falklands 1991–1992 before going to Botswana as the CSM. Canada, two tours in Cyprus and participation in six Troopings of the Colour can be added to the list.

No 2 Company, commanded by Maj TW Jalland, was the next to exercise overseas. In February 1993 the Company spent a month in Cyprus. Live firing at Pyla in Dhekelia Garrison was an unusual experience because the ranges overlooked the Mediterranean. Firing had to be stopped whenever fishing boats encroached into the danger area. The military training culminated in a full-scale company exercise which involved a deployment by landing craft, a forced march, the take-over of Paramali village from where recce and fighting patrols were despatched. Three newspaper reporters accompanied the platoons and seemed to enjoy the compo rations and a late night ambush. They then wrote 'local boy' stories on some of the Guardsmen. The final week was spent on such activities as skiing in Troodos, watersports at Episkopi and three-day boat trips to Egypt and

Israel. As with Botswana and other overseas exercises, it was a great experience for everyone.

The overseas exercises were invariably exciting and well planned. The adventurous training which followed was equally memorable. During exercises in Africa, visiting Mombasa was not permitted due to the danger of catching AIDS and so alternative plans were made such as watching big game on safari.

By now the senior Warrant Officers serving in the 2nd Battalion were WO1 DJ O'Keefe (soon to be replaced by S Swanwick), RQMS S Austin, D/Sgt A Prentice, WO2s L White, SP Milsom, HR Booth, CT Angel, GP Hares, K Walsh, PW Sellors BEM, Clarke and R Dransfield.

However, there was only one major exercise left for the 2nd Battalion. The whole Army was to be drastically reduced following the break up of the Warsaw Pact and the collapse of Communism.

It was with the utmost sadness that it was learnt of the Government's decision to cut the Army by some 30% and to reduce the number of Guards Battalions by three. The Second Battalions of the Grenadiers, Coldstream and Scots Guards were all earmarked to go into suspended animation before the end of 1994.

Many fine infantry battalions were to lose their identity in a mass of amalgamations, regardless, like 2nd Battalion Grenadiers, of their seniority, professionalism or recruiting potential.

The only consolation was that, as a result of many representations to the Ministry of Defence, one company of each of the Guards' Second Battalion's were to remain as a Public Duty increment. This would enable at least two of the five Guards Battalions to be stationed abroad, for example in Germany or on a long operational tour. Hopefully, an excessive proportion of the Household Division would not therefore be confined to Public Duties. Moreover individuals could always volunteer for service on exchange or secondment (as did many named in Chapter 20 who served away from the Regiment).

So many regiments were removed from the Army's order of battle that there was no likelihood of Colonels of Regiments agreeing to the retention of the 2nd Battalion Grenadiers. However, 'hope springs eternal'. In 1993 no solution was in sight to terrorism in Northern Ireland where it was still too dangerous to travel by vehicle on some roads; pressure was mounting on the UK to contribute to more United Nations peacekeeping missions, and some argued that the break up of the Soviet Union would lead to more, rather than less, instability. Had the Government miscalculated? The 2nd Battalion was to be the last to go. Meanwhile an exciting exercise lay ahead of them.

*   *   *

On 27 February 1994 the 2nd Battalion deployed to America on its last major exercise – an excellent challenge in its final year before being placed in suspended animation. The Battalion conducted comprehensive company group training in Washington State on the north-west coast of America, enjoying facilities that are not available in the UK.

By now the Battalion had reduced in strength to two rifle companies, so Lt Col Aubrey-Fletcher had under his command B Company 1st Battalion The King's Regiment, as well as N Battery of 3 Royal Horse Artillery, a troop of 22 Regiment Royal Engineers, two scout helicopters from 658 Squadron Army Air Corps and numerous drivers, cooks and mechanics from the Regular and Territorial Army. The exercise was coordinated by Maj WB Style, Coldstream Guards, the Senior Major. He was assisted by the Operations Officer, Maj PA Rose of the Adjutant General's Corps.

No 2 Company, The Inkerman Company and B Company, 1 KINGS, were put through a five-week training package. It consisted of firing weapons at Fort Lewis, while at Camp Bonneville platoons practised patrolling, ambushes, navigation and fieldcraft. The climax for many was the all-arms training at Yakima where all operations of war were tested.

'Chinook helicopters lifted The Inkerman Company to its dropping-off point from where it embarked on an 18 kms night march across the eerie Yakima Lunar landscape,' wrote Lt JL Davies. 'Each man, regardless of rank, carried 60 lbs. Early the next morning the Company embarked on an advance to contact and then moved to a defensive position.' The Company, commanded by Maj RMT Reames and the company officers (Lieutenants AL Grinling, AB Jowett, JL Davies, 2Lt SGG Witheridge), led their men through a withdrawal over a reserved demolition after fending off counter-attacks during the night.

No 2 Company (Maj TW Jalland, Capt CIT White-Thomson, 2Lts S Courtauld, JA Inglis-Jones, SCE Wade) undertook a similar exercise.

The Mortar Platoon (Capt RT Villiers-Smith) trained in the coordination of fire support with N Battery, Cobra helicopters and American A10 aircraft. The Recce Platoon (Capt SRA Miles) took part independently in several American exercises. Fort Lewis units, including those from the National Guard, assisted the Battalion throughout, making things somewhat easier for Maj DJC Russell-Parsons and Capt FS Acton, commanding Support and Headquarter Company respectively.

The exercise was a welcome change from the routine of London Duties and provided one last opportunity for the Battalion to carry out demanding and realistic training. Maj Reames even arranged for his Company to be shown round an American missile-firing submarine.

The training over, a Battalion review was run by Maj Reames and Drill

Sergeants R Smith and R Sargeant BEM. The Regimental Band accompanied the Battalion to America to do their mandatory training and to play at venues throughout the State. As usual, they were immensely popular.

On return to Caterham on 4 April the Battalion excelled itself for its final seven months together, providing the Escort for the Colour and No 2 Guard. 'Those who witnessed the Queen's Birthday Parade will agree that the standards were as high as ever despite the traumatic reductions in strength of the Household Division.' recorded *The Guards Magazine*.

'The Major General (Robert Corbett) and Bernard Gordon Lennox had fought long and hard to retain the 2nd Battalion and never took "No" for an answer,' recalls Lt Col Aubrey-Fletcher. 'The Battalion's service to the Nation represented the very best the Army had to offer. However, we all made the definite decision not to look back – only to look forward. In 1993 the Battalion's teams in the London District Rifle Meeting had come first, third and fourth out of eleven teams. In the closing months of the Battalion's existence in 1994 it continued to win competitions. The Battalion remained operational right up to its final week – standing by to assist the police at Heathrow airport.'

Finally, on 5 November 1994, Inkerman Day, the 2nd Battalion went into suspended animation. 175 men including The Inkerman Company went to the 1st Battalion, thirty-seven to redundancy over the previous two years, and twenty-seven to serve away from the Regiment, leaving five officers and ninety-eight men, commanded by Maj OP Bartrum MBE, to form the independent Nijmegen Company which proudly carries the 2nd Battalions' Colours.

The Company was named after the courageous capture by the Grenadier Guards of the bridge over the Waal at Nijmegen in September 1944.

The Duke of Cambridge, in Victorian times, was credited with describing the 2nd Battalion as the 'Models'. The 1st Battalion was said to be the grandest, and the 3rd Battalion the maddest, but the men of the 2nd Battalion in his day were the true professionals.

The Nijmegen Company will therefore be representing the spirit, traditions and the standards of the 'Models' in the 21st century and beyond.

CHAPTER 30

NORTHERN IRELAND: CURTAIN DOWN?

1ST BATTALION ARMAGH
SEPTEMBER 1993 – MARCH 1994

On 27 September 1993 the Commanding Officer, Lt Col ET Bolitho, assumed command of the Armagh Roulement Battalion. Apart from the 1st Battalion, he had under command a battery of 26 Regiment, Royal Artillery, at Bessbrook, a Company of the King's Own Scottish Borderers as a Battalion reserve, and a Company of Fusiliers (later the Staffords) at Newry, changing over every four weeks. The Battalion was also reinforced by platoons of the Coldstream, Scots Guards and Royal Anglian Regiment, plus individuals from the Household Cavalry and Irish Guards, bringing the strength up to 830 all ranks – 200 more than usual.

The Queen's Company (Maj NJR Davies) was deployed to Crossmaglen, ten years to the day since its previous tour there. No 2 Company (Majors EF Hobbs then TH Breitmeyer) went to Newtown Hamilton, No 3 Company (Maj FA Wauchope) to Forkhill, and Battalion Headquarters with HQ Company (Maj RD Winstanley) were at Bessbrook. The Adjutant was Capt RL Fanshawe, and the RSM was D Beresford.

The Queen's Company was still taking over from the Duke of Edinburgh's Royal Regiment (DERR) on 23 September when the IRA struck. A Puma helicopter was about to fly out of the base at Crossmaglen while two Lynx helicopters were circling above. (Helicopters flew by day in South Armagh in threes.) The IRA opened fire, hitting a Puma and Lynx, with two 12.7 mm and two GPMG machine guns mounted in the back of a considerable number of vehicles. DERR soldiers in the sangars fired back in different directions but the IRA firing positions were not visible from the sangars. Some friendly forces, including Sgt DI Adkins, were nearly hit by the DERR soldiers. The IRA vehicles drove off chased by the helicopters; the Army Air Corps fired 200 rounds. One helicopter subsequently dropped off six soldiers to intercept a white van.

2Lt GR Denison-Smith, on his first patrol, saw three masked men jump from the van and run behind a house. The constraints of the rules of engagement prevented him from opening fire as the masked men did not have weapons visible. The IRA probably escaped in another vehicle. Patrols were quickly deployed from all the Company locations.

The terrorists left in the van one 12.7 mm machine gun, two general purpose machine guns and an AK 47 rifle. It was a large-scale operation in which over 2,000 rounds were fired. The locals appeared very concerned. Father O'Hanlon, the priest in Crossmaglen, condemned the IRA attack on television.

Less than two weeks later a single shot was fired at a Queen's Company foot patrol led by 2Lt Lord Wrottesley. The firing point may have been 300 metres away on a grass bank near a road. Nobody was hit and no round was fired back as the terrorist was not seen. This attack highlighted the vulnerability of patrols and hence the requirement only to send them out when they had a genuinely worthwhile aim. It also showed the importance of the highest standards of fieldcraft. The terrorists had watched the Army at work for almost twenty-four years.

The attack was useful for it reminded the Grenadiers of the gravity of the situation. There was no complacency.

Lt Col Bolitho had served in Crossmaglen fifteen years earlier. What had changed? 'Tours in Ireland had been extended from four to six months,' he wrote later. 'The Guardsmen are better equipped and more professional, but so are the terrorists. Observation towers had been built in the 1980s to cover much of the Battalion area. These were surrounded by mines, wire and trip flares and provided good surveillance coverage of the area, but they did tie down a company and needed to be resupplied by air. Many of the people on the "IRA wanted list" were the same as they had been fifteen years previously, or their children. Although the RUC has primacy in Northern Ireland, the intractability of the problem in South Armagh and the need to move everywhere by military helicopter means that the Army still has a dominant role to play there. Liaison with the RUC was very close, with most bases co-located and official meetings several times a week, involving myself, the Senior Major (Maj AD Hutchison), the Operations Officer (Capt NPJS Keable) and the Intelligence Officer (Capt CGF James). Visits by senior officers and politicians were very frequent because at that time South Armagh was the only place where British soldiers were regularly being killed.'

50. The Ensign, Second Lieutenant J J R Hunt-Davis carries the Queen's Colour of the 1st Battalion during the 1992 Queen's Birthday Parade with the Guards Memorial in the background.

51. The Queen, accompanied by the Colonel, inspects the parade on the occasion of the Presentation of New Colours to the 1st and 2nd Battalions on Horse Guards Parade, 26th May, 1992.

52. The Regimental Band march from Buckingham Palace to Wellington Barracks.

53. The 2nd Battalion's final Queen's Guard, 6th October 1994. Former 2nd Battalion
Commanding Officers dined on Guard.
Back Row: Captain G J Rocke, Lieutenant S Courtauld.
Centre Row: Colonel D V Fanshawe, Brigadier D H Blundell-Hollinshead-Blundell,
Lieutenant Colonel H M L Smith, Colonel J V E F O'Connell, Colonel A M H Joscelyne,
Lieutenant Colonel G F Lesinski.
Front Row: Lieutenant Colonel J R S Besly, Brigadier P G A Prescott, Major General
The Duke of Norfolk, Lieutenant Colonel R E H Aubrey-Fletcher, Lieutenant General A
A Denison-Smith, Brigadier A N Breitmeyer, Lieutenant Colonel P H Haslett.

On 31 October an RUC policeman was shot in the neck and killed at an RUC vehicle check point in Newry.

On 12 November a tractor towing a silage trailer stopped close to the Borucki sangar which was named after a soldier in the Parachute Regiment who had been killed there. The tractor's driver suddenly started to spray the sangar and its only exit with 1100 gallons of fuel. The sangar is on the edge of Crossmaglen town square, some 75 metres from the base. The terrorist ran from the tractor, out of sight. Seventeen seconds later the fuel was ignited and a huge fireball 30 feet high engulfed the sangar for seven minutes. It was occupied by L/Cpl DR Wells, Gdsm DT Smith, SA Briggs and IR Eachus. Capt CAG Hatherley and CSM RM Dorney in The Queen's Company operations room had been alerted to the approach of the trailer by the closed circuit TV operator. On seeing the flames, a contact report was sent, 999 was rung for the fire brigade, and the quick reaction force, commanded by L/Sgt Elliot, Coldstream Guards, was deployed, followed by Sgt DI Adkins who recalls: 'I stood in the turret of the Saxon armoured vehicle: it was very dark and raining. I screamed at the driver to ram the tractor away from the sangar. I could see locals coming out of the pubs to watch. I was worried about secondary explosive devices, and assumed that the four Grenadiers were all dead – asphyxiated. But then I saw four faces poking out and a thumbs up – all was well.'

The volunteer local fire brigade arrived quickly. Meanwhile a Puma helicopter deployed patrols and cordons were established. Tracker and search dogs failed to detect anything significant. The Ammunition Technical Officer carried out X-rays of the tractor before driving it and the trailer into the Crossmaglen base. With considerable bravery L/Cpl Wells and his team reoccupied the sangar. Although only the seventh attack on the sangar since its establishment in 1977, it was the fifth in eleven months and so was not entirely unexpected. Since the attack the sangar has been replaced with one better designed and protected.

The Regimental Band paid a very welcome visit to Ireland before Christmas and gave concerts in Crossmaglen, Newtown Hamilton, and Bessbrook town hall which were attended by all sections of the community from Sinn Fein councillors to Ulster Unionists.

Also before Christmas a few Grenadiers had lunch with Prime Minister John Major in Armagh. One of them, CQMS RCP Duggan, whose brother had been murdered by the IRA fifteen years previously to the day, made sure that Mr Major's resolve to deal with the IRA was not weakening. (Sgt Johnson, the brother of Gdsm Johnson killed at the same time as Duggan, was also serving in the Battalion.)

On Christmas Day Lt Col Bolitho and the Quartermaster (Capt JA Sandison QGM), who was dressed as Father Christmas, went round all the bases giving a present to every man in the Battalion. They received a glass of port in return which made the day go with a swing. The unusual appearance of Father Christmas in Crossmaglen Square brought squeaks of surprise from local children that 'Santa is a Brit!' The Quartermaster and his team did a magnificent job, resupplying eighteen bases largely by helicopter. At one sangar he found that a defective sewage system was responsible for a year's supply of sewage bubbling to the surface. The smell was very striking.

Lt S Marcham and his 'Buzzard' team also deserve special mention. They were responsible in South Armagh for one third of all RAF and Army helicopter hours worldwide. Lt Marcham inspired the pilots and crews with frequent renditions of Kipling and other poems.

Lt Col AJC Woodrow, a previous Grenadier Company Commander at Crossmaglen, had warned Maj Davies that he must expect to take casualties.

On 30 December at 11 am two patrols each of four men deployed briefly from Crossmaglen's base for the first patrol of the day to check that IRA mortar attacks were not being prepared. Eight minutes later one of them moved into North Street which was full of people. Partly due to a variable cloudbase there was no helicopter cover, but a Gazelle helicopter had flown over the area some minutes earlier.

A terrorist fired one 0.5 inch high-velocity round at the rear left-hand man of L/Sgt KJ Turner's patrol, hitting Gdsm DM Blinco. The internal injuries were huge. He was conscious for only twenty seconds. The Saxon ambulance with L/Cpl KL Edwards was there within four minutes; the incident was handled in the operations room by Capt RAC Duncan.

L/Sgt Turner, next to Blinco, later went into a shop while the cordon was being positioned. He was amazed at the indifference of the locals who simply continued shopping. Blinco was quickly flown to Newry where he died. The probable firing point was 130 metres away. The terrorist may have been lying in wait and fired from a vehicle, the weapon possibly ending up in a house close by. Two other Grenadiers had been covering the rear of the patrol but were on the other side of the road and so not visible to the sniper. A Guardsman's mother who lived in Belfast arranged for some flowers to be placed on the pavement where Blinco was murdered. Some Grenadiers went into the pub nearby and requested that the flowers be left alone, but twenty minutes later the flowers were found kicked down the street.

Blinco was a fine, popular Guardsman. He was buried in Melbourne Church near Derby, his body being borne by those members of the patrol who were with him when he was murdered. The last member of The

Queen's Company killed in action had been Gdsm Lunn in the Cameroons in 1961.

An increase in the number of helicopter flying hours enabled the Battalion to cover all patrols in dangerous areas with armed troop-carrying helicopters. As a result there was no more sniping at patrols for the rest of the tour, otherwise such incidents would have occurred perhaps once a month. The Queen's Company also became much more aware of where they could, and could not, patrol in Crossmaglen during daylight.

On 10 January Gdsm SG Wyatt on sentry saw a flash and heard a bang. He immediately pressed the alarm button a few seconds before a mortar bomb was fired into Crossmaglen's base. It had been fired from a stolen Toyota van 50 yards away, hitting the exact spot of the earlier attack which had just been repaired. The concrete was still drying!

Six hours later, after carrying out all the correct drills, the Ammunition Technical Officer declared the Toyota safe and drove it into the base. While manoeuvring the vehicle into a shed a secondary device, triggered off perhaps by a timer, exploded causing five casualties two of which were serious – the ATO, WO1 McLelland, and Craftsman Pickford REME.

For several years no significant find had materialized from a planned search in Crossmaglen. However, on 16 January Maj Davies organized an ingenious search, led by Capt Hatherley. Patrols with sniper protection deployed to the area at 4.30 am, providing the surprise which helicopter drop-offs at night might have prevented. The sharp eyes of Guardsmen SJ Phillips and M Stanton led to the capture of mortar parts and home-made explosives. The Garda provided cover on their side of the border and had worthwhile finds of their own.

February began with a mortar attack by the IRA at a vehicle checkpoint south of Newry. The mortar bomb landed on an access road causing a crater and severing an overhead power cable but no one was injured. No follow-up was instigated as there was a threat of a further attack on troops deploying. Like those on Heathrow Airport, the mortar was timed to fire after the terrorist had left the scene. In this case he could have strolled into the local pub.

Meanwhile, back at Wellington Barracks, Capt P Harris and Lt D O'Keefe did much to keep the families in touch and cheerful. The visit of Prince Philip, the Colonel, was very welcome. On 22 February he also visited the Battalion in Bessbrook. As always, everyone was delighted to see him. Unfortunately the appalling weather prevented him meeting many from the outstations. On the same day a car bomb in Crossmaglen narrowly missed a patrol. A red Mazda vehicle had just been checked. On receiving 'no trace', the patrol was moving away when the bomb exploded. Despite the last man being only 10 metres away, the only damage was to one man's back pack which was hit by shrapnel. Due to the weather, no helicopter could be used.

Early the next morning the clearance team, led by the Commanding Officer, reached the area by vehicle which was most unusual as, in theory, there is no vehicle movement to Crossmaglen without the route being fully picketed and cleared. It was the first unprotected vehicle move by the Regular Army for some years. A car bomb incident in Crossmaglen had not occurred for over ten years, the last occasion being against The Queen's Company in 1983.

Eight days before the company hand-over to a company of the Worcestershire and Sherwood Foresters the IRA renewed their attack. A Lynx helicopter was flying an underslung load into Crossmaglen's base and dropping off an RUC Constable. The helicopter was struck by a single mortar bomb which severed its tail boom. (The base plate positions were often in dead ground out of sight of the sangars.) The mortar was fired by a collapsing circuit while the firing pack was plugged into the mains. The IRA turned off the mains supply to Crossmaglen, thus firing the round. Therefore when the round fired the village was in darkness.

Lt JJR Hunt-Davis, patrolling less than one mile away to the south, saw an enormous orange flash and feared the worst, particularly as he could not get through on his wireless to the operations room. But the helicopter crew had successfully crash-landed and pulled out the Constable. There were eight other more minor casualties. As The Queen's Company fought the fire, ammunition in the burning helicopter exploded, liberally spraying the base with rounds. Lt BGT Stephens was intrigued that the news announced where the firing party was located; the TV cameras were on the ground very quickly.

As had occurred before when the locals had felt that their lives had been endangered, there was considerable hostility to the IRA. John Fee, the SDLP Councillor for Crossmaglen, was outspoken in his condemnation of the mortaring. He was severely beaten up by three young men with base-ball bats, showing their alarm at the extent of the reaction against the attack.

The main operational incidents described above all concerned The Queen's Company. However, the other Companies most certainly deserve their share of the credit for successful arrests, anticipating IRA action and generally outwitting the terrorists.

Through making best use of the surveillance assets that dominated the Forkhill area, the three Grenadier platoon commanders there (Capt JA McDermott, Lieutenants the Hon James Geddes and AJH Watkins) helped prevent any terrorist attacks in that area. Capt MJ Besly was the company 2IC.

At Newtown Hamilton Capt MA Griffiths was the watchkeeper with Capt RT Maundrell the 2IC and 2Lt TP Barnes Taylor another platoon commander. Captain AS McKie was the Battalion Signals Officer. The Close Observation Platoon (Capt JDMcL Wrench, D/Sgt VJ Overton and C/Sgt DI Grassick) had a particularly successful tour, as did the sniper group (Lt JGR Frost and C/Sgt LJ Herbert).

The reader can compare this tour with the earlier ones (Chapters 14, 17 and 23). Twenty years earlier an IRA sympathizer would have warned the terrorists of the approach of a Grenadier patrol by a hand signal, whereas in 1994 his successor might use a mobile phone. In 1974 Grenadiers could and did search houses indiscriminately, whereas in 1994, as the IRA knew, permission to search a house was very rarely authorized by Special Branch and Brigade HQ. 'You mustn't act like terrorists,' Grenadiers were told by VIPs when this was discussed. 'International opinion wouldn't like it.' And there was no lack of distinguished visitors: the Secretary of State for Northern Ireland, Sir Patrick Mayhew, dined at Borucki sangar which had been engulfed in flame a few months before. He was escorted by the Commanding Officer with a rifle in one hand and a decanter of port in the other.

There were many other notable visitors including the Commissioner of the Metropolitan Police whom Lt Col Bolitho had last seen, similarly dressed, in his smartest uniform on the Queen's Birthday Parade. Others included the Primate of All Ireland and a lady from the Treasury who was so important that she was escorted by the GOC and Lt Gen Sir Thomas Boyd-Carpenter, formerly Scots Guards.

Officers in Battalion Headquarters dined with the Garda in the Officers' Mess – an encouraging indication of greater cooperation. 'There were other frustrations apart from not being allowed to search houses,' noted Maj Wauchope who had fought Mugabe's armed supporters in the Rhodesian guerrilla war before joining the Regiment. 'A terrorist convicted of murder in Northern Ireland would be out of prison within seven years, whereas in England the last attempted murder of a policeman led to a 32-year sentence.'

At last, on 29 March 1994, the long six-month tour came to an end, to everyone's relief. The Battalion had greatly distinguished itself once more. They received a very exceptional number of Honours and Awards as Appendix 2 reveals. They returned to Wellington Barracks for a month's leave – thoroughly deserved.

CHAPTER 31

1995

1995 had an exciting start – the 1st Battalion, commanded by Lt Col ET Bolitho OBE, enjoyed a seven-week exercise in Kenya.

The companies had a fortnight to themselves, rotating around privately owned ranches about an hour from the Battalion's base at Nanyuki. The Queen's Company (Maj RD Winstanley) initially deployed to Dol Dol – an excellent area offering rugged terrain and spectacular scenery. No 2 Company (Maj TH Breitmeyer) went first to Mpala Farm firing all their weapons and working up to a company live firing attack. The Inkerman Company (Maj RMT Reames), meanwhile, started at Solio Ranch amid a private game reserve. Support Company (Capt RT Villiers-Smith) fired their mortars at Archers Post while the Recce Platoon (Lt BGT Stephens) trained at Ol Kangeo.

The Battalion's overall aim in Kenya was to conduct all-arms training up to battle group level, concentrating on aggressive operations. On 20 January the rifle companies began the major exercise with one company advancing to contact firing live ammunition, supported by D Battery 3 Royal Horse Artillery with six 105mm light guns. A second company, meanwhile, rescued a hostage in a night operation against an enemy found by the anti-tank platoon (Capt SG Soskin), while the third undertook vehicle anti-ambush drills and a company night attack with live ammunition, supported by the Mortar Platoon (Capt the Lord Wrottesley). After all the rifle companies had rotated in these tasks, phase two began with an enterprising night attack with overhead fire and mortar illumination. The 'enemy' included Support Company Headquarters, the Corps of Drums and 2 Troop 34 Field Squadron Royal Engineers which had helped prepare the position. This phase concluded with a further battalion attack this time firing live ammunition, supported by the mortars, the Battery, the 94mm light anti-tank weapon (LAW), snipers and the general purpose machine guns (sustained fire).

The exercise then switched to jungle and adventure training. Capt

372

DW Ling at Kathendini spared no effort in hiring a host of locals, with such unlikely names as Francis Awfully-Splendid, to teach survival skills. Everyone learned about trapping and what to eat and drink from within the jungle. Other specialists taught watermanship and jungle-fighting skills.

Adventure Training took place largely at Lakes Naivasha and Baringo, being run by Captains JW Walker and CAG Hatherley respectively. The more orthodox activities, climbing, abseiling and biking were mixed with the more unusual – riding among antelope and giraffe, canoeing alongside hippo and visiting snake farms; some of the Battalion snipers helped in a zebra cull on one of the ranches.

Capt JD McL Wrench, ORQMS MC Elliott and Sgt S Damant discovered that some of the world's top athletes were attempting to run a marathon up Mount Kenya to Point Lenana. On 2 February planes took off, helicopters hovered, the sun came up at dawn and the starting gun went off. Lack of oxygen and total exhaustion reduced the number of runners, but Capt Wrench and ORQMS Elliott surmounted the 1,500 foot scree slope, followed by the glacier, sheer ice and a further vertical climb of 1,000 feet. After seven hours and 40 minutes, having climbed 9,500 feet and run 26 miles, the two of them crossed the finishing line. The international press at the Mount Kenya Safari Club that evening wanted to meet the four servicemen, the only British to enter, who only decided the day before to compete in the hardest, highest marathon in the world against some of the world's finest runners. The soldiers said that the challenge, excitement and achievement would live with them for ever, but enter again? 'Never!'

The Battalion returned to Wellington Barracks between 14–17 February. This overseas exercise was invaluable. The last time the 1st Battalion had undertaken a battalion live-firing attack was in 1989. (There was, of course, much live firing at Battalion level in the Gulf, but not as a Battalion together.) For those 200 Grenadiers who had joined from the former 2nd Battalion, such an attack had not taken place since 1981. With a two-year Northern Ireland tour in prospect, another opportunity may not arise until at least 1999. As well as the benefits of exercising with Gunners, Engineers, the Royal Signals and Army Air Corps, the exercise had a considerable, if unquantifiable, value to retention and motivation within the Battalion.

The new 'London District Activity Plan' greeted the Battalion on its return to Public Duties. In principle one month is to be spent on ceremonial followed by a month free for training or whatever other commitments the Commanding Officer decides. It may prove workable but the Battalion found itself on 2½ solid months of Public Duties. Even so, the Battalion trained for ten days at Longmoor Camp in Hampshire in April and, now commanded by Lt Col JP Hargreaves, in Wales in July. In September the Battalion took on the Spearhead commitment whereby the Grenadiers were placed on short notice to move operationally to anywhere required. This was followed by a five-day field test exercise at Sennybridge in Wales, and the start of the Northern Ireland training prior to the Battalion's deployment there with the families for a residential tour in 1996.

At the end of 1995 the seniors in the Sergeants' Mess were RSM AJ Green, RQMS VJ Overton, D/Sgts SP Milsom and RM Dorney, CSMs SD Ashley, KM Gibbens, P Parker, J Coleman, RM Jolly, CL Jones, SAW Sadler and M Thompson.

An option was considered of the Battalion going to Germany to replace the 1st Battalion Coldstream Guards, who would return early to London District, and this was an attractive possibility for the Regiment. However, after considerable discussion, it was decided that it would be unfair to ask the Coldstream Battalion to return early from Germany, and so the plan for the 1st Battalion to go to Northern Ireland remained unchanged.

An uneasy peace descended upon Northern Ireland when the IRA agreed to a ceasefire in August 1994. A faction within the Republican movement convinced their sceptical comrades that there was more to be gained from a cessation of violence than a bloody stalemate. Traditional sectarian hostilities were greatly reduced and the expectations of ordinary people in Ireland have been transformed. Though essentially suspicious of the Anglo-Irish negotiations, the Unionist Community has not yet upset the peace process. But pro-IRA graffiti and fluttering orange, white and green flags still proclaim political allegiances just as the Orange Order's provocative summer marches endeavour to assert the old Protestant supremacy. By December 1995 more than 3,200 people had been murdered over a quarter of a century of violence, Guardsman Blinco being among the last to be killed in action for the time being. But on 9 February 1996 the IRA announced that they were resuming their 'military operations'. That night an IRA bomb in London shattered the ceasefire.

*   *   *

1995 was the first year that Nijmegen Company operated as the Regiment's Incremental Public Duties Company. Although inaugurated on 3 August 1994, the Company had to assist with the drawdown of the 2nd Battalion and the closure of The Barracks, Caterham. It also went to Nijmegen to take part in the fiftieth anniversary of the City's liberation. In March 1995 the Company, commanded by Maj OP Bartrum MBE, moved to Victoria Barracks, Windsor, before finding Number 5 Guard on the Trooping of the Colour alongside the Incremental Companies of the Coldstream and Scots Guards. (Nijmegen Company's officers initially were Captains SRA Miles and S Courtauld, and Lieutenants JA Inglis-Jones and SGG Witheridge. By August 1995 some of them had been posted, being replaced by Capt MA Griffiths, Lt TP Barnes-Taylor and 2Lt CD Morris. The CSM and CQMS were E O'Keefe and W Orton.)

Between 1 April 1995 and 31 March 1996 the Company spent some forty days on Queen's Guard (St James's Palace, Buckingham Palace and The Tower of London) and twenty-six days on the Windsor Castle Guard. The Company was also fully committed on Guards of Honour, Street Lining for State Visits and the State Opening of Parliament. Despite the emphasis on Public Duties, and the trickle posting of Grenadiers through the Company, it will remain fully trained and operational.

The Company contributed to the training of 1st Battalion The King's Regiment for Northern Ireland in February and March, provided a demonstration platoon at the Army Training Regiment Bassingbourn in July, relieved the Gurkhas at Brecon in the autumn and found the 'hunter force' for 21 SAS selection in October. The Company also trained in 1995 at Thetford and Sennybridge while looking forward to an overseas exercise in Kenya in 1996.

Each Incremental Company is attached to the Battalion with which it normally shares a barracks. F Company Scots Guards at Wellington Barracks was therefore attached to 1st Battalion Grenadier Guards; Nijmegen Company was with the 1st Battalion Scots Guards at Windsor. Sending each Incremental Company to serve with another Regiment was done quite deliberately so that there could be no excuse for not keeping them up to strength. It also leads to healthy competition particularly in the sporting field. Just as F Company won inter-company competitions against the Grenadiers at Wellington Barracks, so Nijmegen Company won the 1st Battalion Scots Guards Athletics and Swimming trophies, and was the winner of the London District Minor units in these events.

Until 1961 the initial training of Grenadiers had been conducted at the Guards Depot at Caterham and at the Guards Training Battalion at Pirbright. From 1961 to 1993 adult recruit training was consolidated at Pirbright. In 1993 all depots were abolished and all basic training was divided into two phases. The first phase consists of the ten-week Common Military Syllabus (Recruit). All Army recruits now complete this course in one of the Army Training Regiments – the Grenadiers remaining at Pirbright, but now training alongside recruits from the Royal Artillery, Royal Logistic Corps, the Royal Electrical and Mechanical Engineers and, of course, the rest of the Household Division.

Cosmetically things have changed. However, the first two Company Commanders of the Guards Company (Maj DJC Russell-Parsons, then RD

Winstanley) have ensured that, within the Company, the standards, culture and ethos of the old Guards Depot have remained. The standard of initial training has always been regarded within the Household Division as of crucial importance, and the Guards Depot in its day invariably enjoyed an exceptional reputation.

On completion of Phase 1 the trainee Guardsmen move to Catterick in North Yorkshire to undertake the 14-week Foot Guards Combat Infantryman course. This is two weeks longer than the rest of the infantry to allow for a drill enhancement package. The training at Catterick is conducted by the Guards Training Company which forms part of 1st Battalion the Infantry Training Centre. The Company is unique within the Centre as it is the only one to maintain Divisional integrity.

Training of junior soldiers for the Brigade of Guards began at Pirbright in 1958, while junior leaders for the infantry were trained at the Infantry Junior Leaders Battalion first at Plymouth, then Oswestry, then at Shorncliffe, and finally after the Battalion was disbanded, at Pirbright. Junior training ceased altogether in 1993, as did the training of potential officers who underwent short voluntary courses at Pirbright prior to sitting the Regular Commissions Board at Westbury.

Although many changes have also been made to the training of potential officers, a spectator at the August 1995 Sovereign's Parade at RMA Sandhurst will have noticed surprisingly few differences to the commissioning parade of, say, 40 years previously. The standard of drill was as high as ever and the form of the parade, culminating in the Guards Adjutant riding up the steps into Old College, has scarcely changed. Inevitably, however, much is different – the presence of female officer cadets and the high proportion of potential officers with university degrees are but two examples.

On the 11 August 1995 parade, the Sovereign's representative, Field Marshal Sir Peter Inge, was full of praise for the officer cadets and those who had trained them. On that day four officers were commissioned into the Regiment (BCP Hancock, RA Harper, TC Nicholson and GEO Stanford) while the Grenadiers on parade were CSM SP Milsom, M Gaunt, P Maher, C/Sgt KM Finney and NA England.

While Sandhurst has lost its own Band, that of the Regiment has survived relatively unscathed compared to most others. Despite a reduction in strength from seventy-eight in 1978 to forty-eight in 1995 the Regimental Band still provides a variety of groups specializing in strings, fanfare trumpets or dance music. In the event of war the Band will provide reinforcements as Regimental Medical Assistants.

The Band's Tercentenary was reached in 1985 and an extremely busy year culminated in a gala concert at the Royal Albert Hall. The Band was joined on stage by the Band of the Dutch Grenadiers and the Royal Choral Society. Lt Col Fred Harris OBE and Lt Col Rodney Bashford OBE returned to conduct as did Sir Harry Mortimer and Iain Sutherland. Later in the year a reunion Dinner was held and attended by all serving members of the Band and their guests plus many former members. The Guest Speaker was Iain Sutherland, BBC conductor and former Band member.

The strength of Regimental Headquarters has been decimated over the years and now consists of only two established military posts. Even so the Headquarters continues to deal as best it can with all Regimental matters including welfare.

The Regiment, through the Grenadier Guards Association, grants financial assistance to needy members or their widows or on behalf of their children, provided their adverse circumstances are of an unavoidable nature and that the funds of the Association permit. Apart from grants to over 100 hardship cases, in 1994 some 600 elderly members received Christmas gifts and forty elderly members and wives or widows were given grants towards holidays. The Regiment also assists serving Warrant and Non-Commissioned Officers with grants towards the boarding school fees of their children. Welfare payments from Regimental and Association Funds in 1994 totalled nearly £62,000.

The reductions in the size of Regimental Headquarters has, seemingly, not prejudiced their planning of 'Grenadier Day', held annually each summer. In 1995 over 4,000 members of the Grenadier Association with their families enjoyed a marvellous day amidst beautiful weather at Pirbright. A carnival atmosphere prevailed. The highlight of the arena events was the re-enactment of the battle of Inkerman, directed by Maj RMT Reames. The Boyton shooting cup was presented to the Nijmegen Company team by Brig EJ Webb-Carter OBE, the Regimental Lieutenant Colonel, who, as already related, succeeded Maj Gen BC Gordon Lennox CB, MBE on 1 April 1995. Traditionally, the Regimental Band concludes the display by marching past the President of the Regimental Association who was Brig MS Bayley MBE in 1995. However the 1995 Grenadier Day marked the last week of service of the Regimental Adjutant, Lt Col TJ Tedder, who had spent seven years in the appointment over two tours. He took the salute instead. Lt Col CJE Seymour has replaced him – the first Retired Officer to become Regimental Adjutant. Another change was the posting of 2/Lt HJR Bond-Gunning to Catterick and the commissioning of Second Lieutenants JMH Bowder, the Hon CL Broughton and WTJ Smiley.

*　　*　　*

During 1995 special events were held in the United Kingdom, the Commonwealth and many parts of the world to mark the fiftieth anniversary of the end of the Second World War. Grenadiers played a full part in England in the Victory in Europe commemorative events in May, and the Victory over Japan celebrations in August. The armed forces of the Crown had paid a high price for victory. The Regiment's casualties in the Second World War were 1,256 killed, and 3,166 wounded. Whether the services of Thanksgiving were held in such places as St Paul's Cathedral, the Church of St Michael and All Angels, Pirbright, or St John's Cathedral in Hong Kong, they had much in common including in all such cases the moving words:

> When you go home,
> tell them of us and say
> 'For your tomorrow
> we gave our today'.

CHAPTER 32

FIFTY YEARS OF PEACE
AND WARS

Since 1945 the Grenadier Guards have played their full part in events both sad and glorious in London – for example providing the bearer parties for King George VI and Sir Winston Churchill – and the Guards of Honour on triumphant occasions, such as the Coronation.

The Regiment has proudly guarded its Sovereign and Colonel-in-Chief at Buckingham Palace and Windsor Castle, and earned the admiration of many at the Queen's Birthday parades and the other great ceremonial events.

It will be seen from this History that the Battalions have also deployed, sometimes at very short notice and at times of crisis, to four continents.

There are few places of consequence where Grenadiers have not served; they have seen action in the jungles of Malaya and the deserts of the Gulf; they have endeavoured to keep the peace amidst the murderous internal struggles in Palestine and Cyprus; they have contributed to the success of lasting peace in West Germany; they have maintained their vigilance and professionalism in Northern Ireland; they have joined other United Nations in peace keeping operations in the Middle East, and much else.

*   *   *

This chapter will not attempt to summarize the overseas tours, nor the lessons learned, but instead it looks briefly at a few of the older Grenadiers, scarcely mentioned hitherto, who have set the standards in peace and war for past, present and future generations. Most of the names which follow will be unfamiliar to Grenadiers today, unlike, hopefully, those who have been referred to in earlier chapters.

Between March and August 1946 the Grenadiers witnessed the deaths of their two Field Marshals, both of whom had been Chiefs of the Imperial General Staff – the professional heads of the Army. Lord Gort VC had become CIGS at the early age of 51. Perhaps his greatest triumph occurred in May 1940. Ignoring orders to join up with the French at Amiens, he subsequently moved the two divisions instead to fill the gap between the British and Belgian armies. 'By doing so,' recorded the official historian, 'he saved the British Expeditionary Force.' Gort's character was upright and honourable; he regulated his conduct by a strict code of duty.

Field Marshal the Earl of Cavan, another Grenadier, also had an enviable reputation as a great fighting commander. Those who knew him were apt to comment on his dependability, imperturbability and friendliness. The Field Marshals, educated at Harrow and Eton respectively, were born with advantages in life. The same cannot be said for many other very distinguished Grenadiers who gave a lifetime of service to the Regiment, rising through the ranks to the pinnacle of their profession.

Lt Col GF Turner OBE DCM was known and beloved by very many Guardsmen, young and old. During the difficult days of the withdrawal to Dunkirk, as Quartermaster of the 3rd Battalion, he never failed to get supplies up to the front line. Thereafter he was largely responsible for the orderly evacuation of a large number of troops from the beaches. But it was after the war that he revealed his greatest talents. As General Secretary of the Grenadier Guards Association for sixteen years and of the Household Brigade Employment Society, he was most successful in looking after everyone's interests.

His successor was the equally popular and capable Lt Col AJ Spratley MBE MM. He had joined the Regiment as a Drummer, and also fought at Dunkirk; thereafter he saw action from Normandy to the Baltic in his Churchill tank, appropriately named "Windsor". Both he and Fred Turner were appointed to a Military Knighthood of Windsor. Both were the most courteous of gentlemen.

Fame can be said to be measured by widespread notoriety, and by the respect, admiration and friendship of a very large number of people. Few will dispute that RSM AJ Brand MVO MBE, and his successor, JC Lord MVO MBE, both Academy Sergeant Majors at Sandhurst, were more than famous in their day. Between 1937 and 1948 30,000 officer cadets were commissioned during RSM Brand's term of office. Since joining The King's Company in 1915 he had struck everyone with his charm and consideration of others.

JC Lord had joined the Grenadiers in 1933. As Academy Sergeant Major he was a counsellor, guide, philosopher, disciplinarian, figurehead and driving force to ensure that every Officer Cadet leaving Sandhurst was well equipped as a soldier and leader to command men. Officers from throughout the Army will remember Brand and Lord with gratitude and esteem.

Many other great personalities, whether they be Guardsmen or Generals, deserve mentioning in these pages, but space is against us and, as already said, hopefully most of them have been referred to in earlier chapters.

One man who knew them all was Maj Gen Sir Allan Adair, the much-loved 24th Colonel. In the long annals of the Regiment, he is one of the most outstanding examples of the saying 'Once a Grenadier, always a Grenadier' – first during his years of active service, then as President of the Regimental Association until 1961 when he became the Colonel of the Regiment for thirteen years, and thereafter as a wise counsellor – in all over seventy years of service to the Regiment he loved.

In 1942 Sir Allan had taken command of the Guards Armoured Division. But that is also the year in which Princess Elizabeth became Colonel of the Regiment. Since then Her Majesty has closely followed the fortunes of her Grenadiers, first as Colonel and since becoming Queen, as Colonel-in-Chief. During these years she has earned the great admiration and affection of all ranks. Wherever she goes, and she goes everywhere, she inspires us by her sense of duty.

The same can be said for Prince Philip, the 25th Colonel of the Regiment, whose informality and encouragement have made him so popular: whether his visit was to the beleaguered Crossmaglen base in South Armagh, or London, or to the wives in Münster waiting anxiously for news from the Gulf War in 1991, or elsewhere, Prince Philip has never failed to raise morale due to his friendliness, easy charm, approachability and interest in his Regiment.

In 1990 both The Queen and Prince Philip were photographed with five former Grenadiers. They were all Knights of the Most Noble Order of the Garter – the oldest order of chivalry in Europe. The five were Lord Carrington who won a Military Cross with the 2nd Battalion at Nijmegen and became High Commissioner in Australia and the Secretary-General of NATO; Lord Cromer had fought in North-West Europe and was later Governor of the Bank of England and Ambassador in Washington. Lord De L'Isle won his Victoria Cross at Anzio in 1944 with the 5th Battalion and was later Governor-General of Australia. The Duke of Norfolk had commanded 1 Division in Germany, before becoming Earl Marshal and Premier Duke. The fifth was the Duke of Grafton. All of them have devoted much of their life to public service.

All these great men are noble examples of those former Grenadiers who have dedicated much of their efforts to serving their country. The same can also be said for those others who have also 'laboured seeking no reward' – for example those who help former Grenadiers – the Branch Secretaries of the Regimental Association. They have ensured that the elderly and sick are looked after, and the widows comforted, let alone organizing the reunions to everyone's great enjoyment.

From the moment recruits join their Depots, whether they be Gunners, Royal Engineers, Green Jackets or Guardsmen, they are presumably taught that they have joined 'the best', something 'special', something 'unique'. A few, when they look back on their service, may ask themselves how their Regiment or Corps achieved its success or otherwise.

Six Grenadiers of different backgrounds were asked, off-the-cuff, why they thought, as they did, that the Regiment was 'special'. The first replied that 'Guardsmen are trained properly from the beginning, and it's a family: they stand by each other, putting their friends first.'

Another, a former NCO, put it differently: 'Everything we do, it's done as well as it can be – the connection with Royalty – the achievements in battle – we are therefore "different".'

A former boy soldier felt that: 'The Regiment is an élite with its inherent spirit due to its traditions, and history, and what people have been through: I was caught up in that spirit myself: it's the sense of belonging to the family – the comradeship which keeps one going in times of real hardship: every Regiment would claim that, wouldn't they? I remember the great characters – the Commanding Officers, the Adjutants, the Warrant Officers and NCOs and all ranks – people who treated you right if you were in trouble – the great sense of humour; it's all a matter of pride, the sense of belonging; most peoples' lives are pretty humdrum in civvy street – but if you look back on your service in the Regiment it's like a beacon – they were happy years.'

A former RSM put the Regiment's success down to: 'The way we were brought up in our recruit and Guardsmen days – the great efficiency and discipline; every effort being made to carry out the traditions; the great respect for the Commanding Officers.'

A serving officer put it thus: 'Why are the Grenadiers and Household Division "different"? There's something special in a Guardsman's swagger; the confidence in belonging to something very special; as my wife once pointed out, the Guardsmen are physically bigger, enabling them to go in and win things. This gives them a certain style. And their extraordinary humour. I saw all sorts when commanding a Brigade; the Guardsmen had different witty retorts, the amusing backchat; the Guardsmen were so good at chatting people up. The attitude at the Guards Depot was so good, as were the Guards instructors at Sandhurst. The young Lance Sergeants, equivalent to Corporals in Line Regiments, getting into the Sergeants' Mess in their early twenties had something to do with it – learning the form from their seniors and Quartermasters.'

Finally, a former Quartermaster put it still differently: 'The strength lies in the man-management – the Guards officers are always loyal; they listen; they rely on their Warrant Officers and NCOs which gives them strength. Our training and history helps – although others call us "wooden tops" or "bloody Guards", inwardly they admire us. When I eventually left my civilian firm, the Directors thanked me for getting so many ex-Guards to join the firm: they'd never had a bad one. It's so sad to lose the 2nd Battalion with very fine officers, Warrant Officers and NCOs who now won't get the fine careers they deserve.'

During the last fifty years the list of those Grenadiers receiving gallantry awards has ever lengthened, while the number of those killed in action has, mercifully, remained relatively few. Just as those killed in the Second World War have never been forgotten, so we now remember those Grenadiers murdered in Palestine, Malaya, Suez, the Cameroons and Northern Ireland. Robert Nairac was the first to be awarded the George Cross for service in Northern Ireland, and Richard Westmacott was the first recipient of the Military Cross ever awarded posthumously – such was their courage.

In 1944 King George VI broadcast his Christmas Day message. To the nation and Empire he offered at last a glimmer of hope: 'We do not know what awaits us when we open the door of 1945. But if we look back to those earlier Christmas days of the war, we can surely say that the darkness grows less and less. The lamps which the Germans put out in 1914 and then in 1939 are slowly being rekindled.'

As this History has shown, there have been few years over the last fifty when there has been peace. There is only one year – 1968 – when a British serviceman has not been killed in action.

Nevertheless, fifty years after King George's broadcast there is now a real sense of optimism which was reflected in The Queen's 1994 Christmas speech: 'In Northern Ireland, peace is gradually taking root . . . and, in the Middle East, long-standing enmities are healing . . . the sight of the happy faces of . . . young people in Russia, in South Africa, where so much has changed with such extraordinary speed in the last year, and in Northern Ireland, where there is real hope of a permanent end to the bitterness of recent years, should be enough to convince even the most hard-hearted that peace is worth striving for.'

This History shows how Grenadiers have proudly kept the peace and what they have achieved over the last fifty years. They look forward to the next fifty with equal determination to do their duty whatever may lie ahead.

ROLL OF HONOUR

OF OFFICERS AND SOLDIERS OF THE REGIMENT KILLED IN ACTION OR ON OPERATIONAL DUTY

Name	Battalion	Date	Place
Gdsm D Roberts	3	2.3.47	Palestine
Gdsm AJ Taylor	3	15.1.48	Palestine
Gdsm FLJ Howlett	1	30.4.48	Palestine
L/Sgt PR Clarke	1	8.5.48	Palestine
Lt JRS Farrer	3	4.3.49	Malaya
L/Cpl JP Chriscoli MM	3	12.3.49	Malaya
Gdsm J Ryan	3	12.3.49	Malaya
Gdsm AE Martin	3	12.3.49	Malaya
Gdsm JR Hall	3	12.3.49	Malaya
Gdsm VT Herrett	3	12.3.49	Malaya
Gdsm FL Smith	3	27.1.52	Egypt
Gdsm AN Smith	3	29.1.52	Egypt
Gdsm JW Lunn	1	10.7.61	Cameroons
Capt RL Nairac, GC	HQ 3 Inf Bde	15.5.77	Northern Ireland
Gdsm GR Duggan	1	21.12.78	Northern Ireland
Gdsm KL Johnson	1	21.12.78	Northern Ireland
Gdsm GN Ling	1	21.12.78	Northern Ireland
Capt HR Westmacott MC	SAS	2.5.80	Northern Ireland
Gdsm DM Blinco	1	30.12.93	Northern Ireland

HONOURS AND AWARDS

GEORGE CROSS

1979	Capt	RL Nairac

THE MOST HONOURABLE ORDER OF THE BATH

Knight Grand Cross
1980	Gen	Sir David Fraser KCB OBE

Knight Commanders
1960	Lt Gen	Sir Rodney Moore KCVO CBE DSO
1973	Lt Gen	DW Fraser OBE

Companions
1949	Maj Gen	JA Gascoigne DSO
1955	Maj Gen	JNR Moore CBE DSO
1959	Maj Gen	GC Gordon Lennox CVO DSO
1960	Maj Gen	CMF Deakin CBE
1965	Maj Gen	EJB Nelson DSO MVO OBE MC
1966	Maj Gen	The Hon MFS Fitzalan Howard CBE MC
1969	Maj Gen	RH Whitworth CBE
1970	Maj Gen	FJC Bowes-Lyon OBE MC
1986	Maj Gen	BC Gordon Lennox MBE

THE MOST DISTINGUISHED ORDER OF ST MICHAEL AND ST GEORGE

KNIGHT COMMANDER
1962	Maj Gen	Sir Julian Gascoigne KCVO CB DSO

THE ROYAL VICTORIAN ORDER

KNIGHT GRAND CROSS
1974	Maj Gen	Sir Allan Adair Bt KCVO CB DSO MC

KNIGHTS COMMANDER
1953	Maj Gen	JA Gascoigne CB DSO
1959	Maj Gen	JNR Moore CB CBE DSO
1966	Maj Gen	EJB Nelson CB DSO MVO OBE MC
1967	Maj Gen	Sir Allan Adair Bt CB CVO DSO MC
1973	Maj Gen	FJC Bowes-Lyon CB OBE MC

COMMANDERS
1952	Col	GC Gordon Lennox DSO
1957	Maj Gen	Sir Allan Adair Bt CB DSO MC
1979	Maj	RAG Courage LVO MBE

MEMBERS 4TH CLASS
(Redesignated Lieutenant on 31st December, 1984)
1952	Maj	AG Heywood MC
	Lt	ADY Naylor-Leyland
1953	Lt Col	VAP Budge OBE
	Maj	HC Hanbury MC
	Lt Col	EJB Nelson DSO OBE MC
1968	Maj	ATW Duncan
1977	Maj	CXS Fenwick

MEMBERS
1948	RSM	AJ Brand MBE
1963	RSM	JC Lord MBE
1969	RSM	JS Bird MBE
1976	RSM	T Taylor MBE
1986	Maj	The Hon AFC Wigram

THE ROYAL VICTORIAN MEDAL (SILVER)

1952	CSM	FJ Clutton MM
	L/Sgt	D McMahon
	L/Cpl	G Baker
	L/Cpl	W Fryer
	L/Cpl	D James
	L/Cpl	J Schofield
	L/Cpl	DR Wilson
	Gdsm	R Dilworth
	Gdsm	NS Perkin
	Gdsm	V Wright
	Gdsm	G Croucher
	Gdsm	CP Snook
	Gdsm	RJ Garner
	Gdsm	EJ Goodall
	Gdsm	J Marsh
	Gdsm	TJ O'Donnell
	Gdsm	AJ Roberts
	Gdsm	FC Saddington
	Gdsm	E Scorer
	Gdsm	A Welding
	Gdsm	GR Whitehead
1953	Sgt	F Brunt

THE MOST EXCELLENT ORDER OF THE BRITISH EMPIRE

KNIGHTS COMMANDER

1964	Lt Gen	GC Gordon Lennox CB CVO DSO
1995	Lt Gen	AA Denison-Smith MBE

COMMANDERS

1956	Brig	CMF Deakin
1960	Brig	The Hon MFS Fitzalan Howard MC
1961	Col	VAP Budge MVO OBE
1964	Col	AG Heywood MVO MC
	Brig	RH Whitworth MBE
1982	Brig	MF Hobbs OBE
1992	Col	ET Hudson OBE
1993	Col	OJM Lindsay

OFFICERS

1949	Lt Col	EJB Nelson
1950	Lt Col	JDC Brownlow
1952	Lt Col	VAP Budge
1954	Lt Col	TP Butler DSO
1959	Col	AMH Gregory-Hood MBE MC
	Maj	FJ Harris
1960	Lt Col	ECWM Penn MC
1961	Lt Col	FJC Bowes-Lyon
1962	Lt Col	DW Fraser
1965	Lt Col	The Lord Freyberg MC
1967	Lt Col	TN Bromage MBE
1974	Lt Col	RB Bashford MBE
1977	Lt Col	DV Fanshawe
1979	Lt Col	MF Hobbs MBE
1983	Col	ATW Duncan MVO
1987	Lt Col	PR Holcroft
	Lt Col	ET Hudson MBE
1988	Lt Col	AJC Woodrow MC QGM
1989	Lt Col	EJ Webb-Carter
1991	Lt Col	ACMcC Mather MBE
	Lt Col	RS Corkran
	Lt Col	EH Houstoun MBE
1994	Lt Col	ET Bolitho

MEMBERS

1950	Capt	TN Bromage
1951	Maj	FJ Harris
	RSM	HE Clarke
1953	Maj	RH Heywood-Lonsdale MC
1954	Maj	AMH Gregory-Hood MC
	Maj	WK Chetwynd
	Lt	RE Butler
1955	Maj	R Steele
	RSM	I Ayling
1956	C/Sgt	LW Mayhew
	RSM	EC Weaver
	Capt	AJ Spratley MM
1957	RSM	A Dickinson
	Maj	GC Hackett
	Maj	LAD Harrod
1958	RSM	JS Bird

388

1959	Capt	AG Everett
1960	Maj	MS Bayley
1962	Maj	PH Haslett
	Capt	LE Burrell
	Lt	FJ Clutton MM RVM
	Lt	LC Drouet
1965	RSM	GW Kirkham
	RSM	D Randell
	Lt	ACMcC Mather
1966	RSM	S Low MM
	Capt	WLA Nash
1967	Capt	ST Felton
	Maj	RAG Courage
1968	Maj	BC Gordon Lennox
	Capt	PWE Parry
1969	Capt	A Dobson
1970	RSM	IDA Thomas
	Maj	RB Bashford
1972	RSM	T Taylor
1973	Capt	LG White
	RSM	RP Huggins
	WO2	GV White
	Maj	AA Denison-Smith
1974	WO2	TJ Farr
	Maj	MF Hobbs
	Maj	J Baskervyle-Glegg
1976	Capt	PA Lewis
1977	Capt	The Hon PJA Sidney
1978	WO2	A Fernyhough
1980	Maj	EH Houstoun
1981	Maj	TH Astill
1982	RSM	DR Rossi
1984	Maj	BT Eastwood
	Maj	ET Hudson
1985	Maj	TH Holbech
	Maj	N Collins
1988	Capt	CE Kitchen
1989	Capt	PT Dunkerley
	WO2	M Dabbs
1990	Maj	JS Lloyd
	Maj	ARK Bagnall
	Maj	WR Clarke

1991	Maj	MJ Joyce
1993	C/Sgt	T Jones
1994	Maj	NJR Davies
	L/Cpl	DR Wells
	Capt	AA Pollock
1995	Capt	JA Sandison QGM
	WO1	DL Cox

BRITISH EMPIRE MEDAL

1949	Sgt	D Ramsbottom
1953	C/Sgt	RA Boyles
	Sgt	WC Goodwin
1954	C/Sgt	PH Croucher
1955	C/Sgt	TAW Leach
1956	C/Sgt	W Price
1960	C/Sgt	B Pyrcroft
	Sgt	T Yardley
	Sgt	EW Fisher
1961	Sgt	DM Creswell
	WO2	J Foster MM
	WO2	ET Weston
1962	Sgt	F Brunt RVM
1963	Sgt	C Tudge
1964	WO2	J Harper
	WO2	RC Page
	WO2	JG Pope
1965	WO2	W Williams
	L/Sgt	L Perkins
	L/Sgt	D Sweeney
	L/Cpl	M Surman
	Gdsm	M Ryan
	Gdsm	J Warner
	Gdsm	A Sharrocks
	Gdsm	N Wright
	Gdsm	D Stoakes
1968	Sgt	DJ Worsfold
1969	C/Sgt	BE Thompson
1972	WO2	PR Taylor
	Sgt	AV Axworthy
1973	C/Sgt	P Hodgkinson

	Sgt	V Goodwin
1974	WO2	D Denness
1976	Sgt	AD Peachey
	C/Sgt	RH Trussler
	C/Sgt	WA Hill
1978	L/Sgt	GS Allen
	Sgt	G Reincke
	C/Sgt	T Dove
1979	Sgt	D Bradley
	C/Sgt	RAJ Phasey
	L/Sgt	KA Reagan
1980	Sgt	T Mann
	L/Sgt	JC Brown
	C/Sgt	JA Steel
1982	C/Sgt	DW Cousins
1983	C/Sgt	S Tuck
1984	Sgt	RTG Warlow
	WO2	M Robinson
1985	Sgt	DE Wilkinson
1987	C/Sgt	LP Gallagher
1988	C/Sgt	RM Carter
	C/Sgt	RJ Le Louet
	Sgt	NG Lawrenson
1990	C/Sgt	AJ West
	Sgt	AT Wedesch
	C/Sgt	SP Milsom
	Sgt	PA Sellors
1991	C/Sgt	R Sargeant
	Sgt	DA Traynor
1992	C/Sgt	RM Jolly
	C/Sgt	PM Ladd
	L/Sgt	G Warner

MILITARY CROSS

1949	Maj	AG Heywood
1979	Maj	AJC Woodrow QGM
1980	Capt	HR Westmacott

MILITARY MEDAL

1949	Sgt	B Clutton
1974	C/Sgt	GR Matthews
1979	Sgt	RG Garmory

THE QUEEN'S GALLANTRY MEDAL

1975	Maj	AJC Woodrow
1981	WO2	JA Sandison

MERITORIOUS SERVICE MEDAL

1957	WO2	WT Garrett
	C/Sgt	PH Croucher BEM
1958	C/Sgt	EC Scott BEM
1972	RSM	RP Huggins

MENTIONED IN DESPATCHES

1949	Lt Col	TFC Winnington MBE
	Lt	JRS Farrer
	L/Sgt	J Buck
1957	Brig	CMF Deakin CBE
1958	Maj	PJC Ratcliffe
	RSM	AL Stevens
	C/Sgt	FW Parker
	C/Sgt	TF Cornall
1959	Lt Col	PC Britten
	Lt	WLA Nash
	Sgt	EM Addison
1972	Lt Col	GW Tufnell
1973	L/Cpl	G Fishwick
	Capt	The Lord Richard Cecil
	Maj	PH Cordle
1974	Maj	HML Smith
	Capt	RG Woodfield
	L/Cpl	JR Knight
1975	Lt Col	BC Gordon Lennox MBE

	2Lt	NWL Hackett Pain
	Sgt	T Mann
	L/Sgt	PS Brown
	Lt Col	DH Blundell-Hollinshead-Blundell
	S/Sgt	JA Sandison
	L/Sgt	TK Cooke
	Sgt	K Watson
1979	Maj	RG Woodfield
	WO2	DW Ling
1980	Maj	RG Cartwright
	Maj	EJ Webb-Carter
1982	Maj	EH Houstoun MBE
1984	Lt Col	A Heroys
	Maj	EJ Webb-Carter
	Capt	JRH Wills
1987	Maj	RG Cartwright
1988	Capt	MCJ Hutchings
	Maj	REH Aubrey-Fletcher
	Capt	FA Wauchope
	Sgt	TE Sentance
1994	Sgt	DI Adkins
	Gdsm	IR Eachus
	Gdsm	SG Wyatt
	L/Cpl	SJ May

QUEEN'S COMMENDATION FOR VALUABLE SERVICE

1994	Capt	JDM Wrench
	Lt	SD Marcham
	WO2	RM Dorney
	L/Sgt	PW Murray
	Gdsm	CS Putley

CERTIFICATES OF MERIT, GALLANTRY, GOOD SERVICE AND OTHER COMMENDATIONS

COMMANDER-IN CHIEF'S CERTIFICATE

1959	WO2	K Pridham (Cyprus)
	WO2	J Downes
	Sgt	N Everitt

QUEEN'S COMMENDATION FOR BRAVE CONDUCT IN CANADA

1973	Sgt	T Mann

GENERAL OFFICER COMMANDING'S (NORTHERN IRELAND) COMMENDATION

1973	Gdsm	RA Sparkes
	Gdsm	S Tuck
1975	WO2	A Davenport
	L/Sgt	RD McLellan
	L/Sgt	TJ Yates
	Gdsm	SP Marriott
	Gdsm	RE Stanley
1978	C/Sgt	DS Randell
	L/Sgt	BH Forsyth
	Gdsm	NE Walker
1980	WO2	T Bingham
	WO2	D Loxton
	WO2	G Dann
	WO2	MK Collins
	Sgt	J Finch
	L/Sgt	MR Pearson
	Capt	WR Clarke
1985	WO2	P Harris
	L/Sgt	BW Lawson
	L/Sgt	MJ Walker
	L/Cpl	PJ Woodgates
1986	L/Sgt	H Moore
1988	C/Sgt	D Hague
	L/Cpl	N Miller
	L/Sgt	GJ Glasspool
	Gdsm	AJ Callender
1991	WO2	V Overton
	C/Sgt	G Couzens

	L/Sgt	DA Ibbotson
	L/Cpl	M Beasley
	Gdsm	W Turner
1995	C/Sgt	LJ Herbert
	L/Sgt	KJ Turner
	Gdsm	DT Smith

COMMANDER BRITISH FORCES HONG KONG COMMENDATION

1976	WO2	J Gowers
	Gdsm	K Winn

COMMANDER-IN CHIEF FLEET COMMENDATION
(Relating to the Falkland Islands)

1982	L/Cpl	J Noble

OTHER GENERAL OFFICER COMMANDING'S COMMENDATIONS

1986	JGdsm	M Foggarty
1990	Gdsm	S Marsland
1991	C/Sgt	M Goodenough
1993	C/Sgt	C Parker

JOINT COMMANDER'S COMMENDATION
(Relating to the Gulf War)

1992	Maj	GJS Hayhoe
	L/Sgt	A Leach

APPENDIX III

SOURCES AND BIBLIOGRAPHY

Incomparably the most important source for this History has been my interviews over five years with many hundreds of serving and retired Grenadiers – too many to name here.

There are few printed sources apart from *The Guards Magazine*, and *The Grenadier Gazette* from 1977 onwards, and the following:

A Guards' General: The Memoirs of Sir Allan Adair, Hamish Hamilton 1986.

Always a Grenadier by Maj Gen Sir John Nelson, published privately.

Brushfire Wars by Michael Dewar.

Withdrawal from Empire by William Jackson, Batsford 1986.

The Regiments Depart by Gregory Blaxland, William Kimber 1971.

I visited the Archives in Singapore on Malaya, in Nicosia on Cyprus, in Hong Kong and UK.

In the sources that follow, RHQ refers to Regimental Headquarters Grenadier Guards, Wellington Barracks, London SW1 where there are files on most of the chapters below; MOD refers to the Ministry of Defence Whitehall Library, 3–5 Great Scotland Yard, London SW1A 2HW.

CHAPTER 1 (GERMANY 1945)

The Grenadier Guards 1939–45 Volume One by Lt P Forbes, Gale and Polden 1949.

The Last Days of the Reich by James Lucas, Guild Publishing Ltd 1986.

CHAPTER 2 (AUSTRIA 1945)

Accounts of 3 GREN GDS 1 March–1 August 1945 Italy and Austria; Report by Capt N Nicolson dated 13 June 1945 in RHQ.
The Grenadier Guards 1939–1945 Volume Two by Capt N Nicolson p 518–23, Gale and Polden 1949.

CHAPTER 3 (PALESTINE 1945–1948)

Search Operations in Palestine by Brig RN Anderson; *The Army Quarterly* 1948; *The Evacuation of Palestine* by Lt Gen GHA MacMillan RUSI Journal November 1948; *The Last Days of the Mandate* – address to Chatham House 22 July 1948 by Gen Sir Alan Cunningham, MOD ref 956–94; *Insurgency and Counter Insurgency in Palestine 1945–1947* by DA Chartres, MOD ref 956–94; *Memoirs*, Field Marshal Montgomery, Collins 1958.

CHAPTER 4 (MALAYA 1948–1949)

Templer, Tiger of Malaya by John Cloake, Harrap 1985 p 188–327; Intelligence Summaries in RHQ; *Communist Struggle in Malaya* by Gent Z Hanrahan, University of Malaya Press 1971; *Hearts and Minds in Guerrilla Warfare*, Oxford University Press 1989; *Repression and Revolt*, Centre for International Studies, Ohio University 1969; *The Malayan Emergency in Retrospect* by RW Korner February 1972, Rand Corporation; *Jungle Green* by Arthur Campbell, George Allan and Unwin 1953; *Guerrilla Communism in Malaya* by Lucian Pye, Princeton University Press 1956; Various documents in National Library, Singapore; Articles in *The Guards Magazine* by 2Lt JGC Wilkinson Summer 1949 and by Col TFC Winnington Winter 1960; Quarterly Historical Reports in RHQ; Newspaper articles including *Daily Telegraph* 25 March 1949.

CHAPTER 5 (GERMANY 1945–1952)

Berlin and the British Ally 1945–1990 by Maj Gen RJS Corbett, Zumm Druck and Satz KG, Berlin 1991.

CHAPTER 7 (MIDDLE EAST 1948–1956)

Mountbatten by Philip Ziegler, Collins 1985; *Memoirs*, Field Marshal Montgomery Collins 1958 pages 420–422; *Full Circle* by Earl of Avon, Cassell 1960; Various *Army Quarterlys* (The Reminiscences of 37 Grenadiers were included in this chapter!)

CHAPTER 8 (CYPRUS 1956–1959)

Island in Revolt by Charles Foley, Longmans Green 1962; *Cyprus Guerrilla* by Doros Alastos, Heinemann 1960; *Riding the Storm* and *At the End of the Day* by Harold Macmillan, published by Macmillan 1973; *Portrait of a Terrorist* by Dudley Baker, Cresset Press 1959; *Memoirs of Gen Grivas* edited by Charles Foley, Longmans Green 1964.

CHAPTER 9 (UK 1958–1961)

For reductions in Battalions see *Festing Field Marshal* by Lyall Wilkes, the Book Guild 1991; and *Tiger of Malaya* p 365; Jeremy Paxman's article on *Changing Face of Britain, Sunday Times* News Review 14 October 1990.

CHAPTER 10 (CAMEROONS 1961)

Reports include 1 GREN GDS S/3/28 dated 10 October 1961 in RHQ, various articles in King's Own Royal Border Regiment journals; *Army Quarterly* April 1970 pages 97–99; *The Guards Magazine* Autumn 1961 pages 197–198.

CHAPTER 11 (BRITISH GUIANA 1963–1964)

Report of a Commission of Inquiry into Disturbances in British Guiana, HMSO 1962; *Guyana and Cyprus, Techniques of Peace Keeping* by Anthony Verrier *RUSI Journal* 1966, MOD 956–43.

CHAPTER 13 (UN IN CYPRUS 1965)

British Army Review No 30 December 1968 p 14–15 and No 41 August 1972 p 46–51.

CHAPTER 14 (NORTHERN IRELAND 1969–1970)

Disturbances in Northern Ireland HMSO September 1969 Cmd 532; *The Guards Magazine* Spring 1976 pages 5–10; *British Army Review* April 1970 pages 42–48; *A Commentary to accompany the Cameron Report* Cmd 534.

CHAPTER 15 (SHARJAH 1968–1970)

Guidance written by British Troops Sharjah to HQ London District in RHQ.

CHAPTER 16 (BRITISH HONDURAS 1971–1972)

Newspaper cuttings and Historical Reports April to August 1972 in RHQ.

CHAPTER 17 (NORTHERN IRELAND 1971–1974)

The British Army in Northern Ireland by Michael Dewar, Arms and Armour Press 1985.

CHAPTER 18 (HONG KONG 1975–1976)

Files in RHQ include 15 April 1975 report on Frontier Duties; 10 September 1975 report to the Major General; Internal Security Seminar papers dated 21 November 1975; *The Army Quarterly* October 1976 pages 423–427; *Life and Death in Shanghai* by Nien Cheng, Grafton Books 1984; *Wild Swans* by Jung Chang, HarperCollins 1991.

CHAPTER 20 (AWAY FROM THE REGIMENT)

A Modern Major General, the memoirs of Maj Gen Sir Julian Gascoigne, published privately. *The Myth of Mau Mau* by Carl Rosberg, FA Praeger 1966; *Mau Mau from Within* by DL Barnett, MacGibbon and Kee 1966.

CHAPTER 21 (ASSISTANCE TO CIVIL MINISTRIES)

Troops in Strikes by Steve Peak, The Cobden Trust 21 Tabard Street SE1 4LA; the Firemens and Ambulances strikes were covered in *The Guards Magazines* of Winter 1977 and Autumn 1990 respectively.

CHAPTER 27 (THE FALKLANDS 1990–1992)

The Falkland Islands: The Facts FCO 1982; *Grenadier Gazette* 1992 pages 22 and 25.

CHAPTER 28 (THE GULF WAR 1991)

Storm Command and *Looking For Trouble* by Gen Sir Peter de la Billière, HarperCollins 1992 and 1994; *It Doesn't Take A Hero* autobiography of Gen H Norman Schwarzkopf, Bantam Press 1992; tape recordings in RHQ; *The Guards Magazines* in 1991 and magazines of the Royal Scots, Staffords and Fusiliers; *The Infantryman* 1991 pages 38–58; *The London Gazette* second supplement 28 June 1991 containing the despatch by Air Chief Marshal Sir Patrick Hine; *Soldier Magazines* late 1990 and 1991.

ACKNOWLEDGEMENTS

I am deeply grateful to those who contributed their reminiscences to this History. (I hope that anyone else who now wishes they had done so will send material to the Regimental archives, or to me, the Editor of *The Guards Magazine*, for possible publication).

I greatly benefited from those who took immense trouble commenting on the manuscript. I must also thank Anthony Mather who checked the Honours and Awards Appendix and Richard Corkran who clarified the British Honduras chapter for me. The librarians at the Prince Consort and Ministry of Defence, Whitehall, libraries were always exceptionally helpful. And then there are the friends of all ranks with whom I soldiered for 35 years. Many are mentioned in the History but far more could not be included without extending these pages endlessly. I thank them all.

However, I must reserve my greatest thanks for my wife, Clare, who has proved to be the most loving of Army wives, and whose patience, during the five years it took me to research and write this book, is truly remarkable. She typed all the chapters several times, helped by our daughter, Fiona.

It has been a privilege to have the opportunity to write the Grenadier History. I hope you enjoy reading it.

Oliver Lindsay

Regimental Headquarters
Grenadier Guards
Wellington Barracks
Birdcage Walk, London SW1E 6HQ

INDEX

Notes:
1. Ranks shown are for those reaching the rank of Brigadier and above.
2. Only those whose names appear in the main text and at Appendix 1 are included here.

403

407

411

412